ARTEMIA: BASIC AND APPLIED BIOLOGY

BIOLOGY OF AQUATIC ORGANISMS

Volume 1

Series editor:
Koen Martens

ARTEMIA: BASIC AND APPLIED BIOLOGY

Edited by

TH. J. ABATZOPOULOS
Aristotle University of Thessaloniki, School of Biology, Greece

J. A. BEARDMORE
University of Wales Swansea, School of Biological Sciences, UK

J. S. CLEGG
University of California, Davis, Bodega Marine Laboratory, USA

and

P. SORGELOOS
Ghent University, Laboratory of Aquaculture & Artemia Reference Center, Belgium

KLUWER ACADEMIC PUBLISHERS
DORDRECHT / LONDON / BOSTON

A C.I.P. Catalogue record for this book is available from the Library of Congress.

ISBN 1-4020-0746-9

Published by Kluwer Academic Publishers,
P.O. Box 17, 3300 AA Dordrecht, The Netherlands.

Sold and distributed in North, Central and South America
by Kluwer Academic Publishers,
101 Philip Drive, Norwell, MA 02061, U.S.A.

In all other countries, sold and distributed
by Kluwer Academic Publishers,
P.O. Box 322, 3300 AH Dordrecht, The Netherlands.

Printed on acid-free paper

All Rights Reserved
@ 2002 Kluwer Academic Publishers
No part of the material protected by this copyright notice may be reproduced or
utilized in any form or by any means, electronic or mechanical,
including photocopying, recording or by any information storage and
retrieval system, without written permission from the copyright owner.

Printed in the Netherlands.

TABLE OF CONTENTS

Preface xi
Editorial Note on Terminology xiii
Acknowledgements xv

Chapter I. *ARTEMIA* MORPHOLOGY AND STRUCTURE
G.R.J. Criel and T.H. Macrae

1. External Form of *Artemia* 1
    1.1. INTRODUCTION 1
    1.2. HEAD 3
    1.3. THORAX 5
    1.4. ABDOMEN 6
2. Internal Anatomy of *Artemia* 6
    2.1. DIGESTIVE SYSTEM 6
    2.2. CIRCULATORY SYSTEM 9
    2.3. EXCRETORY SYSTEM 11
    2.4. NERVOUS SYSTEM 12
    2.5. SENSORY SYSTEM 14
        2.5.1. Compound Eyes 14
        2.5.2. Median Eye (Nauplius Eye) and Ventral Frontal Organs 16
        2.5.3. Dorsal Frontal Organs or Cavity Receptor Organs ... 17
        2.5.4. Other Sense Organs 17
    2.6. ENDOCRINE SYSTEM 18
    2.7. FEMALE REPRODUCTIVE SYSTEM 19
    2.8. MALE REPRODUCTIVE SYSTEM 24
3. *Artemia* Morphology and Structure as Revealed by Immunofluorescent Staining 30
4. References 33

Chapter II. REPRODUCTIVE BIOLOGY OF *ARTEMIA*
G.R.J. Criel and T.H. Macrae

1. Introduction 39
2. Molecular Aspects of Early *Artemia* Development 39

v

3. **Morphology of Early *Artemia* Development** 52
   3.1. OOGENESIS 52
   3.2. SPERMATOGENESIS 64
   3.3. FERTILISATION 82
      3.3.1. Sperm Cell Incorporation and Development of the Male Pronucleus 84
      3.3.2. Oocyte Maturation, Polar Body Formation and Female Pronucleus Development 91
      3.3.3. Migration and Interaction of Pronuclei 97
      3.3.4. Development of the Fertilisation Membrane 100
   3.4. EGG ENVELOPE DEVELOPMENT 103
   3.5. CHANGES IN THE RELEASED CRYPTOBIOTIC CYST 106
4. **Embryonic Development** 107
5. **Development of *Artemia* Larvae** 114
   5.1. DESCRIPTION OF LARVAL STAGES 114
   5.2. INTERNAL DEVELOPMENT OF *ARTEMIA* LARVAE 116
6. **References** 119

Chapter III. PHYSIOLOGICAL AND BIOCHEMICAL ASPECTS OF *ARTEMIA* ECOLOGY
J.S. Clegg and C.N.A. Trotman

1. **Introduction** 129
2. **The Two Paths of Development** 130
3. **The Desiccated Cyst** 132
   3.1. THE WATER REPLACEMENT HYPOTHESIS (WRH) AND VITRIFICATION 132
   3.2. ANTIOXIDANTS AND RELEVANT ENZYMES 134
4. **The Hydrated and Activated Cyst** 135
   4.1. ANOXIA 135
   4.2. THE pH_i SWITCH 136
   4.3. STRESS PROTEINS 137
5. **Thermal Resistance of Cysts, Larvae and Adults** 139
6. **Selected Biochemical Features of Encysted Embryo Development** 142
   6.1. YOLK PLATELETS 143
   6.2. TREHALOSE AND TREHALASE 144
   6.3. NUCLEOTIDES AND THEIR ENZYMES 146
   6.4. PROTEASES AND THEIR INHIBITORS 148
7. **Emergence and Hatching** 149
8. **Emerged Embryos and Instar-I Nauplius Larvae** 150
9. **Adult Osmoregulation** 151
10. **Oxygen Transport and Consumption in Adults** 152

11. Haemoglobins 153
   11.1. PHYSIOLOGY AND CHARACTERIZATION 153
   11.2. SEQUENCE AND STRUCTURE 154
   11.3. GENE EXPRESSION AND REGULATION 155
   11.4. ADDITIONAL ROLES FOR HAEMOGLOBIN? 157
12. Concluding Remarks 158
13. Acknowledgements 159
14. References 159

Chapter IV. ZOOGEOGRAPHY
G. van Stappen

1. Introduction 171
2. Ecological Aspects of *Artemia* Distribution 174
   2.1. INTRODUCTION 174
   2.2. SALINITY 175
   2.3. IONIC COMPOSITION 177
   2.4. TEMPERATURE 177
   2.5. BIOTIC ELEMENTS 179
3. Ecological Isolation of *Artemia* Strains 180
4. Coexistence of Different *Artemia* Strains 181
   4.1. DISTRIBUTION PATTERNS OF *ARTEMIA* SPECIES 181
   4.2. COEXISTENCE OF BISEXUAL STRAINS 182
   4.3. COEXISTENCE OF DIPLOID AND POLYPLOID PARTHENOGENETIC STRAINS 182
   4.4. COEXISTENCE OF PARTHENOGENETIC AND BISEXUAL STRAINS 183
      4.4.1. Competition and Niche Partitioning 183
      4.4.2. The Iberian Peninsula 185
      4.4.3. Lake Urmia, Iran 185
      4.4.4. PR China 186
5. *Artemia* and Birds: Predation and Dispersal 186
6. Australia: *Artemia* versus *Parartemia* 189
   6.1. INTRODUCTION 189
   6.2. MORPHOLOGICAL, PHYSIOLOGICAL AND ECOLOGICAL DIFFERENCES *ARTEMIA-PARARTEMIA* 189
   6.3. DISTRIBUTION OF *ARTEMIA* AND *PARARTEMIA* OVER THE AUSTRALIAN CONTINENT 191
   6.4. BIOCONSERVATION 193
7. 'New' *Artemia* Biotopes: Commercial Exploitation and Fundamental Research 193
   7.1. MEDITERRANEAN BASIN AND SOUTHERN EUROPE . 195
   7.2. PR CHINA 195

7.3. SOUTH AMERICA 196
8. Conservation of *Artemia* Biotopes and their Gene Pool 197
9. Acknowledgements 215
10. References 215

Chapter V. EVOLUTION AND SPECIATION
G. Gajardo, T.J. Abatzopoulos, I. Kappas and J.A. Beardmore

1. Introduction 225
2. *Artemia* Species and Genome Characterization 227
   2.1. KARYOLOGY 228
   2.2. ALLOZYME DIVERGENCE 230
   2.3. DNA MARKERS 231
3. Genetic Variation, Ecology and Evolution 233
   3.1. HETEROZYGOSITY AND LIFE HISTORY TRAITS 233
   3.2. BISEXUAL *VS* PARTHENOGENETIC TYPES 236
4. Population Structure and Pattern of Speciation in *Artemia* 237
   4.1. A CASE STUDY OF A SUPERSPECIES: *A. FRANCISCANA* 240
5. Reproductive Isolation 242
6. Final Remarks 243
7. References 245

Chapter VI. APPLICATIONS OF *ARTEMIA*
J. Dhont and P. Sorgeloos

1. Introduction 251
2. *Artemia* Cyst Supply 251
   2.1. A SHORT HISTORIC OVERVIEW 251
   2.2. GREAT SALT LAKE: ECOLOGICAL ASPECTS 252
   2.3. TRENDS IN SUPPLY AND DEMAND 255
3. *Artemia* and Solar Salt Production 256
   3.1. IMPORTANCE OF SALT 256
   3.2. SOLAR SALT PRODUCTION 256
   3.3. BENEFICIAL EFFECT OF *ARTEMIA* IN SALT PRO-
        DUCTION 257
   3.4. AN EXAMPLE OF *ARTEMIA*, SALT AND SHRIMP
        INTEGRATION: THE BOHAI BAY, PR CHINA 257
4. *Artemia* and Aquaculture 259
   4.1. *ARTEMIA* AS INSTANT LIVE FOOD 259
   4.2. USE OF *ARTEMIA* NAUPLII 259
   4.3. THE MYSTERY OF THE 'UNIDENTIFIED FATTY ACIDS' 261
   4.4. ENRICHMENT OR BIO-ENCAPSULATION TECHNIQUES 265

	4.5. COLD STORAGE OF *ARTEMIA* NAUPLII	267
	4.6. USE OF DECAPSULATED CYSTS	267
	4.7. USE OF JUVENILE AND ADULT *ARTEMIA*	268
5.	**Applications of *Artemia* Pond Culture**	**269**
	5.1. VALORISATION OF UNICELLULAR ALGAE RESOURCES	269
	5.2. INTEGRATED AQUACULTURE OPERATIONS RELYING ON *ARTEMIA* POND CULTURE	269
	5.3. VALORISATION OF ORGANIC WASTE AND USE OF *ARTEMIA* IN EFFLUENT TREATMENT	270
6.	**Use of *Artemia* in Research and Education**	**270**
	6.1. THE CONSEQUENCES OF BEING 'EXTREMO-TOLERANT'	270
	6.2. *ARTEMIA* IN ECOTOXICOLOGY	271
	6.3. *ARTEMIA* AS A DIDACTIC TOOL	271
	6.4. *ARTEMIA* IN PERSONAL, SOCIAL AND SEX EDUCATION	271
7.	**Acknowledgements**	**272**
8.	**References**	**272**
	Index	**279**

PREFACE

The objectives of this volume are to present an up-to-date (literature survey up to 2001) account of the biology of *Artemia* focusing particularly upon the major advances in knowledge and understanding achieved in the last fifteen or so years and emphasising the operational and functional linkage between the biological phenomena described and the ability of this unusual animal to thrive in extreme environments.

Artemia is a genus of anostracan crustaceans, popularly known as brine shrimps. These animals are inhabitants of saline environments which are too extreme for the many species which readily predate them if opportunity offers. They are, thus, effectively inhabitants of extreme (hypersaline) habitats, but at the same time are able to tolerate physiologically large changes in salinity, ionic composition, temperature and oxygen tension. Brine shrimp are generally thought of as tropical and subtropical, but are also found in regions where temperatures are very low for substantial periods such as Tibet, Siberia and the Atacama desert. They have, thus, great powers of adaptation and are of interest for this capacity alone.

The earliest scientific reference to brine shrimp is in 1756, when Schlösser reported their existence in the saltpans of Lymington, England. These saltpans no longer exist and brine shrimp are not found in Britain today. Later, Linnaeus named the brine shrimp *Cancer salinus* and later still, Leach used the name *Artemia salina*. The strong effect which the salinity of the medium exerts on the morphological development of *Artemia* is now widely recognised. Unfortunately, biologists in the late nineteenth and early twentieth centuries did not realise the effect of salinity on the morphology of *Artemia*. This resulted in a proliferation of apparent species, subspecies and varieties described for the genus, a solution which has only relatively recently been tidied up.

Studies on the cytology and reproductive mode of the genus were initiated by Artom in the late nineteenth and early twentieth centuries when the existence of the two distinct, stable reproductive states of bisexuality and parthenogenesis were recognised.

Since those early days much work has been done on many aspects of *Artemia* biology and much practical use has been made of the animal. For example, Seale in 1933 started to use nauplii for feeding early stages of marine fish and commercial production of cysts from the coastal saltworks in San Francisco Bay (California, USA) started in the 1950s to be followed slightly later by material from the Great Salt Lake of Utah (USA). *Artemia* found favour too as a 'standard' organism in toxicological assays, although with the recognition that it is too robust organism to be a sensitive indicator species; this interest has now diminished.

Co-ordination of international research activity on *Artemia* took a major step forward in 1976 with the establishment, through the initiative of Patrick Sorgeloos, of the International Study of Artemia (ISA) a group embracing a number of laboratories based on a variety of disciplines and sharing the brine shrimp as a central organism of study. In the intervening years many useful working relationships have been established, research on *Artemia* had blossomed through the multidisciplinary approaches fostered by ISA and many publications in a variety of journals have resulted. It is interesting to look back at the reasons why ISA was established. In the mid 70s many workers using *Artemia* regarded it in much the same light as the classic white rat, *i.e.* a standard and standardised organism. In fact, *Artemia* of many different origins were used on the assumption that all brine shrimp are equal. Unfortunately, some are more equal than others, so that attempts at intercalibration of results were attended by inconsistencies and frustration. There was a recognition, too, that the cyst forming ability of *Artemia* represented a faculty leading to evolutionary capital composed of different year classes much as with the seed banks of many flowering plants. However, because cyst composition has a significant degree of environmental determination (depending on maternal nutrition and other factors), cyst variation in nutritional value for aquaculture was considerable. There was, therefore, a clear need for international co-operation in characterizing strains and developing a unanimously recognised standard. ISA developed a systematic multi-disciplinary programme of research which led to resolution of many of the practical problems attending the use of *Artemia* in aquaculture and opened up a variety of questions of fundamental biological significance. We believe that this book would not have been possible without the existence of ISA.

The editors hope that this book will be useful to several constituencies. In the first place, it should be helpful to graduate students and other young biologists who wish to know more about the life of the brine shrimp. It will also, we trust, indicate a number of areas in which opportunities exist for analysis of some fascinating and fundamental biological problems such as the control of diapause and the population and ecological dynamics of complex multiclonal parthenogenetic populations to name but two. Finally, it is intended to be helpful in developing the interface and information flow between basic biology and aquaculture – an area of applied biology which, for a variety of reasons, is expanding rapidly in most areas of the world.

The editors hope that the users of the volume will find in it material which will both assist them in their work and also inspire within them the same spirit of curiosity and interest in the biology of a unique and fascinating group of organisms as they themselves feel.

The Editors

EDITORIAL NOTE ON TERMINOLOGY

In many different publications extending over a considerable period the early stages in development of *Artemia* have been referred to as 'instars'. In the strict sense some may argue that this is incorrect as the word instar refers to an inter-moult stage of development applicable to all Arthropods but not beyond that phylum. Nevertheless, as the usage is so deeply embedded in the *Artemia* literature we have elected to perpetuate the usage in the interests of clarity.

The population of Mono Lake (California) is very similar in the composition of its gene pool to the many populations of *A. franciscana* in N. America. The genetic distance between the Mono population and such *franciscana* populations is characteristic of the distances found between populations of one species and a significant number of *Artemia* biologists consider the Mono population to be *A. franciscana*. Nevertheless from the viewpoint of the biological species concept it is clear that the exchange of genes between the Mono population and other populations is extremely small if it exists at all. The emphasis in this volume is on the essential biology of *Artemia* and for this reason we have elected to use the binomen *A. monica* for the Mono population.

For European bisexual populations the name used in this volume is *A. salina* which is the binomen most supported by modern genetical evidence. It must be noted, however, that there has been a considerable amount of misuse of the name *A. salina* in the literature in past decades.

We have chosen to refer to the many parthenogenetic forms of *Artemia* as populations. Despite the proposal by Barigozzi to term all of these *A. parthenogenetica*, respect for the biological species concept strongly suggests that obligate parthenogens existing in numerous clones at several different levels of ploidy cannot readily be considered as belonging to a single species.

ACKNOWLEDGEMENTS

We thank Mr **A. Baxevanis** for his devoted help, efficient technical contribution and critical reading. The technical help of Mr **C. Vasdekis** is also acknowledged. We would like to express our gratitude to all authors for their contributions and patience. Last but not least, we acknowledge the cooperation with Dr **Koen Martens** and the people from Kluwer.

The Editors

CHAPTER I

ARTEMIA MORPHOLOGY AND STRUCTURE

GODELIEVE R.J. CRIEL
Department of Anatomy, Embryology and Histology
University of Ghent
Godshuizenlaan 4
B-9000 Gent
Belgium

THOMAS H. MACRAE
Department of Biology
Dalhousie University
Halifax, NS B3H 4J1
Canada

1. External Form of *Artemia*

1.1. INTRODUCTION

Artemia is a typical primitive arthropod with a segmented body to which are attached broad leaf-like appendages that greatly increase the apparent size of the animal. The total length is usually about 8–10 mm for the adult male and 10–12 mm for the female, but the width of both sexes, including the legs, is about 4 mm. The body is divided into head, thorax, and abdomen. The head consists of one prostomial and five metameric segments, which bear in order, the median and compound eyes and *labrum*, first antennae, second antennae, mandibles, first maxillae or *maxillulae*, and second maxillae or maxillae. The thorax is constructed of eleven segments, each provided with a pair of swimming legs, while the abdomen is composed of eight segments. The anterior two abdominal segments are often referred to as the genital segments and of these, the first bears the gonopods, either the ovisac of the female or the paired penes of the male. Abdominal segments two through seven lack appendages. The eighth or last abdominal segment possesses the cercopods, also called the *furca* or telson (Cassel, 1937).

The entire body is covered with a thin, flexible exoskeleton of chitin to which muscles are attached internally. The exoskeleton is shed periodically and, in females a moult precedes every ovulation, although in the male, a correlation between moulting and reproduction has not been observed. In larval stages, the cuticle is 0.3–1.0 μm thick and is thought to be similar in all regions of the integument (Freeman, 1989). Specifically, the cuticle is divided into

Th. J. Abatzopoulos et al. (eds.), Artemia: *Basic and Applied Biology,* 1–37.
© 2002 *Kluwer Academic Publishers. Printed in the Netherlands.*

an outer epicuticle and an inner fibrous procuticle without a differentiated exo- or endocuticle. A cuticle of variable thickness covers adults (Criel and Walgraeve, 1989) (Figure 1). For example, the cuticle covering the male clasper reaches 7 µm in width, whereas it is only 1–1.5 µm in trunk and thoracopod exopodites. In most regions, the cuticle is composed of a thin, three-layered epicuticle, a laminated pre-exuvial exocuticle and a fibrous post-exuvial endocuticle. However, the post-exuvial endocuticle is lacking on the medially oriented inner surface of the exopodite, and it is also absent from the cuticle covering the gills or metepipodites (Copeland, 1967). Differences between the cuticle of larvae and of adult *Artemia* probably indicate modification of

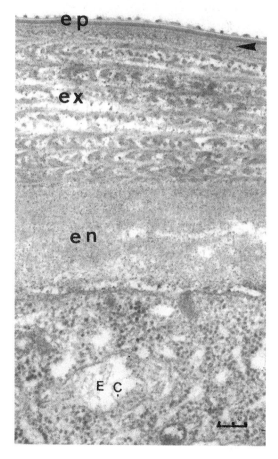

Figure 1. Electron micrograph of the cuticle in the trunk region showing the three-layered epicuticle (ep), the exocuticle (ex), with a thin homogeneous outer layer (arrowhead) and a broad laminated inner layer. (en) the endocuticle; (EC) epidermal cell; bar: 200 nm. (From Criel, 1991)

the mechanism responsible for its synthesis during the larval period (Freeman, 1989). Moreover, the nauplii have a diecdysic type of moult cycle, with a predominant pre-moult stage and a greatly shortened inter-moult period (Freeman, 1989), while in adults, the inter-moult period predominates (Criel and Walgraeve, 1989). Horst (1989) studied ultrastructural aspects of the larval epidermis in relation to cuticle formation, but adult epidermal cells have not been fully investigated in this regard.

The only tegumental glands discernable in *Artemia* are the proximal thoracopodal glands (Dornesco and Steopoe, 1958), although Benesch (1969) reports a similar pair of glands in the abdominal region of the first thoracic segment. Three to four duct cells link the proximal thoracopodal gland cells to the outside, where they open at the base of a protoendite spine (Benesch, 1969). Claus (1886) proposed a sensory function for these glands, each of which consists of one large and two small cells.

1.2. HEAD

The six segments of the head represent the most specialized region of the body, as indicated by the fusion of segments, specialization of the appendages, extensive development of sensory and nervous tissues and formation of complex muscular and skeletal systems. Excretory organs and a part of the digestive tract are located in the head. The prostomial segment is without appendages. The stalks of the compound eye, which contains the many *ommatidia* of the eye, arise laterally from it. The *labrum* forms ventrally between the bases of the compound eyes, serving to hold food in position for mastication and swallowing. The *labrum* is relatively large, elastic and muscular in early developmental stages. Additionally, the median eye and the frontal organ are located in this segment.

The first metameric segment of the head, characterized by paired appendages termed the first antennae or *antennulae*, is fused with the prostomial and second metameric segments. These antennae are cylindrical tubes with very flexible walls, capable of bending in any direction. The central cavity of each antenna is a blood sinus, penetrated by two nerves, at least one of which passes to the tip of the antenna and ends in a mass of ganglion cells. Through the use of scanning electron microscopy (Tyson, 1980; Tyson and Sullivan, 1978, 1979a) found that the antennae of adults and nauplii support two types of innervated *sensillae*. Segment two of the head exhibits a pair of jointed appendages, the second antennae, which differ considerably in males and females. Male antennae, termed claspers, are very large with a joint separating the heavy, rectangular, basal protopodite from the triangular, flattened exopodite. The claspers fit tightly around the female, just anterior to the ovisacs, and on the medial border of the clasper base is a knob covered with sensory spines. The protopodite is thickened medially near its articulation, forming a hinge upon which the exopodite swings, and is very muscular with a generous supply of nerves, while the exopodite is practically without either. The frontal knobs of the male clasper contain two types of processes, the numerous small non-cellular spines and the rarer but larger setae protruding from a broad basal elevation (Figure 2) (Tyson and Sullivan, 1979b, 1980b; Wolfe, 1980). Only the setae are innervated (Wolfe, 1980). The second antennae of the female are similar to those of the male, except they extend anterio-laterally and are much smaller. The distal exopodite of the second antennae may articulate with the protopodite or be entirely fused, depending on the individual. Three sets of sensory spines located on the anterior border of the protopodite correspond to those of the male, and a fine nerve strand leads to a spine at the tip of the exopodite. Muscles are not developed in the second antennae of females, being replaced by a large sinus and fat cells, the latter filling most of the internal space (Cassel, 1937)

Segment three is more sharply defined than the anterior segments just described, due to the bulging of the head on either side that is required for attachment to the strong mandibular muscles. The paired appendages of the segment are represented by the C-shaped mandibles, which lie on either side of the head and meet below. Each mandible possesses peripheral incisors and blunt-crowned teeth used for mechanical processing of filtered food particles. The dentitions of mandible molar regions are mirror images of one another (Tyson and Sullivan, 1981a, b; Schrehardt, 1987), although variations occur in their outer aspects in different *Artemia* species (Mura and del Caldo, 1992). The mandibles move laterally, and in an anterio-posterior direction, to grind

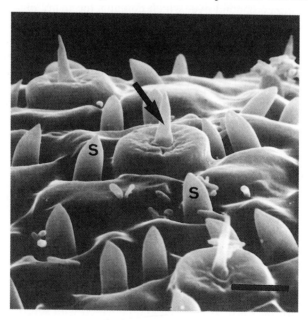

Figure 2. Scanning electron micrograph of male frontal knob showing uninnervated spines (s) arising from depressions in the exoskeleton and sensory setae (arrow) arising from a dome shaped supporting cell, bar: 1 μm. (From Criel, 1991)

the food. The first maxillae, paired appendages on the fourth head segment which extend inward to meet on the medial line, are endowed medially and ventrally with a flattened lobe which arises from the protopodite. Posteriorly, each of these protopodites has a palp provided with 12 to 14 spines in a vertical row. The setose spines are moved by a strong muscle attached to the palp, thus allowing food material to be received from the ventral groove. Segment four is also enlarged dorsally and ventrally to accommodate the maxillary glands. Segment five is almost completely fused with segment four, and its appendages are the paired second maxillae. Each limb consists of a basal portion (protopodite) with an inner blunt, cone-shaped lobe and a lateral tubular lobe, respectively called the gnathobase and the exopodite. The median lobe supports three strong setose spines that project into the ventral groove and are usually covered by the median flaps of the first maxilla. These spines aid the first maxillae in passing food particles from the ventral groove into the mouth. The lateral tubular lobes of the second maxillae do contain the distal part of the maxillary gland ducts. The openings of these ducts are at the extreme tip of each lobe and the rims surrounding the pores bear one or more spines. In the living animal, this lobe always projects laterally so that the pores discharge near the lateral border of the first maxilla.

1.3. THORAX

The thorax consists of eleven segments, each with a pair of flattened leaf-like jointed appendages, the swimming legs or phyllopods, used for locomotion, osmoregulation, respiration and nutrition. Phyllopods of the first and eleventh segments are the smallest, with both size and length increasing toward the middle of the thorax. The tips of the legs, therefore, form an arc on either side of the body. The thoracic limbs define a mid-ventral channel, termed the food groove. Spines and setae on the phyllopods, by extending in the same planes as the endites, considerably increase the functional area of the legs and aid greatly in swimming. The protopodial and endite setae of each limb carry *setulae* about 3 μm in length, arranged on their borders and forming an effective food filtering apparatus with an inter-setular distance of about 500 nm. Cuticular setae are found on all thoracic segments (Tyson and Sullivan, 1980a), but no information is available about their innervation nor on their anatomical connection with the central nervous system.

Swimming and filter-feeding have been examined (Barlow and Sleigh, 1980; Schrehardt, 1987), revealing that *Artemia* larvae use antennae and mandibles as filtering organs, whereas in adults the food groove serves for this function. Setae and *setulae* of the protopodites and endites assume the filter-feeding function in adults. That is, the particle flow passes the setal filter, with suspended bacteria and algae eventually adhering to the *setulae*. The swimming legs of *Artemia* do not all move simultaneously in the same direction. In this context, the phase difference changes such that limbs 11 and 1 beat in phase, although this difference can shorten in faster beating animals, thus synchronizing legs 11 and 2. The largest limbs travel about 120° during the effective stroke, with this metachronal rhythm producing a continuous water flow which draws food into the filter system and creates two propulsive water streams at the rear of the animal. During the recovery stroke, the protopodial and endite setae of one pair of thoracopods comb the setae of the adjacent posterior extremities, collecting the filtered food organisms and concentrating them with their filtrate during the next effective stroke. As a result, particles are transported within the food groove to the mouth. The frontal limit of the food groove is marked by the first and second maxilla, whose feathery setae carry particles through the paragnath channel to the mandibles, the latter covered by the *labrum* (Schrehardt, 1987). The exites do not seem to have a major role in swimming, but probably function as respiratory devices.

1.4. ABDOMEN

The abdomen lies behind the thorax and consists of eight annular segments. The gonopods, either the paired penes of the male or the ovisac of the female, are specialized segmental appendages of the first abdominal segment. The ovisacs are united into a single structure, hanging from abdominal segments one and two, whereas the penes are paired. Abdominal segments two to seven lack appendages. Segment eight is long, possibly representing two fused segments, and it ends posteriorly in a pair of cercopods, also called furcal *rami* or telson. The cercopods are thought by some to be an additional body segment and by others to be the jointed appendages of the last segment. The cercopods possess a variable number of setose spines, a characteristic that may be influenced by environmental factors.

2. Internal Anatomy of *Artemia*

2.1. DIGESTIVE SYSTEM

The alimentary tract of *Artemia* lies freely in the haemocoel and is bathed in haemolymph. Multicellular digestive glands are lacking but the alimentary tract is divided histologically into three regions. These include a short vertical *stomodeum*, followed by a foregut or oesophagus leading up from the mouth and opening at right angles into the mesenteron or midgut, which joins a short *proctodeum*, or hindgut. In front of the entrance to the oesophagus, the midgut expands into a pair of pouches called the gastric *caeca*. At the oesophagus and gastric *caeca* junction, a fine granular matrix is secreted between the cell apices and the cuticle by the oesophagus cells. Also in this region the gastric *caeca* cells contain electron dense granules of unknown function (Schrehardt, 1987).

The histological differences between the regions of the alimentary tract have an embryological foundation, with the oesophagus and hindgut of ectodermal origin (Benesch, 1969), but the midgut of endodermal origin. The oesophagus and hindgut are surrounded by circular and longitudinal muscles and a dilatator muscle, whereas the midgut is surrounded by only a circular muscle layer. The midgut epithelium consists of cylindrical to cubical cells with a conspicuous brush border, and it is separated by a basement membrane from the underlying circular muscle layer. Light microscopic observations indicate that secretion predominates in the anterior midgut and absorption in the posterior part (Frenzel, 1893). The secretion remains to be identified, but examination by light microscopy indicated both holocrine (Frenzel, 1893) and apocrine (Kuenen, 1939) characteristics. Additionally, electron microscopic studies have not resolved this issue. However, fixation artefacts may be the source of some cell differences (Kikuchi, 1972; Schrehardt, 1987). Midgut cells are lined with lòng *microvilli* coated with a fine fibrous material, probably mucopolysaccharides. Apical membranes of these cells are associated with pits and vesicles, and the phagolysosomal cytoplasmic body in every cell suggests the capacity for intracellular digestion (Kikuchi, 1972) (Figure 3).

Under the epithelial cells lies a basement membrane that, in adults, is composed of two layers (Kikuchi, 1972; Schrehardt, 1987). A thick, often

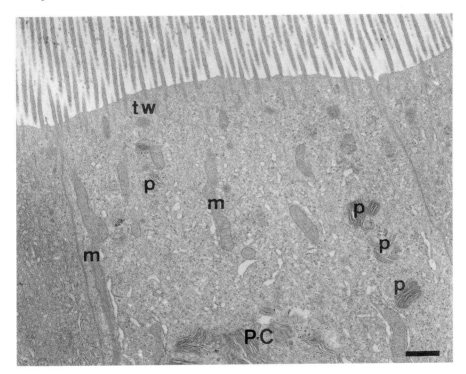

Figure 3. Electron micrograph of the apical region of a midgut epithelial cell. Under the terminal web (tw) long mitochondria (m) are oriented in the microvillar axis, while small phagolysosomes (p) coalesce to a large phagolysosomal cytoplasmic body (PC), bar: 1 µm. (From Criel, 1991)

striated inner layer adjacent to the epithelial cells is densely packed with rough granules that are mixed with amorphous material. An outer layer of the basement membrane facing the haemocoel is thin, loose and finely granular. It does not penetrate into mitochondria-associated basal infoldings of the cell membranes found in the posterior part of the midgut (Figure 4).

The internal faecal pellets of *Artemia* are surrounded by a peritrophic membrane, as occurs in insects (Reeve, 1963). Hansen and Peters (1998) determined that the peritrophic membranes of metanauplii are fully permeable for latex beads with a diameter of 70 nm, less permeable for those of 130 nm and completely impermeable for beads with a diameter of 327 nm. Adults have nearly the same effective pore size of the peritrophic membranes as the metanauplii.

The naupliar and adult midguts are similar (Hootman and Conte, 1974). A glycocalyx layer, seen on the *microvilli* in other regions of the gut, seems to be absent from the naupliar midgut, but a peritrophic membrane possibly arising from numerous clear vesicles found between the *microvilli* is present. These vesicles might, however, represent digestive enzymes. No phagocy-

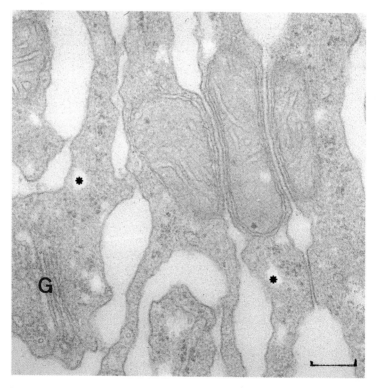

Figure 4. Electron micrograph of the basal region of an epithelial cell in the distal part of the midgut showing the special association between mitochondria and infoldings of the plasma membrane. (G) Golgi apparatus; (*) clear spaces due to elution of glycogen during fixation; bar: 200 nm. (From Criel, 1991)

tolysosomes are found in the nauplii gut, but basal infoldings associated with mitochondria are present, and the naupliar gut is rich in Na^+K^+-activated ATPase, suggesting an active Na^+ exchange. There is only one ultrastructural study of the oesophagus and hindgut in nauplii (Schrehardt, 1987), and the many tendon cells involved in attachment to dilatator muscles (Cassel, 1937) are not mentioned in this work. Both epithelial layers were found to be cuboidal rather than cylindrical, and they are covered by a thin cuticle (Schrehardt, 1987). The midgut cells store small amounts of glycogen and lipid. Many myelin figures are found in the nauplius hindgut, and the basement membrane is a monolayer.

2.2. CIRCULATORY SYSTEM

The fundamental plan for blood circulation in *Artemia* was shown as early as 1840 by Joly (1840), to be an open lacunar system, as it is characteristic of crustaceans. The heart is a simple longitudinal tube suspended dorsally to the alimentary tract in the body cavity. The anterior end of the heart opens at the base of the antennae, and at the posterior end is a slit called the caudal *ostium*. Segmentally arranged *ostia* open over the entire length along both sides of the heart. Other descriptions of the circulatory system were published: Leydig (1851), Claus (1886), Greene (1924), Cassel (1937), Vehstedt (1940), and Benesch (1969). Of these, the most comprehensive is by Vehstedt (1940), who demonstrated that the pericardial septum and its side openings channel the blood in a steady stream in the thorax and abdomen. On the other hand, connective tissue fibres, muscles and nerves fulfil the same function in the head region. There is some uncertainty surrounding the cranial and caudal openings of the heart, the number of *ostia* and the cranial extent of the pericardial septum. However, ultrastructural analysis indicated that the heart wall consists of a single layered myocardium, arranged as a tube posteriorly (Økland *et al.* 1982). In contrast, the anterior portion of the heart is open dorsally, forming a trough with upper edges attached firmly to the basement membrane of the epidermis.

The blood cell forming organs of *Artemia* were described extensively by Cassel (1937) and by Lochhead and Lochhead (1941), who proposed that cell groups in thin walled pockets on both sides of the leg muscles produced the haemocytes. The blood cell forming organs are small, variably sized nodules surrounded by a thin discontinuous membrane and found at the base of each trunk limb. The nodules consist of independent, rounded cells, differing from each other in size and mitotic stage, although resting cells within nodules exhibit uniform nuclear and cytoplasmic content. Toward the periphery of nodules, the cytoplasmic content of haemocytes increases, and there is a decrease in the nucleus. The circulating blood cells have an even more extensive cytoplasm and usually contain numerous inclusions. Intermediate stages, between fully developed haemocytes and undeveloped nodular cells are found in spaces adjacent to the blood forming organs, but unlike the situation in

Malacostraca, no stroma cells are present. According to Benesch (1969), the first blood cells can be distinguished posteriorly and dorsally from the first antennal mesoderm in nauplius stage Na-5. At this stage, the primary haemocytes, as they are called, look like fat storage cells and contain yolk droplets. They appear before the mandibular and thoracal blood organs supply the secondary and tertiary haemocytes, respectively. Upon emergence of the prenauplius from the cyst, however, the primary blood cells are round and free of yolk.

Blood cells (haemocytes) were described by Claus (1886), Cassel (1937), and Lochhead and Lochhead, (1941) and more recently by Martin *et al.* (1999) and Day *et al.* (2000). The circulating blood cells of *Artemia* are amoeboid, containing a relatively small nucleus and cytoplasm filled with reflecting granules. Morphologically, the *Artemia* haemocytes are similar to the granulocytes found in most crustaceans, although their granules are generally larger than those in other species (Martin *et al.* 1999). All blood cells are of the same type, but vary considerably in size and shape (Martin *et al.* 1999; Day *et al.* 2000). Electron micrographs sometimes show the granules as an electron dense crescent deposited against a surrounding membrane (Figure 5), but this may be a fixation artefact (Martin *et al.* 1999). Although several cyto-

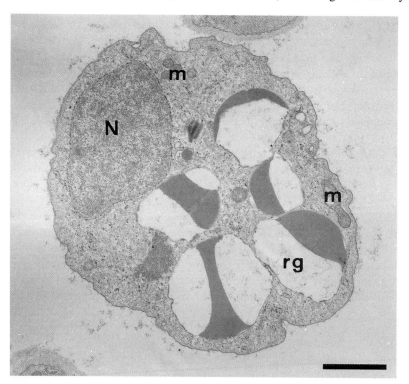

Figure 5. Electron micrograph of a blood cell. (N) nucleus; (m) mitochondria; (rg) refringent granules; bar: 2 μm.

chemical techniques were used, the chemical composition of the refringent granules was initially undetermined (Lochhead and Lochhead, 1941). However, it has since been shown that the granules contain acid phosphatase and that they react with L-DOPA, suggesting an involvement in degradation of ingested material (Martin *et al.* 1999). The phenoloxidase system is also present in blood cells from *Artemia*. Release of granules has been induced and the haemocytes demonstrated to phagocytize bacteria (Martin *et al.* 1999), however, this capability needs more extensive study. The haemocytes have a major function in clotting and wound healing (Lochhead and Lochhead, 1941; Steopoe and Dornesco, 1945; Martin *et al.* 1999) and in females they may play a role in vitellogenesis (Lochhead and Lochhead, 1941). Immunofluorescent staining revealed, for the first time in crustaceans, the distribution of post-translationally modified tubulins in *Artemia* haemocytes (Day *et al.* 2000). Acetylated tubulin was found in only a subset of haemocyte microtubules, perhaps indicating those microtubules are stable. Staining for detyrosinated tubulin yielded a ring of fluorescence associated with the nuclear membrane, a very unusual observation. Morphological change of *Artemia* haemocytes entailed a major spatial rearrangment of microtubules, and this may reflect the types of changes that occur during phagocytosis by these cells.

2.3. EXCRETORY SYSTEM

Excretion in adults is probably performed by the maxillary glands, while the antennal glands, occurring only as rudiments in adults, operate in early developmental stages (Warren, 1938; Lochhead, 1950). Both glands have a similar structure (Claus, 1886; Cassel, 1937; Warren, 1938; Lochhead, 1950). The larval antennal glands are located laterally on the head, on either side of the digestive tract where the second antenna arises, opening on the lateral side of the posterior portion of the protopodite. The antennal gland is present by the time of hatching, but continues to grow and reach its greatest development in the 6th instar. Thereafter, the antennal gland degenerates, and only remnants are found in 9th and 10th instars. The maxillary gland develops in the nauplius and its formation is complete in the 6th instar. Thus, during the 6th and 7th instar, both excretory glands are fully developed (Warren, 1938) and presumably functional. Gross morphology and fine structure of the maxillary glands are known in some detail (Tyson, 1966, 1968, 1969a, b). The maxillary glands are surrounded by haemocoelic spaces and protrude from the sides of the body, just behind the mandibles. Each gland consists of a blind, centrally located end sac, around which an efferent duct makes three loops before continuing as a terminal duct and opening via a small aperture on the second maxilla (Figure 6). The epithelium of the end sac has a striking resemblance to the podocytes of Bowman's capsule of the vertebrate nephron; that is, the cells possess regularly arranged foot processes and junctional elements similar to 'filtration slit membranes'. On the basis of these ultrastructural findings, it is postulated that formation of primary urine involves ultrafiltra-

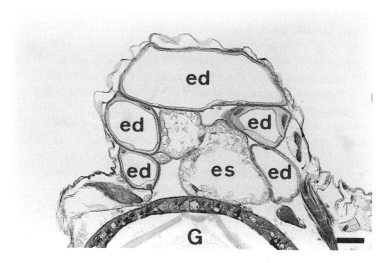

Figure 6. Light micrograph of a section through the maxillary gland. The centrally located end sac (es) is surrounded by coils of the efferent duct (ed). (G) gut; bar: 50 μm. (From Criel, 1991)

tion. The efferent duct of the maxillary gland is divided into two morphologically distinct regions, an efferent tubule and a terminal duct (excretory duct) (Warren, 1938). In the efferent tubule, an apical microvillous border and associations of the basal membrane with mitochondria suggest a role in modifying the lumenal contents. The terminal duct is lined with a cuticle and probably functions as a conduit for the final urine.

2.4. NERVOUS SYSTEM

The typical primitive Branchiopod-type (Hanström, 1928) nervous system of *Artemia*, fully described but scarcely illustrated (Leydig, 1851; Claus, 1886; Spencer, 1902; Warren, 1930; Cassel, 1937; Benesch, 1969), consists of a dorsal brain, or supra-oesophageal ganglion, circum-oesophageal connectives and a double ventral row of segmental paired ganglia. The latter are united longitudinally by connectives and laterally by a large anterior and a smaller posterior commissure. Two types of neurons, the larger 'ganglia cells' characterized by a well-developed cytoplasm and a nucleus with little chromatin, and the smaller '*globuli* cells' have been reported (Benesch, 1969). The latter were described by Hanström (1928) as unipolar associative neurons, possessing little cytoplasm and a relatively large nucleus full of chromatin.

The *protocerebrum* of the nervous system receives nerves from both the complex and nauplius eyes and the frontal organs. The *deuterocerebrum*, clearly separated from the *protocerebrum*, is the region where the right and left ganglia of the first antenna, linked by a commissure, are found (Hanström,

1928). The course of the antennal nerve has been described (Cassel, 1937), and it contains sensory and motor neurons (Benesch, 1969).

The circum-oesophageal connectives extend tailward from the posterior surfaces of the brain, on either side of the gut, connecting the brain to the ventral nerve cord. The cell bodies which constitute the *tritocerebrum* are found on these connectives. Close to the brain, the second antenna ganglion (Warren, 1930) innervates sensory receptors in both males and females (Cassel, 1937). More caudally, the post-oesophageal ganglia on the connectives have been described (Warren, 1930; Cassel, 1937) although others do not mention them (Claus, 1886; Hanström, 1928; Benesch, 1969). The post-oesophageal ganglia consist of a large dorsal lobe and a smaller ventral mass (Cassel, 1937), with both right and left lobes linked by a commissure. The dorsal lobe gives off connectives to the mandibular ganglia. The ventral commissure is the origin of the stomatogastric visceral nervous system. A large ventral branch, which meets its homologue to form the circum-oral ring arises at the ventral commissure, passing ventrally and laterally from the oesophagus into the *labrum*, where it connects to the labral ganglion (Warren, 1930) or oesophageal ganglion (Benesch, 1969). Here, three nerves arise (Cassel, 1937; Benesch, 1969) one connecting to the anterior part of the oesophagus, and innervating the circular muscles of the oesophagus and probably its dilatator muscles. The other two are labral nerves, responsible for innervating the muscle at the tip of the *labrum* and supplying the sensory areas of the mouth.

The ventral nerve cord has been extensively described (Warren, 1930; Cassel, 1937; Benesch, 1969), and all innervated muscles named (Benesch, 1969). The mandibular ganglia are linked to the dorsal lobe of the post-oesophageal ganglia (Benesch, 1969). A large anterior nerve leading to the mandibular muscles (Warren, 1930; Cassel, 1937) contains two motor and one sensory branch and may be comparable to the nerves of the thoracic segments (Benesch, 1969). A smaller posterior motor nerve leads to the longitudinal body muscles (Tyson, 1980). In the metanaupliar region, a dorsal nerve arises from each ganglion, and innervates the muscles of the adjacent segment. The total number of nerves varies up to the first thoracic segment. The initial pair of metanaupliar ganglia are the first maxillary ganglia. These ganglia deliver a large branched nerve to the muscles of the metanaupliar region, a smaller nerve to the muscles of the first maxilla, and possibly a nerve to the maxillary glands (Warren, 1930). A maxillulary ganglion or its rudiments have been described (Warren, 1930; Cassel, 1937), with Benesch (1969) finding two nerves arising from the maxillulary ganglion, a ventral mixed and a dorsal motor type.

The eleven pairs of thoracic ganglia, each linked by a larger anterior and a small posterior commissure, follow the mandibular ganglia. The nerves arising from their ganglia are similar in structure along the thoracic segments (Leydig, 1851; Benesch, 1969), with three nerves generated from each thoracic ganglion, including a cranial motor, medial sensory and caudal motor nerve (Leydig, 1851). The elaborate structure of the thoracal peripheral nerves may be required to coordinate swimming movements, which involves the motor and sensory components in these nerves (Kane, 1975). The first and second pairs of genital ganglia are almost fused into a pair of large nervous tissue masses. The innervation of the male genital tract has been described (Cassel, 1937). No ganglia are found in the abdomen, but from the posterior end of the genital ganglia, one (Warren, 1930) or several (Cassel, 1937) nerves pass posteriorly. One of these nerves gives off a branch in each segment, which innervates the longitudinal body muscle of the following segment. Nerves in this region of the animal also contact the heart, midgut and muscles of the hindgut.

2.5. SENSORY SYSTEM

2.5.1. Compound Eyes

Artemia has two widely separated, compound eyes mounted on unjointed but flexible stalks (Figure 7). These apposition-type eyes have a distal layer of about 300 *ommatidia*, separated from the optic lobe by a basement membrane and linked to the lateral protocerebral lobe by an optic peduncle (Hanström, 1931; Horridge, 1965a, b). An unspecialised cuticle covers the compound eye. Earlier work on ommatidial structure (Leydig, 1851; Claus, 1886; Nowikoff, 1905; Warren, 1930; Cassel, 1937) was extended by the comprehensive study of Debaisieux (1944), and by electron microscopic approaches (Elofsson and Odselius, 1975; Hertel, 1980; Nilsson and Odselius, 1981). The *ommatidia* (Figure 8) are built from crystalline cone and *retinula* cells,

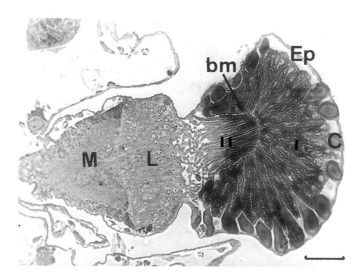

Figure 7. Light micrograph of a 2 µm section through an epon-embedded eyestalk showing the eye and the optic lobe. (bm) basement membrane; (C) cristal cone; (Ep) epidermis and cuticle; (I) zone of the cone cell roots and the retinula cells; (II) zone of the photoreceptor axons; (L) *lamina ganglionaris*; (M) *medulla* of the optic lobe; bar: 50 µm. (From Criel, 1991)

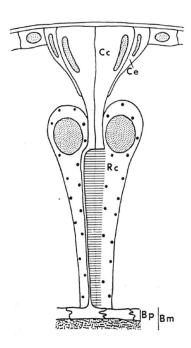

Figure 8. Schematic drawing of an *ommatidium* in the compound eye of *Artemia*. (Bm) basement membrane; (Bp) basal plate; (Cc) cristalline cone; (Ce) juxta-cristalline epidermal cell; (Rc) *retinula* cell. (From Elofsson, R., and Odselius, 1975)

with a pronounced regional difference in the morphology of the crystalline cones important for light/dark adaptation (Nilsson and Odselius, 1981). The top of each cone is enveloped by two juxta-crystalline epidermal cells, surrounded by a clear intercellular space often containing blood cells. The crystalline cone is composed of four tightly apposed cells, although sometimes three, five or six occur (Debaisieux, 1944). The cone cells consist of a main body, a crystalline tract abutting the rhabdom, and a 'cone cell root' extending between the *retinula* cells to the basement membrane. Suspended within the crystalline tract of each cone cell is a lens-like structure, the highly refractive glycogen body (Elofson and Odselius, 1975). Each *ommatidium* also contains six *retinula* cells, which contribute to a fused rhabdom. The retinula cells constitute the only pigmented cells of the eye and pigment granules are found over the entire *retinula* cell body. Axons penetrate the basement membrane of the compound eye, becoming thinner and losing pigment during this process. Below the basement membrane, axons from each *ommatidium* unite in bundles of six (Nässel *et al.* 1978). With dark adaptation, the crystalline tract shortens, accompanied by a change in focal length of the glycogen body (Debaisieux, 1944). The rhabdom simultaneously lengthens, while its distal part becomes thicker, but there is no adaptation movement of pigment in the *retinula* cells. Two neuropiles are found in the eyestalk within the optic lobe (Hanström, 1928; Horridge, 1965b; Elofsson and Dahl 1970; Elofsson and Klemm, 1972; Amarant and Elofsson, 1976; Nässel *et al.* 1978), these being the distal *lamina ganglionaris* and the proximal medulla. As in other Branchiopoda, no chiasma exists between these neuropiles (Horridge, 1965b; Elofsson and Dahl, 1970).

2.5.2. Median Eye (Nauplius Eye) and Ventral Frontal Organs

Early morphological work on these structures (Claus, 1891; Nowikoff, 1906; Moroff, 1912; Warren, 1930; Debaisieux, 1944; Vaissière, 1956, 1961; Elofsson, 1966) has been extended by more recent studies (Rasmussen, 1971; Anadon and Anadon, 1980).

The median eye, situated anteriorly in the centre of the head is the only functional optical sense organ in the nauplius. Covered by an unmodified cuticle and epidermis, the median eye persists throughout the various instars, enlarging slightly but maintaining its general structure until the adult stage (Warren, 1930). Posteriorly and dorsally, the median eye is separated from the gut diverticles by a basement membrane. Ventrally, the eye resides close to the *protocerebrum* (Rasmussen, 1971), which may extend in front of the eye (Anadon and Anadon, 1980). The median eye is pear-shaped and surrounded by a wide haemocoel, except where it is adjacent to the epidermis, the cavity receptor organ (described later) and the cerebrum. The thinnest part of the eye is pressed against the epidermis, and its thickest part faces the brain. In larvae, the eye may even protrude into the brain (Moroff, 1912). The most obvious feature of the median eye are two pigment cells arranged as an inverted Y, establishing two latero-dorsal and one medial-ventral cup containing the sensory *retinula* cells (Claus, 1891) (Figure 9). The lateral cups are deep and, depending on the specimen, contain from 25 to 75 cells each; the same range as occurs in the ventral cup (Elofsson, 1966). Anadon and Anadon (1980) obtained a better insight into the three-dimensional structure of the median eye by examination of serial ultrathin sections, and concluded that organization of the median eye does not permit visualization of form. Also, the three groups of *retinula* cells were shown to have different orientations and to receive luminous rays proceeding from different points, though some rays may reach more than one *ocellus*.

The sensory cell axons of each cup form a nerve running to the nauplius eye centre (Elofsson, 1966; Anadon and Anadon, 1980) that is clearly separated from the rest of the brain.

Closely related to the median eye are the ventral frontal organs, thought to have a photosensory function (Hanström, 1931; Elofsson, 1966; Rasmussen, 1971; Anadon and Anadon, 1980). *Retinula* cells of the ventral frontal organs are similar to those of the median eye, but no pigment cells are found, nor do they contain pigment granules (Rasmussen, 1971). Moreover, although there are similarities, the ventral frontal organs of *Artemia* are not homologous of Malacostraca frontal organs and, together with three naupliar *ocelli* they may represent five ancestral frontal eyes (Elofsson, 1966).

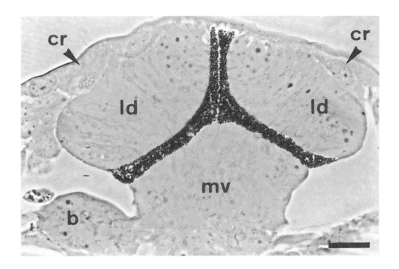

Figure 9. Light micrograph of a 2-μm thick epon-section through the nauplius eye and cavity receptor organ. Two pigment cells forming an inverted Y separate the two latero-dorsal (ld) and the medial ventral (mv) cup of sensory cells. (b) brain; (cr) cavity receptor organ; bar: 20 μm. (From Criel, 1991)

2.5.3. Dorsal Frontal Organs or Cavity Receptor Organs

The dorsal frontal organs, earlier termed the X-organ (Elofsson, 1966), terminate in the epidermis laterally, and slightly dorsally, to the lateral cups of the median eye (Claus, 1886; Spencer, 1902; Nowikoff, 1906; Warren, 1930; Hanström, 1931; Elofsson, 1966; Rasmussen, 1971). As shown by electron microscopic study, the dorsal frontal organs consist of neurons, which protrude into a cavity beneath the cuticle (Elofsson and Lake, 1971). Dendrites of these neurons penetrate a giant accompanying cell, as well as epidermal cells, before entering the cavity. In contrast, axons run on both sides of the nauplius eye, straight to the *medulla terminalis*, but they do not enter the nauplius eye centre. Previously, the cavity receptor organ was considered neurosecretory, although it is now regarded as sensory, but a double function cannot be excluded. Monoaminergic bodies were found in the cavity receptor organ (Amarant and Elofsson, 1976), providing the first example of monoaminergic sensory cells in Arthropods, if indeed these cells are sensory.

2.5.4. Other Sense Organs

Sense organs of the cuticle are discussed in the section dealing with external form. These sense organs have not yet been described in detail, although they may be associated with subcuticular ganglion cells (Claus, 1886). Mandibular and maxillar sense organs have also been mentioned previously by Warren (1930), but they have not been characterized.

2.6. ENDOCRINE SYSTEM

Investigations of the neurosecretory system in *Artemia* have yielded conflicting results, perhaps due to the diversity in histological techniques and the lack of physiological data available on the animals under study (van den Bosch de Aguilar, 1979). However, it is known that the primitive neurosecretory system lacks a clearly defined storage organ. The fact that the presumed neurohaemal organ of the eyestalk can be confused with the zone of *ommatidia* proliferation and growth (Debaisieux, 1944, 1952; Elofsson and Dahl, 1970; van den Bosch de Aguilar, 1979) is confirmed by an electron microscopic study wherein neurosecretory elements were not reported (Nässel *et al.* 1978). Thirty-five to forty putative neurosecretory cells were observed in the anterior part of the brain, and three cells on each ventro-posterior side (Lochhead and Resner, 1958). No neurosecretory structures were found in the eyestalks, nor did eyestalk removal cause physiological disturbance.

Neurosecretory cells were classified into three groups based on their secretions (Baid and Ramaswami, 1965), and the neurosecretory systems of *Chirocephalus* and *Artemia* were compared by Hentschel (1963, 1965). Fewer neurosecretory cells are seen in *Artemia*, and they are difficult to trace due to their small size. Furthermore, several neurosecretory cells were found in the *proto-*, *deutero-*, and *trito-cerebrum*, and on the periphery of each ganglion of the ventral nerve cord (Hentschel, 1965), whereas van den Bosch de Aguilar (1979) reported the presence of only four protocerebral neurosecretory cells. He writes ". . . before labelling a structure as an endocrine organ, it is essential to define its function in the physiology of the animal." Van den Bosch de Aguilar (1976) did this and showed the two median protocerebral neurosecretory cells to be stimulated in hypotonic media, and an osmoregulatory function was suggested because they are related to the cavity receptor organ (Elofsson and Lake, 1971). Activity of the two lateral protocerebral neurosecretory cells was maximal during reproduction. Hentschel (1963, 1965) did not, however, describe the function of neurosecretory cells found separately, or in small groups, in the *deutero-* and *trito-cerebrum*. These cells are often associated with the roots of the *antennula* and antenna nerves, and in the ventral nerve cord. Demonstration of neurosecretory cells by use of an immunocytochemical technique to trace neuropeptide-like material has been attempted (Van Beek, 1988). FMRF-amide and neuropeptide Y occurred in a few median protocerebral cells and in two cell bodies of the sub-oesophageal ganglion. Some small cells in the *protocerebrum* stained with neuropeptide Y antiserum, while leuk- and met-enkephalins were localized in the same median protocerebral cells. Additionally, some protocerebral cells were immunoreactive for only one of these enkephalins.

2.7. FEMALE REPRODUCTIVE SYSTEM

The female reproductive system consists of ovaries and oviducts leading into a single, median ovisac wherein several clusters of shell glands open (Figure 10). The ovaries are paired tubular structures, which extend from the 11th thoracic segment, pass through the genital segments, and end in the abdominal segments (Cassel, 1937). Adult female *Artemia* ovulate approximately every 140 hours, depending on rearing conditions (also, strain specific) and whether development of embryos occurs oviparously or ovoviviparously. An accurate description of the sequence of macroscopical events in oogenesis is given by Metalli and Ballardin (1970). The ovaries are translucent during previtellogenesis and show cyclic changes, reflecting the events of vitellogenesis. As vitellogenesis progresses, small opaque dots initially scattered in the ovary increase in number and enlarge. The ovary gradually acquires a beaded appearance and develops into a knotty rod. Endocrine control of oogenesis, through the action of ecdysteroids or moulting hormones has been examined (Adiyodi, 1985), and a relationship has been shown between ecdysteroid levels and vitellogenesis (Van Beek *et al.* 1987, Walgraeve *et al.* 1988).

Like most malacostracans and entomostracans, *Artemia* moult after puberty, and in females, spawning is followed by a moult where after animals ovulate (Bowen, 1962). The major morphological transformations of oogenesis occur during inter-moult. These include the differentiation of oogonia into either primary oocytes, which undergo previtellogenesis and vitellogenesis, or nurse

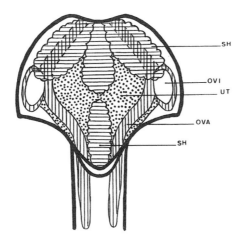

Figure 10. Schematic drawing of a ventral view of the female reproductive system. The uterus (UT) whose volume is determined by the presence or absence of developing cycts or nauplii is a pouch within the ovisac. Two antero-lateral and a lobed medial group of shell glands (SH) open into it. The lateral ovaries (OVA) stretch more caudally along the ovisac, and open into the oviducts (OVI) which are found in the lateral horns of the ovisac. The oviducts which temporarily store the eggs between ovulation and fertilization in lateral pouches are separated from the ovisac by a shutter. (From Criel, 1991)

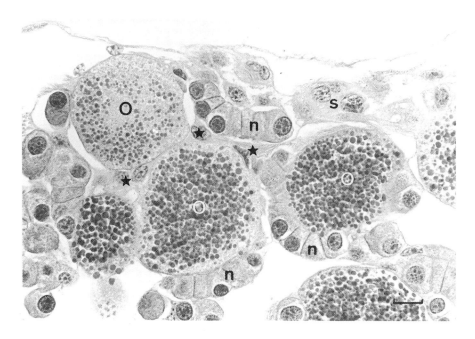

Figure 11. Light micrograph of the ovary during vitellogenesis. The maturing oocytes (O) are surrounded by nurse cells (n). Somatic cells separate the groups of oocytes and nurse cells (*) and form a lateral more or less continuous layer (s); bar: 50 μm. (From Criel, 1991)

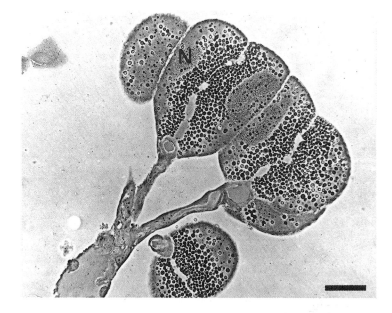

Figure 12. Light micrograph of a 2-μm epon-section of a brown shell gland. Note the cytoplasm filled with homogeneous dark granules. (N) nucleus; bar: 25 μm. (From Criel, 1991)

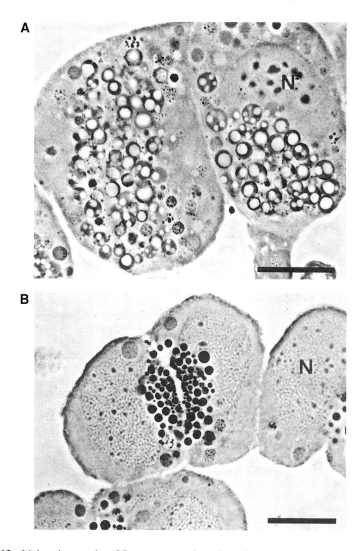

Figure 13. Light micrographs of 2-μm epon-sections through transparent or slightly coloured shell glands. **(a)** cells of a gland presumed to provide material for the nauplius developing *in utero*; note the cytoplasm filled with large granules of irregular density; (N) nucleus; bar: 25μ; **(b)** cells of a gland presumed to provide non- diapausing cysts with a thin shell; note the cytoplasm only partly filled with homogene dense granules, while some heterogeneous larger granules are also seen in the periphery; (N) nucleus; bar: 25 μm (From Criel, 1991)

cells, which are eventually phagocytosed by somatic cells (Figure 11). Formation of the oocyte-nurse cell complex is one of the most intriguing processes of oogenesis, and the role of the nurse cells in *Artemia* is not settled, although in insects they transport compounds to the oocyte (de Cuevas and Spradling, 1998). The last phases of vitellogenesis and spawning occur in the pre-moult period.

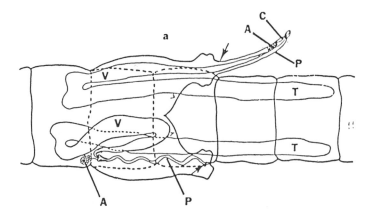

Figure 14. Schematic drawing of ventral view of the male reproductive system, showing an everted penis on one side (a, arrow) while the other penis is in normal retracted condition. (A) accessory gland; (C) collecting duct; (P) eversible penis; (T) testis; (V) *vas deferens* (Modified from Wolfe, 1971)

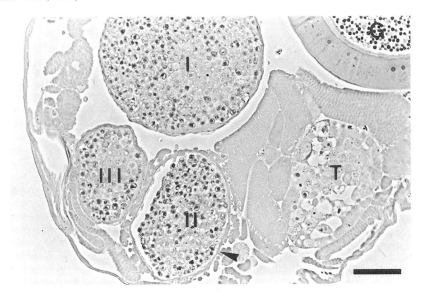

Figure 15. Light micrograph of a semi-thin transverse epon-section through the male genital system. (G) gut; (I, II, III) section through the proximal, medial, and distal part of the *vas deferens* respectively, all filled with sperm; (T) testis; (arrowhead) well developed muscle layer of the intermediate part of the *vas deferens*; bar: 50 μm. (From Criel, 1991)

The oviducts emerge from the ovaries near the anterior part of the third abdominal segment (Cassel, 1937). Each oviduct leads outward, and forward through the lateral areas of the ovisac, emptying into the antero-lateral border of the ovisac. The secretory epithelium of the oviducts is surrounded by longitudinal and circular muscle fibres. The expanded oviducts or lateral pouches

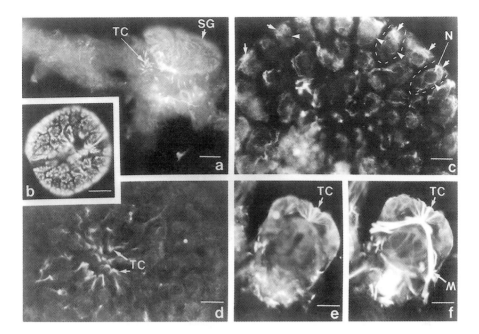

Figure 16. Immunofluorescent staining of the *Artemia* larval salt gland. **a:** A low-magnification micrograph showing a salt gland (SG) attached to a larva. **b:** A detached salt gland seen from the top with the microscope focussed on the apical portion of the cells. **c:** Lateral view of peripheral salt gland cells revealing the enrichment of tubulin in the apical region of the cells (arrows) and the decreased staining of the midcell (arrowheads) and basal regions. Nonfluorescent regions demarcating the lateral borders of cells are indicated by the dashed lines in c. **d:** A view of the lightly staining basal region of the salt gland with associated tendon cells (TC). The primary antibodies were 6-11B-1 (a, c), and YL1/2 (b, d). **e, f:** The same salt gland at the same level of focus as revealed by TAT, a general anti α-tubulin antibody (e), and phalloidin, a stain for filamentous actin (f). (M) muscle; (N) nucleus. The bar represents 50 µm in b, e, f, 25 µm in a, and 20 µm in c, d. (From MacRae *et al.* 1991)

are separated by a shutter from the median and frontal parts of the ovisac (Criel, 1980a). During the time between copulation and fertilization the median part of the ovisac functions as a seminal receptacle, wherein stored sperm mature and either acquire peripheral arms or disintegrate. The secretory cycle of the oviduct epithelium was studied (Criel, 1980b).

The shell glands, which vary from dark brown (Figure 12) to white or even colourless as reproductive cycles differ, consist of several cell clusters. Whether shell glands have a role in determination of ovoviviparity and oviparity is not yet elucidated, although the brown shell glands have been studied extensively at light and electron microscopic levels (Anderson *et al.* 1970; Fautrez and Fautrez-Firlefyn, 1971; De Maeyer-Criel, 1978). Additionally, two types of white shell glands have been distinguished on the basis of secretory granule appearance (De Maeyer-Criel, 1978; Criel, 1980a) (Figure 13). One type is thought to occur when swimming nauplii are released

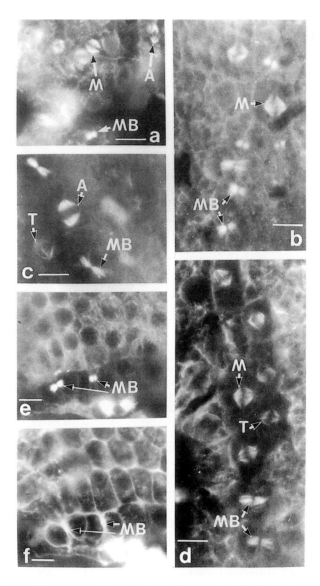

Figure 17. Immunofluorescent staining of mitotic figures and midbodies in developing *Artemia* larvae. The primary antibodies were **a**, DM1A; **b**, KMX; **c**, YL1/2; **d**, 6-11B-1. **e, f:** The same epidermal region reacted with 6-11B-1 (e) and phalloidin (f). Clearly shown are metaphase (M), anaphase (A), and telophase (T) spindles along with midbodies (MB). The bar represents

from females, while the other is postulated to form the thin shell of embryos that hatch shortly after oviposition.

2.8. MALE REPRODUCTIVE SYSTEM

The male reproductive system, consisting of paired testes, *vasa deferentia*, accessory glands and penes (Figures 14 and 15), extends from the first genital

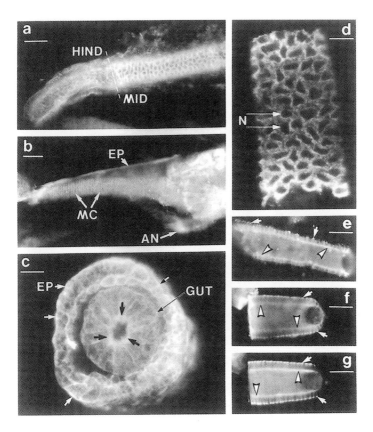

Figure 18. Immunofluorescent staining of *Artemia* endodermal and mesodermal cells. **a, b:** Low-magnification micrographs showing, as viewed from above but at different focal planes, the midgut (MID) and the hindgut (HIND) surrounded by simple muscle cells (MC). The organisms were stained with 6-11B-1 (a) and anti-Glu (b). **c:** Cross-section of the thorax, stained with TAT, showing the epidermal cell layer and the gut cells. The mesenchymal cell layer is sandwiched between the epidermis and gut (GUT). The unlabeled arrows indicate apical cell regions enriched in tubulin. **d:** Apical region of gut epithelial cells, stained with TU01, and showing rings of fluorescence surrounding unstained nuclei (N). **e–g:** Gut viewed from the top with focus about midway though the tubes. The primary antibodies were YL1/2 (e) and DM1A (f): phalloidin (g). Staining in the apical region of gut epithelial cells is indicated by arrowheads while stained muscle cells encircling the gut are indicated by arrows. AN, antenna; EP, epidermal cell layer. The bar represents 50 μm in e–g, 25 μm in a, b, and 20 μm in c, d. (From MacRae *et al.* 1991)

segment to the second, third or fourth abdominal segment (Cassel, 1937; Wolfe, 1971). Although the testis is described as a tube, its lumen is not straight nor is it continuous. The testis is surrounded by a thin basement membrane and contains both somatic and germ cells. The supporting somatic cells are located immediately beneath the testicular membrane, and they form a continuous sheet along the entire testes length. Although the function of supporting cells has not yet been investigated, similarity with vertebrate Sertoli cells is striking. Spermatogonia and early primary spermatocytes lie near the periphery

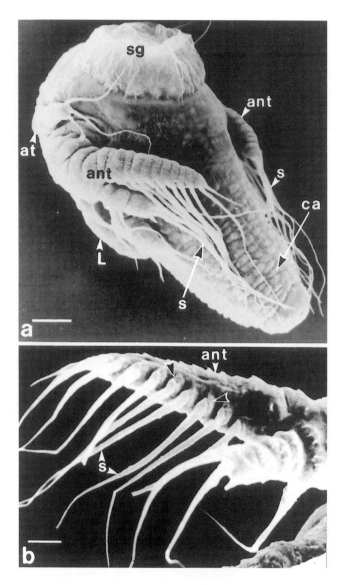

Figure 19. Scanning electron micrographs of antennae and setae of *Artemia* instar-I nauplius (a). (b) An enlarged view of an antenna, showing the region (unlabelled arrowheads) where setae are attached. *ant*, antenna; *at*, antennulla; *ca*, cell array; *L*, labrum; *s*, setae; *sg*, salt gland; bars: 30 μm in (a), 20 μm in (b). (Adapted from MacRae and Freeman, 1995)

of the testis, whereas cells in other developmental stages occur in clusters. Within a cluster all germ cells are at approximately the same stage of differentiation. The caudal end of the testis contains mostly clusters of cells in different meiotic stages, with each cluster reaching about 50 μm in length. Groups of secondary spermatocytes and spermatids prevail in more cephad

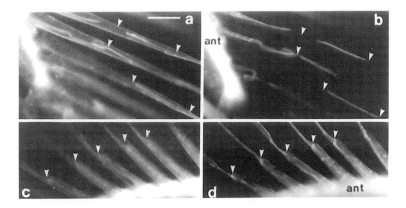

Figure 20. Co-localization of microtubules and microfilaments in setal cells. *Artemia* larvae were fixed with paraformaldehyde and fluorescently stained with either DM1A (**a, c**) for tubulin or rhodamine-phalloidin (**b, d**) for microfilaments. The unlabelled white arrowheads demarcate co-localized tubulin and actin. *ant*, antenna; bar: 20 μm and the magnification is the same in all figures. (Adapted from MacRae and Freeman, 1995)

areas, and these clusters of cells may extend over 100 μm in length. The relative extent of both clusters differs between animals and between the right and left testis of the same animal, but the variations do not seem to be related to copulation or moult phase. Few mature spermatozoa are housed in the testicular lumen, but sperm presumed to be mature occur in the *vas deferens*. The mature sperm are ovoid or spherical and lacking typical characteristics such as flagella, acrosomes and condensed chromatin in the nuclei (Brown, 1966, 1970; Wingstrand, 1978). The *vasa deferentia* join the testes at the ventral surface, near the anterior end, form a loop and terminate at the top of both penes in a somewhat distended distal section that is called the seminal vesicle (Cassel, 1937). Each *vas deferens* leads into a penis. The histology of the penis is similar throughout its entire length, consisting of squamous epithelial cells surrounded by a thin layer of longitudinal muscle and a thick layer of circular muscle. The epithelial cells contain a neutral polysaccharide that is secreted into the lumen and constitutes the major portion of the seminal fluid. Sperm cells are stored along the entire length of the *vas deferens*, wherein rhythmic muscle contractions keep the sperm well mixed.

Artemia has a double penis, which consists of a non-eversible and an eversible portion. The non-eversible section is single but bifurcated, with a cuticle similar to that found on the general body surface covering its epithelium. The eversible part consists of a tortuous tube connecting the *vas deferens* to the outside. In the retracted state, the eversible portion of the penis is lined with a cuticle and an epithelium surrounded by several bands of muscles. The lining of the lumen becomes the outer covering of the penis in the everted state, and the *vas deferens* is the distal part of the reproductive tract. In some species three to five spines are located near the junction of the eversible and

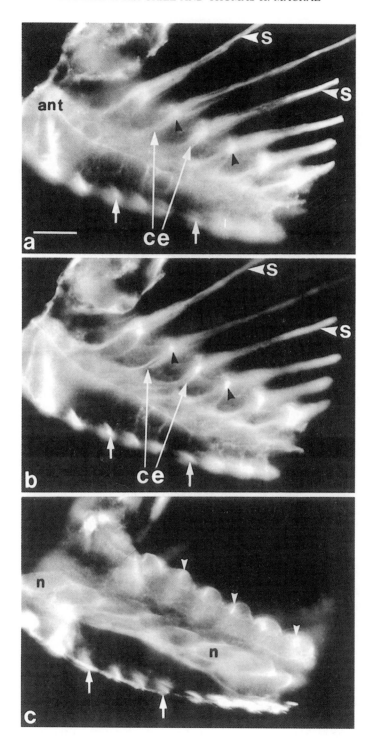

Figure 21. Relationship between tubulin foci and cytoskeletal elements in the antennae. *Artemia* larvae were fixed with paraformaldehyde and immunofluorescently stained with the anti-tubulin antibody TAT. The same antenna at three focal planes is show in **a–c**. The black arrowheads point to brightly stained tubulin foci that appeared to interact with cytoskeletal elements originating in the setal cell apical process, while the unlabelled with arrows indicate tendon cells. *ant*, antenna; *ce*, cytoskeletal element; *n*, nucleus; *s*, setae; bar: 20 μm; magnification is the same in both figures. (Adapted from MacRae and Freeman, 1995)

the non-eversible sections of the penis. Also, accessory glands composed of approximately twenty pairs of cells are located near the junction of the *vas deferens* and the eversible penis. Their secretion consists of mucoproteins and mucopolysaccharides (Bruggeman and Wolfe, 1996).

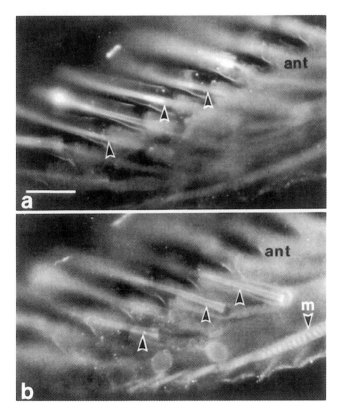

Figure 22. Retraction of setal cell cytoskeletal elements. *Artemia* larvae, fixed with paraformaldehyde and stained with DM1A, were photographed at different focal planes (shown in a and b). The black arrowheads with white borders point to microtubules partially retracted into the basal region of the seta. *ant*, antenna; *m*, muscle; bar: 20 μm; magnification is the same in all figures. (Adapted from MacRae and Freeman, 1995)

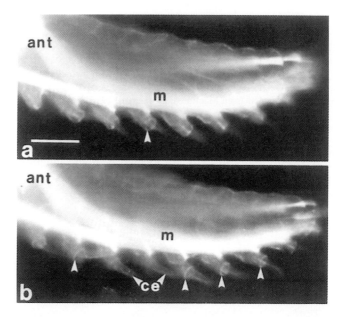

Figure 23. Setal cell morphology during premoult. *Artemia* larvae were fixed with paraformaldehyde and stained with rhodamine-phalloidin. The same antenna, photographed at different focal planes, is shown in **a** and **b**. The unlabelled white arrowheads indicate the region of a premoult setal cell distal to the antenna and enriched in microfilaments. *ant*, antenna; *m*, muscle; *ce*, cytoskeletal element; bar: 20 μm and the magnification is the same in both figures. (From MacRae and Freeman, 1995)

3. *Artemia* Morphology and Structure as Revealed by Immunofluorescent Staining

Antibodies have been used in conjunction with fluorescence and confocal microscopy to examine morphological aspects of structure within *Artemia*. The

Figure 24. Anastral mitotic spindles in epidermal cell layers of *Artemia* lack γ-tubulin. Early instar-II larvae were fixed in paraformaldehyde, dissected and immunofluorescently stained with antibodies to α- and γ-tubulin. **a)** is a low magnification extended focus, constructed by combining several optical sections of a segment of thoracic epidermis visualized by laser scanning confocal microscopy, and **b)** is an enlargement of the 2 spindles enclosed in the box. Examination of individual optical sections demonstrated that a minor foci of γ-tubulin staining near the pole of one spindle in the box of panel a was associated with a portion of a cell slightly above the focal plane of the spindle. Thus, this staining is not seen in b). The primary antibodies were antibody 9 and TAT. **c)** is an extended focus of a malformed bipolar spindle stained with an antibody raised to the γ-tubulin fusion protein and TAT, showing γ-tubulin at one pole only. **d)** is a low magnification extended focus of a segment of thoracic epidermis visualized by laser scanning confocal microscopy, and **e)** is an enlargement of the 4 spindles enclosed in the box. The primary antibodies were TU-30 and Anti-Y. (M) metaphase spindle; (A) anaphase spindle; (MB) midbody. Bar in d = 23 μm and it is the same magnification as a. Bar in b = 11 μm and it is the same magnification as c) and e). (From Walling et al. 1998)

31

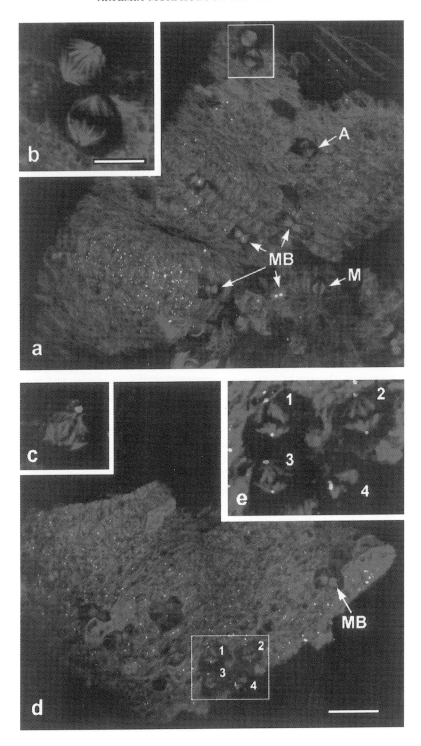

approach has proven very useful, because it provides a way to visualize cells fixed *in situ* rather than after growth in culture. Moreover, the results complement those obtained by other microscopic techniques as just described in this chapter. As one example, tubulin-specific antibodies were used to study the spatial distribution and function of microtubules in *Artemia* larvae (Langdon *et al.* 1991; MacRae *et al.* 1991; Freeman *et al.* 1992, 1995; MacRae and Freeman, 1995; Xiang and MacRae, 1995; Walling *et al.* 1998). Microtubules in several polarized cell types of *Artemia* larvae were shown by immunofluorescence microscopy to possess tyrosinated, detyrosinated and acetylated tubulins (MacRae *et al.* 1991), all known to arise posttranslationally (MacRae, 1997). Included in this survey were epidermally-derived cells such as those of the salt gland and associated tendons (Figure 16). Epidermal layers of the thorax afforded brightly stained mitotic spindles and midbodies when incubated with the antibodies just mentioned (Figure 17) (Langdon *et al.* 1991; MacRae *et al.* 1991; Xiang and MacRae, 1995). Endodermal and mesodermal layers of the gut, including the surrounding muscle cells, were also defined by immunofluorescent labelling of microtubules (Figure 18).

Spatial rearrangement of setal cell microtubules and microfilaments during the moult cycle has been demonstrated by fluorescent staining of *Artemia* larvae (MacRae and Freeman, 1995). Setae contain polarized epidermal cells of the antennae characterized by an elongated apical process containing co-localized microtubules and microfilaments (Figures 19 and 20). The setal cell extensions shorten as larvae enter pre-moult; microtubules and microfilaments recede into the antennae, perhaps through the action of tubulin foci located at the base of the seta (Figure 21). Concurrently, the setal cell shaft telescopes into the antennae (Figure 22), a process accompanied by the appearance of a microfilament ring in a distal fold of the retracting membrane (Figure 23). How the microtubule/microfilament bundle enters antennae and the mechanism of membrane retrieval, two key events of setal cell telescoping, are unknown. They are, however, almost certain to be related processes which employ molecular motors such as cytoplasmic dynein, kinesin and/or myosin, the first two being effectors of microtubule-based mobility and the latter generating movement along microfilaments.

Artemia epidermal cell layers were labelled with antibodies to γ-tubulin, the isotype responsible for nucleation of tubulin assembly at microtubule-organizing centres (Figure 24) (Walling *et al.* 1998). Incubation of interphase epidermal cells with antibodies to γ- and α-tubulin gave two or more foci of γ-tubulin as well as microtubules. However, γ-tubulin was only rarely found at mitotic spindle poles, the normal position for this protein in dividing cells. Interestingly, the mitotic apparatus in epidermal cells of *Artemia* lacks asters suggesting, in concert with the absence of γ-tubulin, an unusual mechanism of spindle construction.

4. References

Adiyodi., R.G. (1985) Reproduction and its control. In D.E. Bliss and L.H. Mantel (eds.), *The Biology of Crustacea*, Vol. 9, Academic Press, pp. 147–215.

Amarant, R. and Elofsson, R. (1976) Distribution of monoaminergic neurons in the nervous system of non-malacostran crustaceans. *Cell and Tissue Research* **166**, 1–24.

Anadon, A. and Anadon, E. (1980) Nauplius eye and adjacent organs of adult *Artemia*. In G. Persoone, P. Sorgeloos, O. Roels and E. Jaspers (eds.), *The Brine Shrimp Artemia*, Vol. 1, Universa Press, Wetteren, Belgium, pp. 41–60.

Anderson, E., Lochhead, J.H., Lochhead, M.S. and Huebner, E. (1970) The origin and structure of the tertiary envelope in the thick-shelled eggs of the brine shrimp, *Artemia. Journal of Ultrastructure Research* **32**, 497–525.

Baid, I.C. and Ramaswami, L.S. (1965) Neurosecretory cells in *Artemia salina* L. *Experientia* **21**, 528–529.

Barlow, D.I. and Sleigh, M.A. (1980) The propulsion and use of water currents for swimming and feeding in larval and adult *Artemia*. In G. Persoone, P. Sorgeloos, O. Roels, and E. Jaspers (eds.), *The Brine Shrimp Artemia*, Vol. 1, Universa Press, Wetteren, Belgium, pp. 61–73.

Benesch, R. (1969) Zur Ontogenie und Morphologie von *Artemia salina* L. *Zoologische Jahrbucher Abteilung für Anatomie und Ontogenie der Tiere* **86**, 307–458.

Bowen, S.T. (1962) The genetics of *Artemia salina* I. The reproductive cycle. *Biological Bulletin* **122**, 25–32.

Brown, G.G. (1966) Ultrastructural studies on crustacean spermatozoa and fertilization, Ph.D. Thesis, Coral Gables, Florida, USA.

Brown, G.G. (1970) Some ultrastructural aspects of spermatogenesis and sperm morphology in the brine shrimp *Artemia salina* Leach (Crustacea: Branchiopoda). *Proceedings of the Iowa Academy of Sciences* **76**, 473–485.

Bruggeman, R.D., and Wolfe, A.F. (1996) A study of the ultrastructure and the secretions of the male accessory gland of *Artemia* (Crustacea Branchiopoda). *Journal of the Pennsylvania Academy of Science* **70**, 40–45.

Cassel, J.D. (1937) The morphology of *Artemia salina* (Linnaeus), M.A. Thesis, Leland Stanford Junior University, California, USA.

Claus, C. (1886) Untersuchungen über die Organisation und Entwicklung von *Branchipus* und *Artemia* nebst vergleichenden Bemerkungen über andere Phyllopoden. *Arbeiten aus dem Zoologische Institute Wien* **6**, 267–370.

Claus, C. (1891) Das Medianauge des Crustaceen. *Arbeiten aus dem Zoologische Institute Wien* **9**, 225–266.

Copeland, D.E. (1967) A study of salt secreting cells in the brine shrimp (*Artemia salina*). *Protoplasma* **63**, 363–384.

Criel, G. (1980a) Morphology of the female genital apparatus of *Artemia*: a review, in G. Persoone, P. Sorgeloos, O. Roels, and E. Jaspers (eds.), The Brine Shrimp *Artemia*, Vol. 1, Universa Press, Wetteren, Belgium, pp. 75–86.

Criel, G. (1980b) Ultrastructural observations on the oviduct of *Artemia*. In G. Persoone, P. Sorgeloos, O. Roels, and E. Jaspers (eds.), *The Brine Shrimp Artemia*, Vol. 1, Universa Press, Wetteren, Belgium, pp. 87–95.

Criel, G.R.J. (1991) Morphology of *Artemia*. In R.A. Browne, P. Sorgeloos and C.N.A. Trotman (eds.), *Artemia Biology*, CRC Press, Boca Raton, Florida, pp. 119–153.

Criel, G.R.J. and Walgraeve, H.R.M.A. (1989) Molt staging in *Artemia* adapted to Drach's system. *Journal of Morphology* **199**, 41–52.

Day, R., Criel, G.R.J., Walling, M.A. and MacRae, T.H. (2000) Posttranslationally modified tubulins and microtubule organization in hemocytes of the brine shrimp. *Artemia franciscana, Journal of Morphology* **244**, 153–166.

de Cuevas M. and Spradling A.C. (1998) Morphogenesis of the *Drosophila* fusome and its implication for oocyte specification. *Development* **125**, 2781–2789.

Debaisieux, P. (1944) Les yeux des Crustacés. Structure, développement, réactions à l'éclairement. *La Cellule* **50**, 9–137.
Debaisieux, P. (1952) Histologie et histogénèse chez *Chirocephalus diaphanus* Prév. (Phyllopode, Anostracé). *La Cellule* **54**, 251–294.
De Maeyer-Criel, G. (1978) Elektronenmicroscopisch onderzoek naar de ultrastructuur tijdens de secretiecyclus van de schaalklier bij *Artemia salina*. *Verhandelingen van de Koninklijke Academie voor Wetenschappen, Letteren en Schone Kunsten van België, Klasse der Wetenschappen* **40**, nr. 145.
Dornesco, G.T. and Steopoe, J. (1958) Les glandes tégumentaires des phyllopodes anostracés. *Annales des Sciences Naturelles Zoologie* **20**, 29–69.
Elofsson, R. (1966) The nauplius eye and frontal organs of the non-malacostraca (Crustacea). *Sarsia* **25**, 1–128.
Elofsson, R. and Dahl, E. (1970) The optic neuropiles of Crustacea. *Zeitschrift für Zellforschung und Mikroskopische Anatomie* **107**, 343–360.
Elofsson, R. and Lake, P.S. (1971) On the cavity receptor organ (X-organ or organ of Bellonci) of *Artemia salina* (Crustacea: Anostraca). *Zeitschrift für Zellforschung und Mikroskopische Anatomie* **121**, 319–326.
Elofsson, R. and Klemm, N. (1972) Monoamine-containing neurons in the optic ganglia of crustaceans and insects. *Zeitschrift für Zellforschung und Mikroskopische Anatomie* **133**, 475–499.
Elofsson, R. and Odselius, R. (1975) The anostracan rhabdom and the basement membrane. An ultrastructural study of the *Artemia* compound eye (Crustacea). *Acta Zoologica* **56**, 141–153.
Fautrez, J. and Fautrez-Firlefyn, N. (1971) Contribution à l'étude des glandes coquilières et des coques de l'oeuf d'*Artemia salina*. *Archives de Biologie (Liège)* **82**, 41–83.
Freeman, J.A. (1989) The integument of *Artemia* during early development. In T.H. MacRae, J.C. Bagshaw and A.H. Warner (eds.), *Biochemistry and Cell Biology of Artemia*, CRC Press, Boca Raton, Florida, USA, pp. 233–256.
Freeman, J.A., Cheshire, L.B. and MacRae, T.H. (1992) Epithelial morphogenesis in developing *Artemia*: the role of cell replication, cell shape change and the cytoskeleton. *Developmental Biology* **152**, 279–292.
Freeman, J.A., Whittington, C. and MacRae, T.H. (1995) Relative growth of the tendinal cell and muscle in larval *Artemia*. *Invertebrate Reproduction & Development* **28**, 205–210.
Frenzel, J. (1893) The mid-gut of *Artemia salina*. *Journal of the Royal Microsc. Society, London* **11**, 37.
Greene, W. (1924) The circulatory system of the brine shrimp. *Science* **60**, 411–412.
Hansen, U. and Peters, W. (1998) Structure and permeability of the peritrophic membranes of some small crustaceans. *Zoologischer Anzeiger* **236**, 103–108.
Hanström, B. (1928) *Vergleichende Anatomie des Nervensystems des wirbellosen Tiere*, Kapitel 22 Crustacea, Julius Springer Verlag, Berlin.
Hanström, B. (1931) Neue Untersuchungen über Sinnesorgane und Nervensystem der Crustaceen, I. *Zeitschrift für Morphologie und Ökologie der Tiere* **23**, 80–236.
Hentschel, E. (1963) Zum neurosekretorischen System der Anostraca, Crustacea (*Artemia salina* Leach und *Chirocephalus grubei* Dybowsci). *Zoologischer Anzeiger* **170**, 187–190.
Hentschel, E. (1965) Neurosekretion und Neurohämalorgane bei *Chirocephalus grubei* Dybowci und *Artemia salina* Leach (Anostraca, Crustacea). *Zeitschrift für Wissenschaftliche Zoologie* **171**, 44–79.
Hertel, H. (1980) The compound eye of *Artemia salina* (Crustacea). I. Fine structure when light and dark adapted. *Zoologische Jahrbucher-Abteilung für Allgemeine Zoologie und Physiologie der Tiere* **84**, 1–14.
Hootman, R. and Conte, F.P. (1974) Fine structure and function of the alimentary canal in *Artemia salina* nauplii. *Cell and Tissue Research* **155**, 423–436.
Horridge, G.A. (1965a) Arthropoda: General Anatomy. In T.H. Bullock and G.A. Horridge (eds.),

Structure and Function in the Nervous System of Invertebrates, Freeman W.H. and Company, San Francisco and London, p. 810.
Horridge, G.A. (1965b) Arthropoda: Receptors for light and optic lobe. In T.H. Bullock and G.A. Horridge (eds.), *Structure and Function in the Nervous System of Invertebrates*, Freeman W.H. and Company, San Francisco and London, p. 1063.
Horst, M.N. (1989) Molecular and cellular aspects of chitin synthesis in larval *Artemia*. In A.H. Warner, T.H. MacRae, J.C. Bagshaw (eds.), *Cell and Molecular Biology of Artemia Development*, NATO ASI Series, Series A: Life Sciences, 174, Plenum Publishing Corporation, pp. 59–76.
Joly, M. (1840) Histoire d'un petit crustacé (*Artemia salina* Leach), auquel on a faussement attribué la coloration en rouge des marais salans méditerrenéens, suivie de recherches sur la cause réelle de cette coloration. *Annales des Sciences Naturelles (Paris) Zoologie et Biologie Animale* **13**, 225–290.
Kane, E.X.C. (1975) Neuroanatomical clues to peripheral locomotor control in small crustaceans (*Artemia salina*). *American Journal of Anatomy* **142**, 485–500.
Kikuchi, S. (1972) The fine structure of the alimentary canal of the brine shrimp, *Artemia salina*: the midgut. *Annual Report of Iwate Medical University, School of Liberal Arts and Sciences* **7**, 15–47.
Kuenen, J.D. (1939) Systematical and physiological notes on the brine shrimp, *Artemia*. *Archives Néerlandaises de Zoologie* **3**, 365–445.
Langdon, C.M., Freeman, J.A. and MacRae, T.H. (1991) Post-translationally modified tubulins in *Artemia*: prelarval development in the absence of detyrosinated tubulin. *Developmental Biology* **148**, 147–155.
Leydig, F. (1851) Über *Artemia salina* und *Branchipus stagnalis*. *Zeitschrift für Wissenschaftliche Zoologie* **3**, 280–307.
Lochhead, J.H. (1950) *Artemia*. In F.A. Brown (ed.), *Selected Invertebrate Types*, Wiley and Sons, New York, pp. 394–399.
Lochhead, J.H. and Lochhead, M.S. (1941) Studies on the blood and related tissues in *Artemia* (Crustacea, Anostraca). *Journal of Morphology* **68**, 593–632.
Lochhead, J.H. and Resner, R. (1958) Functions of the eye and neurosecretion in Crustacea Anostraca. *15th International Congress on Zoology*, London, pp. 397–399.
MacRae, T.H. (1997) Tubulin post-translational modifications. Enzymes and their mechanisms of action. *European Journal of Biochemistry* **244**, 265–278.
MacRae, T.H. and Freeman, J.A. (1995) Organization of the cytoskeleton in brine shrimp setal cells is molt-dependent. *Canadian Journal of Zoology* **73**, 765–774.
MacRae, T.H., Langdon, C.M. and Freeman, J.A. (1991) Spatial distribution of posttranslationally modified tubulins in polarized cells of developing *Artemia*. *Cell Motility and the Cytoskeleton* **18**, 189–203.
Martin, G.G., Lin, H.M.J. and Luc, C. (1999) Reexamination of hemocytes in brine shrimp (Crustacea, Branchiopoda). *Journal of Morphology* **242**, 283–294.
Metalli, P. and Ballardin, E. (1970) Radiobiology of *Artemia*: radiation effects and ploidy. *Current Topics in Radiation Research Quarterly* **7**, 181–240.
Moroff, T. (1912) Entwicklung und phylogenetische Bedeutung des Medianauges bei Crustaceen. *Zoologischer Anzeiger* **40**, 11–25.
Mura, G. and Del Caldo, L. (1992) Scanning electron microscopic observations on the molar surface of mandibles in species of *Artemia* (Anostraca). *Crustaceana* **62**, 193–200.
Nässel, D.R., Elofsson, R. and Odselius, R. (1978) Neuronal connectivity patterns in the compound eyes of *Artemia salina* and *Daphnia magna*. *Cell and Tissue Research* **190**, 435–457.
Nilsson, D.E. and Odselius, R. (1981) A new mechanism for light-dark adaptation in the *Artemia* compound eye (Anostraca, Crustacea). *Journal of Comparative Physiology A* **143**, 389–399.
Nowikoff, M. (1905) Über die Augen und die Frontalorgane der Branchiopoden. *Zeitschrift für Wissenschaftliche Zoologie* **79**, 432–464.

Nowikoff, M. (1906) Einige Bemerkungen über das Medianauge und die Frontalorgane von *Artemia salina*. *Zeitschrift für Wissenschaftliche Zoologie* **81**, 691–698.

Økland, S., Tjønnenland, A., Larsen, L.N. and Nylund, A. (1982) Heart ultrastructure in *Branchinecta paludosa*, *Artemia salina*, *Branchipus schaefferi*, and *Streptocephalus* sp. (Crustacea, Anostraca). *Zoomorphology* **101**, 71–81.

Rasmussen, S. (1971) Die Feinstruktur des Mittelauges und des ventralen Frontalorgans von *Artemia salina* L. (Crustacea: Anostraca). *Zeitschrift für Zellforschung* **117**, 576–597.

Reeve, M.R. (1963) The filter-feeding of *Artemia*, III. Faecal pellets and their associated membranes. *Journal of Experimental Biology* **40**, 215–221.

Schrehardt, A. (1987) Ultrastructural investigations of the filter-feeding apparatus and the alimentary canal of *Artemia*. In P. Sorgeloos, D.A. Bengtson, W. Decleir and E. Jaspers (eds.) *Artemia Research and its Applications*, Vol. 1, Universa Press, Wetteren, Belgium, pp. 33–52.

Spencer, W.K. (1902) Zur Morphologie des Centralnervensystems des Phyllopoden, nebst Bemerkungen über deren Frontalorgane. *Zeitschrift für Wissenschaftliche Zoologie* **71**, 508–524.

Steopoe, J. and Dornesco, G.T. (1945) La cytologie et les propriétés des éléments sanguins des phyllopodes anostracés. *Bulletin de la Section Scientifique de l' Académie Roumaine* **28**, 220–225.

Tyson, G.E. (1966) Gross morphology and fine structure of the maxillary gland of the brine shrimp, *Artemia salina*. *American Zoologist* **6**, 555.

Tyson, G.E. (1968) The fine structure of the maxillary gland of the brine shrimp, *Artemia salina*: the end-sac. *Zeitschrift für Zellforschung und Mikroskopische Anatomie* **86**, 129–138.

Tyson, G.E. (1969a) Intercoil connections of the kidney of the brine shrimp, *Artemia salina*. *Zeitschrift für Zellforschung und Mikroskopische Anatomie* **100**, 54–59.

Tyson, G.E. (1969b) The fine structure of the maxillary gland of the brine shrimp, *Artemia salina*: the efferent duct. *Zeitschrift für Zellforschung und Mikroskopische Anatomie* **93**, 151–163.

Tyson, G.E. (1980) Fine structure of the type 2 antennular sensillum of the brine shrimp. *American Zoologist* **20**, 816.

Tyson, G.E. and Sullivan, M.L. (1978) Scanning electron microscopy of the antennular sensilla of the brine shrimp. *American Zoologist* **18**, 632.

Tyson, G.E. and Sullivan, M.L. (1979a) Antennular sensilla of the brine shrimp, *Artemia salina*. *Biological Bulletin* **156**, 382–392.

Tyson, G.E. and Sullivan, M.L. (1979b) Frontal knobs of the male brine shrimp: scanning electron microscopy. *American Zoologist* **19**, 891.

Tyson, G.E. and Sullivan, M.L. (1980a) Scanning electron microscopy of cuticular sensilla of *Artemia*: setae of the adult trunk segments. In G. Persoone, P. Sorgeloos, O. Roels, and E. Jaspers (eds.), *The Brine Shrimp Artemia*, Vol. 1, Universa Press, Wetteren, Belgium, pp. 99–106.

Tyson, G.E. and Sullivan, M.L. (1980b) Scanning electron microscopy of the frontal knob of the male brine shrimp. *Transactions of the American Microscopical Society* **99**, 167–172.

Tyson, G.E. and Sullivan, M.L. (1981a) A scanning electron microscopic study of the molar surfaces of the mandibles of the brine shrimp (Cl.Branchiopoda: O.Anostraca). *Journal of Morphology* **170**, 239–251.

Tyson, G.E. and Sullivan, M.L. (1981b) Scanning electron microscopy of the molar surfaces of the mandibles of the brine shrimp. *American Zoologist* **21**, 957.

Vaissière, R. (1956) Evolution de l'oeil d' *Artemia salina* Leach (Crustacé, Branchiopode, Phyllopode) au cours de ses stades post-embryonnaires. *Comptes Rendus de l' Academie des Science* **242**, 2051–2054.

Vaissière, R. (1961) Morphologie et histologie comparées des Crustacés Copépodes. *Archives de Zoologie Experimentale et Génerale* **100**, 1–126.

Van Beek, E. (1988) Endocrinogical aspects of vitellogenesis in the brine shrimp *Artemia* sp. (Crustacea: Anostraca), Ph.D. Thesis, Catholic University of Leuven, Belgium.

Van Beek, E., Van Brussel, M., Criel, G. and De Loof, A. (1987) A possible extra-ovarian site for synthesis of lipovitellin during vitellogenesis in *Artemia* sp. (Crustacea; Anostraca). *International Journal of Invertebrate Reproduction and Development* **12**, 227–240.

Van den Bosch de Aguilar, P. (1976) Neurosécrétion et régulation hydroélectrolytique chez *Artemia salina*. *Experientia* **32**, 228–229.

Van den Bosch de Aguilar, P. (1979) Neurosecretion in the entomostraca crustaceans. *La Cellule* **73**, 1–22.

Vehstedt, R. (1940) Über Bau, Tätigkeit und Entwicklung des Rückengefässes und des lacunären System von *Artemia salina* var. arieta. *Zeitschrift für Wissenschaftliche Zoologie* **154**, 1–39.

Walgraeve, H.R., Criel, G.R., Sorgeloos, P. and De Leenheer, A.P. (1988) Determination of ecdysteroids during the moult cycle of adult *Artemia*. *Journal of Insect Physiology* **34**, 597–602.

Walling, M.A., Criel, G.R.J. and MacRae, T.H. (1998) Characterization of γ-tubulin in *Artemia*: isoform composition and spatial distribution in polarized cells of the larval epidermis. *Cell Motility and the Cytoskeleton* **40**, 331–341.

Warren, H.S. (1930) The central nervous system of the adult *Artemia*. *Transactions of the American Microscopical Society* **49**, 189–209.

Warren, H.S. (1938) The segmental excretory glands of *Artemia salina*, Lin., var. Principalis Simon, the brine shrimp. *Journal of Morphology* **62**, 263–289.

Wingstrand, K.G. (1978) Comparative spermatology of the Crustacea Entomostraca 1. Subclass Branchiopoda. *Biologiske Skrifter* **22**, 1–66.

Wolfe, A.F. (1971) A histological and histochemical study of the male reproductive system of *Artemia* (Crustacea, Branchiopoda). *Journal of Morphology* **135**, 51–71.

Wolfe, A.F. (1980) A light and electron microscopic study of the frontal knob of *Artemia* (Crustacea, Branchiopoda). In G. Persoone, P. Sorgeloos, O. Roels, and E. Jaspers (eds.), *The Brine Shrimp Artemia*, Vol.1, Universa Press, Wetteren Belgium, pp. 117–130.

Xiang, H. and MacRae, T.H. (1995) Production and utilization of detyrosinated tubulin in developing *Artemia* larvae: evidence for a tubulin-reactive carboxypeptidase. *Biochemistry and Cell Biology* **73**, 673-685.

CHAPTER II

REPRODUCTIVE BIOLOGY OF *ARTEMIA*

GODELIEVE R.J. CRIEL
Department of Anatomy, Embryology and Histology
University of Ghent
Godshuizenlaan 4
B-9000 Gent
Belgium

THOMAS H. MACRAE
Department of Biology
Dalhousie University
Halifax, NS B3H 4J1
Canada

1. Introduction

The genus, *Artemia*, consists of bisexual and parthenogenetic strains. In North America, only bisexual species are found, whereas in Europe, Asia and Africa, bisexual and parthenogenetic populations occur. Females of most *Artemia* strains reproduce ovoviviparously and oviparously, releasing either nauplius larvae or encysted embryos, respectively (Jackson and Clegg, 1996; Liang and MacRae, 1999) (Figure 1). Reproductive mode may switch, in good rearing conditions *Artemia* tend to reproduce ovoviviparously, whereas under adverse situations they reproduce oviparously. Females differ in their genetic tendency to reproduce either ovoviviparously or oviparously.

2. Molecular Aspects of Early Artemia Development

Within the five days from oocyte fertilisation to release of encysted embryos, a small heat shock/α-crystallin protein termed p26 is synthesized, but only if the oviparous pathway is followed (Clegg *et al.* 1994, 1995, 1999; Jackson and Clegg, 1996; Liang *et al.* 1997a, b; MacRae and Liang, 1998; Liang and MacRae, 1999). Embryos exhibit p26 mRNA on the second day following fertilisation, and the message peaks at day four post-fertilisation, before discharge of cysts from females (Figure 2a) (Liang and MacRae, 1999). p26 protein initially appears on day three after fertilisation and is present in near maximum amounts at the time of cyst release (Figure 2b) (Jackson and Clegg, 1996; Liang and MacRae, 1999). Embryos undergoing ovoviviparous devel-

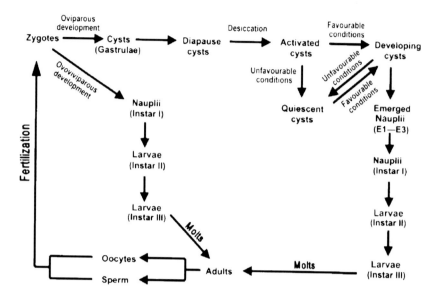

Figure 1. Alternative developmental modes of *A. franciscana*. Fertilised *Artemia* oocytes may undergo ovoviviparous development, wherein free-swimming nauplii (instar-I larvae) are produced (left side of diagram). Alternatively, oviparous development occurs and gastrulae encyst, leave the female and enter diapause, after which they are often desiccated (top of diagram). Upon activation, cysts develop into larvae, but if unfavourable conditions prevail, growth is arrested and cysts become quiescent (right side of diagram). (From Liang and MacRae, 1999)

opment, on the other hand, synthesize neither p26 mRNA nor contain the protein (Figure 2a, b). Both nuclei and cytoplasm of oviparous *Artemia* embryos possess p26, which begins to move into nuclei three days post-fertilisation (Figure 3), a process that persists after cyst release (Jackson and Clegg, 1996; Liang and MacRae, 1999). Although p26 has not been precisely localised in nuclei of diapause embryos, it is sometimes found in discrete intranuclear compartments or foci of activated cysts (Figure 4) (Liang et al. 1997a). Moreover, p26 is reversibly translocated into nuclei under conditions of stress such as anoxia, heat shock and diapause, perhaps in response to reduced intracellular pH (Clegg et al. 1994, 1995, 1999). Nuclear localisation suggests that p26 affects transcription, thereby contributing to metabolic inhibition as cysts enter diapause. Another possibility is that p26 prevents DNA replication, mitosis and cytokinesis, none of which occur during encysted growth, although all resume upon emergence (Nakanishi et al. 1962; Olson and Clegg, 1978), once p26 has undergone a major quantitative reduction. Disappearance of p26 mRNA is well underway five hours after activated cysts resume development, and the message is all but gone when nauplii emerge (Figure 5a) (Liang and MacRae, 1999). p26 is detectable on western blots of cell-free extracts from *Artemia* up to and including instar-I larvae, but not in extracts from instar-II (Figure 5b) (Clegg et al. 1994; Liang and MacRae,

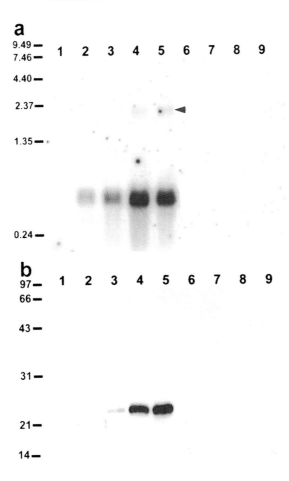

Figure 2. Synthesis of p26 during ovoviviparous and oviparous development. Approximately 200 *Artemia* embryos or larvae, collected at 24-h intervals for 5 days postfertilisation, were homogenised in Trizol solution. (a) Northern blot with 2 mg of total RNA from each developmental stage hybridised to ^{32}P-labelled p26-3-6-3. (b) Equal volumes of cell-free extract from each developmental stage were transferred to nitrocellulose after electrophoresis in SDS-polyacrylamide gels and probed with antibody to p26 by the ECL procedure. Lanes 1–5 received samples from cyst-destined embryos developed 1–5 days, respectively, while lanes 6–9 contain samples from nauplius-destined embryos 2–5 days post fertilisation, respectively. The arrowhead in (a) indicates an mRNA band of about 1.9 kb. Size markers in kb × 10^{-3} (a) and molecular weight × 10^{-3} (b) are to the left. (From Liang and MacRae, 1999)

1999). Immunofluorescent staining reveals a progressive loss of p26 from *Artemia* larvae, with the protein last detected in a subset of salt gland nuclei (Figures 6 and 7) (Liang and MacRae, 1999). The significance of this latter observation is not clear, but the salt gland cells containing p26 do not divide. The salt gland undergoes developmentally programmed disappearance and is missing from adult brine shrimp. p26 may contribute to salt gland loss, perhaps

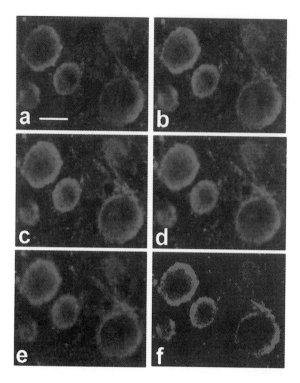

Figure 3. p26 entered nuclei of encysting embryos soon after it was synthesized. Nuclei from cyst-destined embryos 3 days after fertilisation were stained with antibody to p26 and examined with the confocal microscope. **(a–e)** Individual optical sections of stained nuclei; **(f)** composite image obtained by superimposing images a–e. The bar in (a) represents 5 μm and all images are the same magnification. (From Liang and MacRae, 1999)

by modulating transcription and/or DNA replication, or it may linger because these cells are facing apoptotic stress. However, in other examples, the presence of small heat shock/α-crystallin proteins in cells indicates that differentiation and growth, rather than apoptosis, will occur (Arrigo, 1998; Mehlen *et al.* 1999; Wagstaff *et al.* 1999).

Molecular analyses of embryo development are limited, but post-diapause development leading to excystment has been examined extensively. This process is reversibly halted by dehydration (Morris, 1971) and anoxia, with some cysts surviving in the absence of oxygen for as long as 6 years (Clegg, 1997; Clegg *et al.* 1999). A role for p26 can be imagined during either desiccation and/or anoxia, especially as resistance to these stresses is gained and disappears more or less coincidentally with the gain and loss of p26, respectively. Additionally, metabolic cessation upon exposure to anoxic conditions, and its reversal upon aeration, is modulated by changes in cyst intracellular pH (Busa *et al.* 1982; Carpenter and Hand, 1986; Hand and Gnaiger, 1988). Similar modifications to intracellular pH characterise the

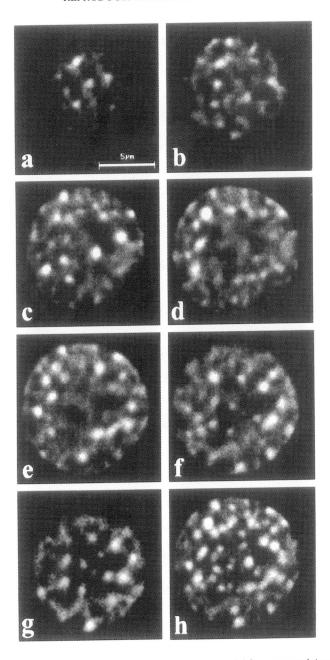

Figure 4. p26 is found in discrete compartments throughout nuclei from encysted *Artemia*. Nuclei purified from *Artemia* cysts were stained with antibody to p26 and examined in the confocal microscope. (**a–g**) represent a continuous series of 1 μm optical sections through a single nucleus. The top and bottom sections of the nucleus are not shown because they contained limited information. (**h**) is a three-dimensional reconstruction of the stained nucleus. The bar in (a) represents 5 μm, and the magnification is the same for all figures. (From Liang *et al.* 1997a)

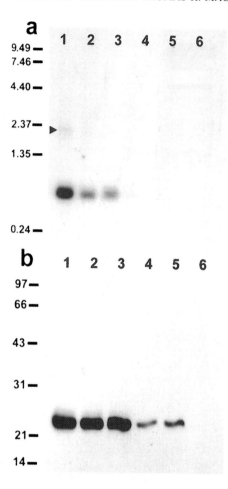

Figure 5. p26 mRNA and protein disappeared during postgastrula development of *Artemia*. Total RNA (a) and protein (b) from encysted *Artemia* incubated for 0 (1), 5 (2) and 10 h (3) and from emerged nauplii (4), instar-I larvae (5), and instar-II larvae (6). (a) Northern blots with 2 mg of total RNA from each developmental stage were hybridised to ^{32}P-labelled p26-3-6-3. (b) Western blots containing equal volumes of cell-free extract from each developmental stage were probed with antibody to p26 by the ECL procedure. The arrowhead in (a) indicates an mRNA band of about 1.9 kb. Size markers in kb × 10^{-3} and molecular weight × 10-3 are to the left of (a) and (b), respectively. (Adapted from Liang and MacRae, 1999)

switch from dormancy to growth of *Artemia* cysts (Busa and Crowe, 1983) indicating that pH is a major regulator of development. (Clegg and Trotman discuss this pH switch further in Chapter III of this volume). Interestingly, the nuclear translocation of p26 is strongly pH-dependent *in vitro*, in a fashion that reflects the *in vivo* situation (Clegg *et al.* 1995).

The initiation of post-diapause development is accompanied by activation of transcription and protein synthesis (Golub and Clegg, 1968; Clegg and

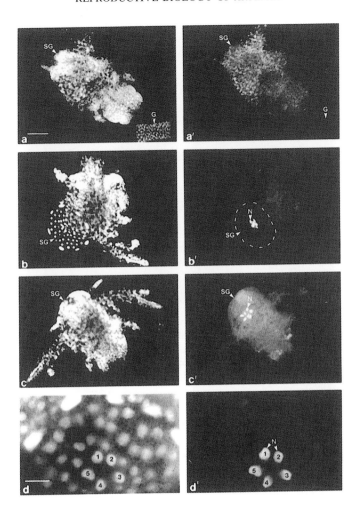

Figure 6. Localisation of p26 in *Artemia* larvae. Instar-I (**a, a′**) and late instar-II (**b, b′–d, d′**) *Artemia* larvae, fixed in paraformaldehyde, were double stained with DAPI (a–d) and antibody to p26 (a′–d′). The location of the salt gland in b′ is indicated by the dashed lines. (d and d′) A salt gland with corresponding nuclei (N) indicated by the same number (1–5). G, gut; SG, salt gland. The bar in (a) represents 100 μm and (a), (a′–c), (c′) are the same magnification. The bar in (d) represents 15 μm and it is the same magnification as (d′). (From Liang and MacRae, 1999)

Golub, 1969; Finamore and Clegg, 1969; Wahba and Woodley, 1984; Drinkwater and Clegg, 1991; Tate and Marshall, 1991), but not by DNA synthesis and cell division, at least not until very late in pre-emergence development (Nakanishi *et al.* 1962; Finamore and Clegg, 1969; Olson and Clegg, 1976, 1978). Thus, brine shrimps are unusual because they undergo development, from encysted gastrulae to swimming nauplii, without an increase in cell number. Few organisms are capable of this, and it has prompted many

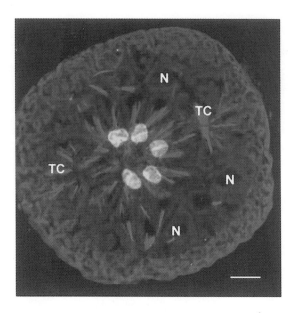

Figure 7. p26 was last observed in a subset of salt gland nuclei in instar-II *Artemia* larvae. A salt gland from an instar-II larva double stained with rhodamine phalloidin (red) and antibody to p26 (green). The image was constructed by superimposing optical sections obtained by confocal microscopy. N, unstained nucleus; TC, tendon cell. The bar represents 10 μm. (From Liang and MacRae, 1999)

studies, including examination of tubulin synthesis during post-diapause development. The possibility that changes seen in tubulin are functionally related to development, rather than mitosis and cell division, offers a rare opportunity to study these processes separate from one another.

Artemia tubulin, differing from mammalian neural tubulin in electrophoretic mobility and assembly properties, consists of three α- and two β-tubulins, as revealed by Coomassie blue staining of two-dimensional gels (Figure 8) (MacRae and Ludueña, 1984; Rafiee *et al.* 1986a; Langdon *et al.* 1990). The isoform patterns of α- and β-tubulins do not change during the first 24 hours of post-diapause development (Rafiee *et al.* 1986a), but posttranslationally generated detyrosinated tubulin is initially found in instar-I larvae, this representing the only change in tubulin isoforms demonstrated for the species (Figures 8 and 9) (Langdon *et al.* 1991a; Xiang and MacRae, 1995). Detyrosination, the reversible removal of the carboxy-terminal tyrosine from α-tubulin (MacRae, 1997), depends in *Artemia* on the developmentally regulated synthesis and/or activation of a tubulin-reactive carboxypeptidase (Figure 10) (Xiang and MacRae, 1995). These findings are interesting because they establish that encysted *Artemia* differentiate in the absence of isotubulin modification and without detyrosinated tubulin.

Hybridisation of α- and β-tubulin specific probes to northern blots demon-

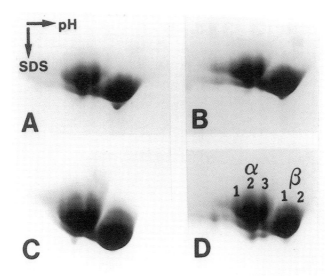

Figure 8. Two-dimensional gels of *Artemia* tubulin induced to assemble with taxol. Tubulin was from organisms allowed to develop 0 (**A**), 15 (**B**), and 24 (**C**) h before homogenisation and subsequent taxol-induced assembly. Biochemically purified tubulin from 15 h organisms (**D**). Migration in isoelectric focusing gels was from left to right and SDS-polyacrylamide gel electrophoresis was from top to bottom as indicated in (A). Only the tubulin regions of the gels are shown and results from 5 and 10 h embryos, which had the same isotubulin populations as at the other time intervals, are not included. (From Rafiee *et al.* 1986a)

strates similar amounts of tubulin mRNA in extracts from *Artemia* cysts through to instar-I larvae, none of which is detectable in polysomal mRNA (Figure 11) (Langdon *et al.* 1990, 1991b). Although incubation of cysts in ^{14}C-bicarbonate reveals a very low level of tubulin synthesis during post-diapause development, non-polysomal tubulin mRNA is in large excess, suggesting an operative translational control mechanism. Additionally, *in vitro* translation of *Artemia* mRNA from all developmental stages examined, gives one isoform each of α- and β-tubulin (Figure 12) (Langdon *et al.* 1991b). Although there are limitations to isoform analysis in two-dimensional gels, the evidence indicates two major tubulin gene products in *Artemia*, one each for α- and β-tubulin, and these are unchanged during development (Figure 12). Isoform diversity observed in two-dimensional gels of *Artemia* tubulin is most likely the result of post-translational modifications. In this context, acetylated α-tubulin was identified in cell-free extracts of *Artemia* cysts by staining western blots with an isoform-specific antibody (Figure 13) (Langdon *et al.* 1990). Clearly, *Artemia* is a very useful model system for the study of tubulin molecular biology.

Once the pre-nauplius has developed within the cyst as a consequence of re-established metabolic and biosynthetic activities, it emerges. The shell enclosing the cyst, laid down during pre-diapause oviparous development, is

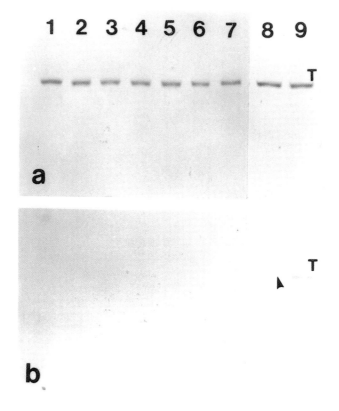

Figure 9. Detyrosinated tubulin was first synthesized during growth of instar-I *Artemia* larvae. Cell-free supernatant were prepared from developing *Artemia*, electrophoresed in SDS-polyacrylamide gels, blotted to nitrocellulose, and immunostained with (a) anti-Y and (b) anti-E. Incubation times for lanes 1-9 were, respectively, 0, 4, 8, 12, 16, 20, 24, 28, and 32 h. Each lane in panel (a) received 25 µg protein, whereas lanes in (b) were each loaded with 50 µg cell-free extract protein. Developing shrimp, synchronised at 16 h (instar-I), were incubated for an additional 16 h to yield instar-II larvae. T, tubulin; arrowheads, location of detyrosinated tubulin visible on blots but not reproduced photographically. (Adapted from Xiang and MacRae, 1995)

impermeable to all non-volatile solutes and structurally complex (Morris and Afzelius, 1967; Anderson *et al.* 1970; Trotman *et al.* 1980, 1987; Drinkwater and Clegg, 1991; Trotman, 1991). The outer two layers of the tertiary envelope are products of the maternal shell gland (Anderson *et al.* 1970), and are thought to serve a protective function. Immediately below the tertiary envelope, and exterior to the embryonic cuticle, is the outer cuticular membrane. This membrane, of uncertain origin, constitutes the permeability barrier of the cyst shell (Morris and Afzelius, 1967; Anderson *et al.* 1970; De Chaffoy *et al.*, 1978; Drinkwater and Clegg, 1991). Achieving impermeability is a prerequisite for normal development of embryos and the nauplii (De Chaffoy *et al.* 1978). Chapter III of this volume contains an electron micrograph of the

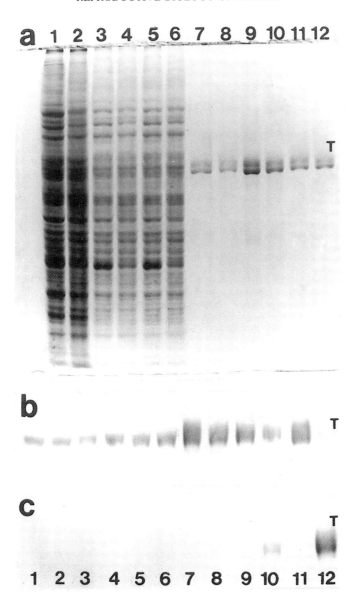

Figure 10. Evidence for a tubulin-reactive carboxypeptidase in developing *Artemia*. Protein fractions obtained during purification of tubulin from *Artemia* developed for either 12 h (1, 3, 5, 7, 9, 11) or 24 h (2, 4, 6, 8, 10, 12) were electrophoresed in SDS-polyacrylamide gels and stained with (**a**) Coomassie blue or blotted to nitrocellulose for reaction with (**b**) anti-Y and (**c**) anti-E. The lanes contained: 100 μg cell-free extract (lanes 1, 2); 50 μg Phosphocellulose P-11 fraction (lanes 3, 4); 50 μg Phosphocellulose P-11 fraction after it was frozen, thawed, and centrifuged, as occurred routinely during preparation (lanes 5, 6); 8.4 μg DEAE-cellulose fraction (lanes 7, 8); 5 μg $(NH_4)_2SO_4$ fraction (lanes 9, 10); 5 μg tubulin after assembly-disassembly (lanes 11, 12). T, tubulin. (From Xiang and MacRae, 1995)

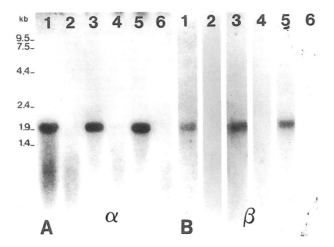

Figure 11. Comparison of *Artemia* α- and β-tubulin mRNA in total and polysomal poly(A)$^+$ mRNA. Five micrograms of total poly(A)$^+$ mRNA was applied to lanes numbered 1, 3, and 5 while 5 mg of polysomal poly(A)$^+$ mRNA was applied to lanes numbered 2, 4, and 6. The RNA was prepared from organisms developed 0 hr (lanes 1 and 2), 15 hr (lanes 3 and 4), and 24 hr (lanes 5 and 6). The RNA was electrophoresed on 1.5% denaturing agarose gels, transferred to nitrocellulose, and hybridised with ^{32}P-labelled pDmTα1 (**A**) and DTB2 (**B**). Size markers in kb are shown on the left side of the figure. (Adapted from Langdon et al. 1991b)

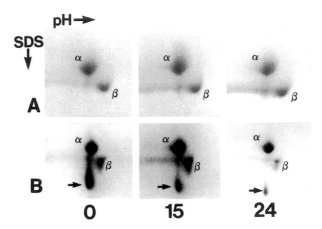

Figure 12. Two-dimensional analysis of *Artemia* tubulin translated *in vitro*. Five micrograms of poly(A)$^+$ mRNA from organisms developed 0, 15, and 24 hr (as labelled) was translated *in vitro* with [^{35}S]methionine in rabbit reticulocyte lysate, coassembled with unlabeled 15 hr *Artemia* tubulin, and then analysed in one direction by isoelectric focusing (pH) and in the second direction by SDS-polyacrylamide gel electrophoresis (SDS). **A**, Coomassie blue-stained gels; **B**, corresponding fluorograms. The arrows in B point to an unidentified translation product. Only the tubulin region of the gels is shown as there were no other spots on the fluorograms. (From Langdon *et al.* 1991b)

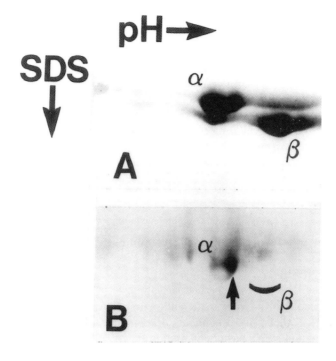

Figure 13. Immunoblotting of *Artemia* tubulin electrophoresed in two-dimensional gels. Purified *Artemia* tubulin was resolved into isoforms by isoelectric focusing in one direction (pH) and SDS-polyacrylamide gel electrophoresis in the other direction (SDS), blotted to nitrocellulose and probed with antitubulin antibodies using the immunoperoxidase method. **a**, Coomassie blue-stained gel; **b**, Blot probed with 6-11B-1 (anti-acetylated α-tubulin). The locations of α- and β-isotubulin are indicated on the figures. (Adapted from Langdon *et al.* 1990)

cyst shell, and documents the importance of shell impermeability to encysted embryo survival.

Escape of the pre-nauplius from within the shell (emergence) has been described morphologically several times (Wheeler *et al.* 1979; Rafiee *et al.* 1986b; Go *et al.* 1990; Rosowski *et al.* 1997). Mechanistically, however, the emergence of nauplii is less clear, although the process is inhibited by heavy metals (Rafiee *et al.* 1986b; Go *et al.* 1990; MacRae and Pandey, 1991; Pandey and MacRae, 1991), bicarbonate deficiency (Trotman *et al.* 1987; Trotman, 1991) and metabolic perturbation (Trotman *et al.* 1980). Metabolism of the disaccharide trehalose (1-α-D-glucopyranosyl 1,1-D-α-glucopyranoside), which increases in concentration for several days after diapause cysts leave females and is massively abundant in cysts (15% of dry weight), may mediate emergence (Clegg, 1964, 1965; Drinkwater and Clegg, 1991; Clegg and Jackson, 1992, 1998). Trehalose is undoubtedly used as an energy source during cyst development and its utilisation is matched by increases in glycerol and glycogen (Clegg, 1964). As glycerol builds up, internal osmotic pressure rises and water is taken in, eventually leading to rupture of the shell and release of membrane enclosed nauplii (Clegg, 1964), but other processes might be involved (Clegg and Conte, 1980). Although not proven, this mechanism is supported by several results, not the least of which are enhanced accumulation of glycerol in response to greater external osmotic pressure and the deposition of glycerol exterior to the embryo but enclosed by the shell (Clegg, 1964). This glycerol-based process may be supplemented by ionic gradients, formed as ATP becomes available to the cyst (Trotman et al. 1987; Trotman, 1991). Trehalose is also an important protective solute (Clegg and Conte, 1980; Crowe et al. 1998), preserving membranes against desiccation damage and proteins against thermal denaturation and aggregation. The participation of trehalose in the protection of cysts against damage due to desiccation and temperature extremes is considered further in chapter III of this volume.

3. Morphology of Early *Artemia* Development

3.1. OOGENESIS

The fundamentally different somatic and germ cells are discernible in the ovary of *Artemia* (Lochhead and Lochhead, 1967). The somatic cells form a continuous layer under the basement membrane facing the gut, and their well developed network of indented plasma membranes suggests a role in nutrient transport. Somatic cells separate the oocyte-nurse cell complexes, and at the end of the vitellogenic cycle they phagocytose residual nurse cells. Germ cells include oogonia and primary oocytes, most of which differentiate into nurse cells. Oogonia are the most lateral elements of the ovary, residing along the entire length of previtellogenic ovaries. In vitellogenic ovaries, on the other hand, oogonia make up islands between the oocytes, as occurs in the decapod, *Gecarcinus* (Weitzman, 1966). The proliferation of oogonia continues after sexual maturity as in many crustaceans (Adiyodi and Subramoniam, 1983). Maturing germ cells lie between oogonia and somatic cells, settling close to the gut as they mature. Premeiotic oocytes cluster, linked to remnants of the mitotic spindle and comprising a structure that looks like a fusome (Figure 14), a vesiculated cytoplasmic organelle found in insects (King et al. 1982). The insect fusome contains several membrane-associated cytoskeletal proteins (Lin et al. 1994), and dynein (de Cuevas and Spradling, 1998). Fusomes are thought to control critical aspects of oogenesis (Lin et al. 1994), including synchronisation of gonial divisions and differentiation of one germ cell into an oocyte. Further investigation of this organelle in *Artemia* will yield interesting information.

Synaptonemal complexes (Figure 15) develop in all cells of the gonial cluster indicating that differentiation between oocytes and nurse cells has not taken place yet. In comparison, most nurse cells described previously either have incomplete synaptonemal complexes or lack those structures (Cassidy and King, 1972; Choi and Nagl 1976; Sabelli Scanabissi and Trentini, 1979;

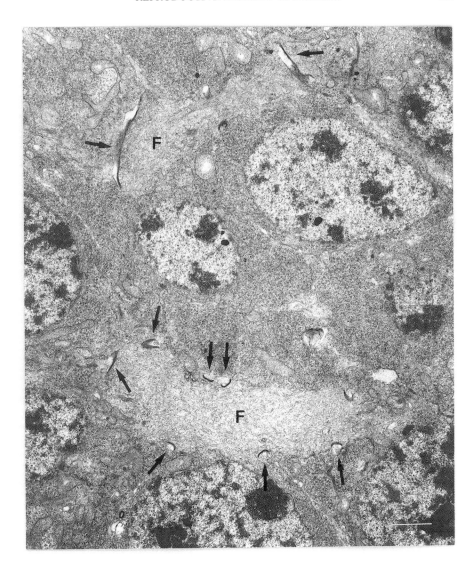

Figure 14. Two fusomes (F) between young oocytes. Arrows, bridge rims. Bar: 1 µm. (From Criel, 1989)

King *et al.* 1982). The primary function of the synaptonemal complex may be to provide a structural framework, ensuring that homologous DNA regions are brought into proximity for crossing-over, but its actual function is still in doubt (Santos, 1999). At the end of pachytene, or in early diplotene, fusomal material disappears and oocytes align into a ribbon, linked by cytoplasmic bridges delineated by electron-dense rims and filled with cytoplasm. Transitory cytoplasmic bridges are described for male and female germ cells in verte-

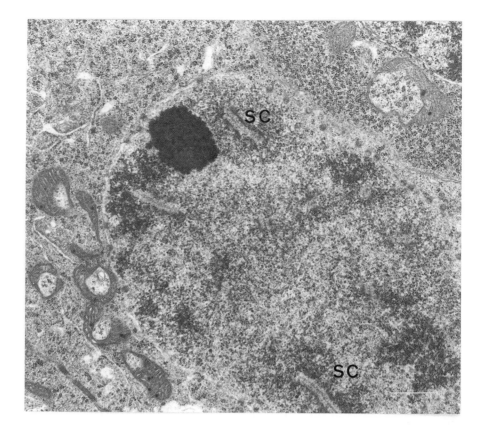

Figure 15. Nucleus of an oocyte in pachytene. Synaptonemal complexes (SC) surrounded by some chromatin. Bar: 1 μm. (From Criel, 1989)

brates and invertebrates, perhaps allowing germ cells to differentiate in a coordinated manner. After pachytene, cells prepare for synthesis and storage of reserve materials, and cytoplasmic volume increases, as is typical during previtellogenesis. This increase is found initially in all cells of the ribbon, due mainly to a greater number of ribosomes and, to a lesser extent, more endoplasmic reticulum. In late previtellogenesis, nuclei enlarge to approximately 12 μm, whereas cell height increases to 25 μm. Nuclei move from a central position to the convex side of the now curved oocyte ribbon, and the large central nucleolus in each cell is highly ramified. The nuclear pores appear to be sites for nucleocytoplasmic exchange, a function indicated by the electron-dense material associated with pores. The term 'nuage' (Dohmen, 1985) is used for material in the perinuclear region of germ line cells, and for cells later in development, where it is often associated with annulate lamellae (Figure 16). Kessel (1989) provided evidence for the progressive transformation of annulate lamellae into rough endoplasmic reticulum (RER) with intermediate smooth endoplasmic reticulum (SER) connections. 'Nuage'

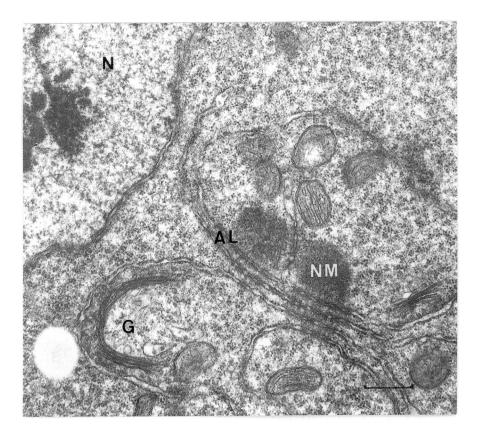

Figure 16. 'Nuage' material (NM) associated with annulate lamellae (AL) in late previtellogenic oocyte. The cytoplasm is filled with polyribosomes. N, nucleus; G, Golgi apparatus. Bar: 0.5 µm. (From Criel, 1989)

on the cytoplasmic side of the nuclear membrane was first described in *Artemia* by Anteunis *et al.* (1968). Transfer of nuclear material to the ooplasm is common for oocytes of various animal groups, and is seen as a prelude to endogenous yolk synthesis (Adiyodi and Subramoniam, 1983). At the end of previtellogenesis, ribosomes and polyribosomes constitute the bulk of the cytoplasm, but increasing numbers of mitochondria appear. RER cisternae arise and develop intracisternal granules, but Golgi systems are sparse. Small lipid and protein droplets appear, together with vesicular yolk bodies, the latter no greater than 2 µm in diameter. Oocyte-nurse cell differentiation marks the end of previtellogenesis wherein cytoplasmic increase occurs. Only oocytes undertake vitellogenesis, they round up, become covered with microvilli, and protrude above sister cells. The remaining cells of the ribbon become nurse cells. Oocyte nuclei transform into germinal vesicles at the end of previtellogenesis.

Incorporation of labelled thymidine demonstrated that DNA synthesis is restricted to the nuclei of nurse and somatic cells (Iwasaki, 1970). In meroistic insects, oocytes stop RNA synthesis as soon as the germinal vesicle is formed, whereas this activity continues in nurse cells (Bier, 1965). Nurse cells are connected to the oocyte and to each other by intercellular bridges arising from incomplete cytodieresis, providing a transport route for materials to the oocyte. *Artemia* oocytes grow at the expense of nurse cell activities, and the cytoplasm transferred from adjacent nurse cells contains mainly free ribosomes, mitochondria, lipid droplets, endoplasmic reticulum and annulate lamellae. Even at this early stage apoptosis is observed in nurse cells located at ribbon ends (Figure 17).

Accessory nuclei, apparently produced by blebbing of the germinal vesicle, as soon as it is formed disperse over the oocyte surface. Each accessory nucleus often contains a small nucleoid (Figure 18) and is covered by a double envelope containing annuli similar in size to those in the nuclei. Although, accessory nuclei are frequently found during vitellogenesis in insects, the only other crustacean where accessory nuclei are found is *Siphonophanes grubei* (Kubrakiewicz *et al.* 1991), another anostracan. At the same time small granules are formed in some endoplasmic reticulum cisternae and the lipid droplets and the clear vesicular yolk bodies increase to about 10 μm, but the small protein droplets remain unchanged.

As in most crustaceans an initial phase of endogenous yolk formation precedes exogenous uptake (Adiyodi and Subramoniam, 1983; Meusy and

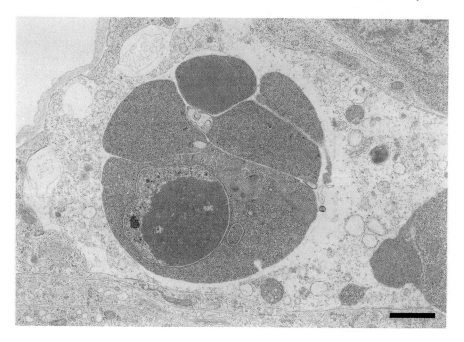

Figure 17. Apoptotic germ cell from an ovary in previtellogenesis. Bar: 1 μm.

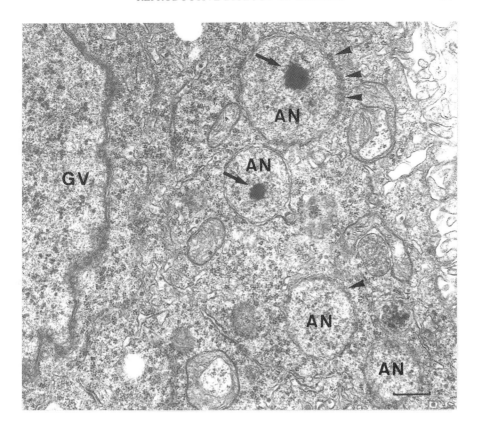

Figure 18. Accessory nuclei (AN) originate between the germinal vesicle (GV) and the oocyte surface at the onset of vitellogenesis. Arrows point to the nucleoids in the accessory nuclei, arrowheads point to their nuclear pores. Bar: 0.5 µm. (From Criel, 1989)

Charniaux-Cotton, 1984; Charniaux-Cotton, 1985; Cuoc *et al.* 1993). A so-called yolk nucleus appears in the centre of the oocyte (Anteunis *et al.* 1964). The yolk nucleus, composed of small vesicles, multivesicular and dense bodies and surrounded by 'nuage' material and dictyosomes, may contain acid phosphatase (Roels, 1968). The first accumulation of larger yolk products occurs as a layer midway between the yolk nucleus and the oocyte surface (Figure 19). Because of this location a major influence of the yolk nucleus on endogenous yolk formation is questionable. After ovulation, the yolk nucleus moves to the oocyte surface, and even though it has been traced until mitosis begins, it remains an enigmatic structure.

The layer of endogenous yolk products is composed of electron-dense protein, yolk platelets, lipid yolk droplets and clear vesicular yolk bodies. How yolk platelets arise is not clear, but newly formed endogenous yolk platelets often show a vesicular periphery, suggesting they have arisen in multivesicular bodies. The small yolk platelets seem to increase in size by fusion, as

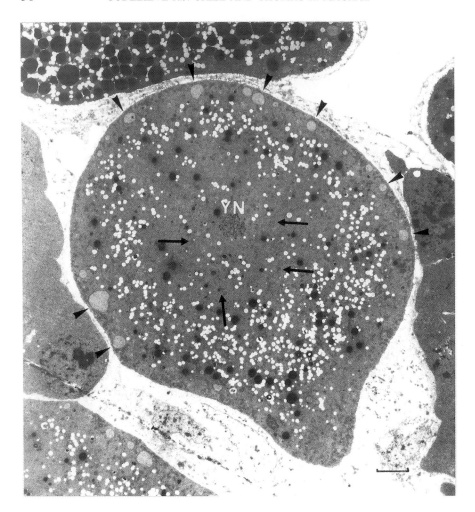

Figure 19. Very low magnification of an oocyte in early vitellogenesis: the yolk nucleus (YN) lies in the centre of the oocyte. The eccentric germinal vesicle is not in the plane of this section. Many accessory nuclei (arrowheads) lie at the periphery of the oocyte. Yolk products accumulate midway between the yolk nucleus and the oocyte surface. The clear granules are the lipid yolk droplets and the vesicular yolk bodies which cannot be discriminated at this low magnification, the dense granules are the protein yolk platelets. Arrows point to the 'nuage' material surrounding the yolk nucleus. Bar: 0.5 μm.

suggested by the presence of many irregular immature granules (Figure 20). As in *Orchestia* the lipid droplets lack a definite relationship with another oocyte organelle (Zerbib, 1980). The clear vesicular yolk bodies arise as dilated cisternae of the endoplasmic reticulum. Clear vesicular yolk bodies reside near the surface of mature eggs, but they do not represent typical cortical granules because they persist past the second cleavage. They were shown to contain peroxidase; Roels (1970) suggested that clear vesicular yolk bodies are per-

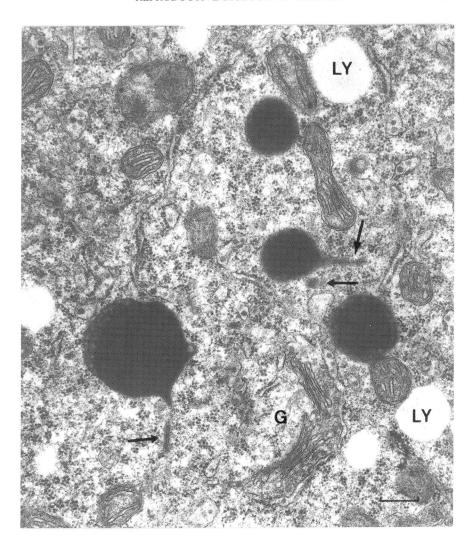

Figure 20. Onset of endogenous yolk synthesis. Immature yolk platelets show a vesiculated periphery. Fusion with dense vesicles and tubules (arrows) is seen. An endoplasmic reticulum cisterna with intracisternal granules lays at the forming face of a Golgi apparatus (G). LY, lipid yolk droplet. Bar: 0.5 µm. (From Criel, 1989)

oxisomes; Roels and Wisse (1973) suggested that they might play a role in shell hardening. Throughout the oocyte, granules accumulate within the endoplasmic reticulum, and vesicles containing several intracisternal granules pinch off from cisternae ends prior to ribosome loss (Figure 21). The relation of intracisternal granules to yolk platelets has yet to emerge. As in most crustaceans studied, the Golgi complex seems to play a minor role in *Artemia* yolk synthesis.

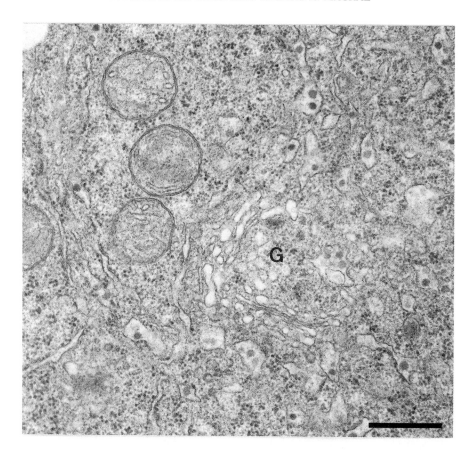

Figure 21. Cytoplasm at the onset of vitellogenesis: endoplasmic reticulum cisternae containing intracisternal granules in the neighbourhood of a Golgi apparatus (G). Bar: 0.5 μm.

Exogenous yolk formation starts forty to sixty hours after onset of the vitellogenic cycle. Intense endocytosis is evident, with coated pits at the bases of microvilli on the side of the oocyte directed toward the ovarial surface (Figure 22). Accessory nuclei concentrate in this region and several stages in the uptake of material are evident. The exogenous yolk product ingested by pinocytosis is probably vitellogenin. Its presence in the blood of vitellogenic female *Artemia* was demonstrated by Van Beek *et al.* (1987), who showed with immunocytochemical techniques that vitellogenin is also synthesized in fat storing cells. The possibility of protein uptake by the ovary was demonstrated by incubation with horseradish peroxidase. However, peroxidase was not transported to yolk platelets but to vacuoles, another finding that needs further study. In early stages, yolk platelets constitute a rim between the yolk nucleus and oocyte surface, indicating that mature yolk products arise by confluence of endogenous material transported from the oocyte centre to the periphery, with exogenous material moving in the opposite direction. Exo-

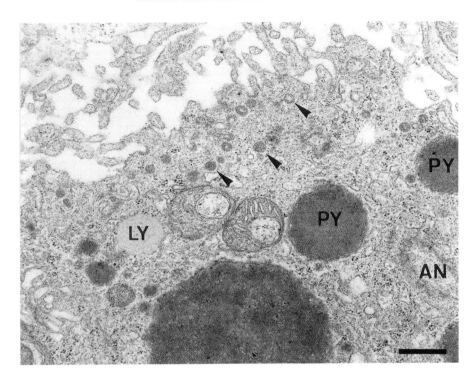

Figure 22. Exogenous yolk synthesis: coated vesicles (arrowheads) are pinched off between the villi and are transported to the protein yolk platelets (PY). AN, tangential section through an accessory nucleus; LY, lipid yolk droplets. Bar: 0.5 µm. (From Criel, 1989)

genous yolk platelets are morphologically indistinguishable from those formed by endogenous synthesis, but it is uncertain if they fuse with one another. Gradually, the entire oocyte fills with yolk. Four different yolk products are discerned: electron-dense protein yolk platelets, clear lipid yolk droplets, vesicular yolk bodies surrounded by an undulating membrane, and small intracisternal granules (Figure 23). Lack of any relationship between mitochondria and yolk platelets and new biochemical evidence allowed Warner *et al.* (2002) to refute the assumption that mitochondria are stored in yolk platelets (Marco and Vallejo, 1976; Vallejo and Marco, 1976a, b; Vallejo *et al.* 1979, 1980; Marco *et al.* 1980, 1981, 1983; Vallego and Perona, 1982).

Nurse cells enlarge and round up during vitellogenesis, caused in part by an increase in nucleus volume. Nuclei become polyploid, increasing rDNA content (Barigozzi, 1942; Lison and Fautrez-Firlefyn, 1950; Lochhead and Lochhead, 1967), a requirement for massive accumulation of rRNA in oocytes (Cave, 1982). Each nurse cell nucleus contains multiple, extremely ramified, nucleolar masses and many small chromatin blocks (Figure 24). Unlike oocytes, diffusion of 'nuage' continues in nurse cells until vitellogenesis ends. Many mitochondria, frequently of irregular shape and appearing as circles or dumbbells in section, occur in the sparse nurse cell cytoplasm. No yolk products

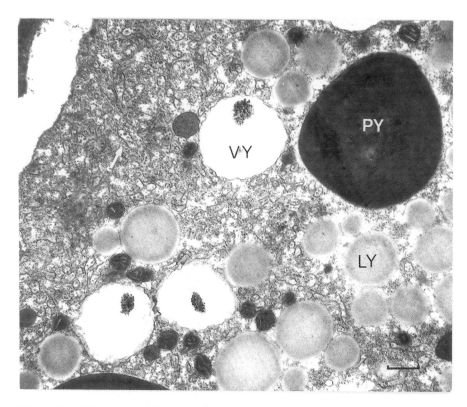

Figure 23. Yolk products of *Artemia* shown in an unfertilised egg. PY, proteinaceous yolk platelets; LY, lipid yolk droplets; VY, vesicular yolk bodies; arrow, endoplasmic reticulum cisternae with granular material. Bar: 1 µm. (From Criel, 1989)

are formed in *Artemia* nurse cells, although there are often a few lipid droplets associated with mitochondria. *Artemia* nurse cells exhibit neither pinocytotic activity nor microvilli extending toward somatic cells, suggesting exogenous material is not incorporated. *Artemia* nurse cells remain linked to each other and to the oocyte by cytoplasmic bridges until the end of vitellogenesis (Figure 25). Intercellular bridges in *Artemia* do not preclude asynchronous nuclear differentiation and endomitotic divisions of nurse cell chromatin occur, even though they are connected with the oocyte.

The nurse cell ribbon detaches from the oocyte upon termination of vitellogenesis and it is surrounded by somatic cells (Figure 26). Nurse cells then separate from each other, their nuclei show characteristics of apoptosis (Zheng *et al.* 1991), and they are phagocytosed by somatic cells. Degeneration of *Artemia* nurse cells was described previously (Lison and Fautrez-Firlefyn, 1950; Lochhead and Lochhead, 1967), but not related to termination of their role in vitellogenesis. After ovulation, large zones of somatic cells filled with degenerating nurse cells occur in some ovary regions (Figure 27), however, the fate of degeneration products is uncertain.

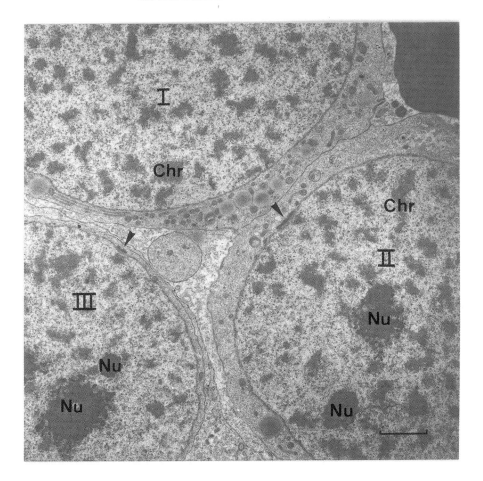

Figure 24. Parts of three nurse cells (I, II, III). The large nucleus contains several nucleoli (Nu) and many lumps of chromatin (Chr). Diffusion of 'nuage' material continues (arrowheads). Note the many lipid yolk droplets and mitochondria. Bar: 2 μm. (From Criel, 1989)

Female moulting happens during the final stage of the vitellogenic cycle when oocytes are covered with small villi while pinocytosis has ended. Vesicular yolk bodies concentrate in cortical layers, persisting subsequent to fertilisation. Accessory nuclei concentrate after moulting (Figure 28), and the oocyte initiates release from surrounding somatic cells, ultimately lying free in an irregular lumen of the ovary. The germinal vesicle breaks down, a tangential metaphasic spindle appears (Figure 29), and meiotic resumption of prophase I-arrested oocytes occurs in the ovary. Germinal vesicle breakdown and resumption of meiosis at metaphase I in *Artemia* ovaries is acknowledged by some authors (Brauer, 1893, 1894) and rejected by others (Gross, 1935; Barigozzi, 1944; Fautrez-Firlefyn, 1951). This difference may arise because germinal vesicle breakdown occurs after the moult, which imme-

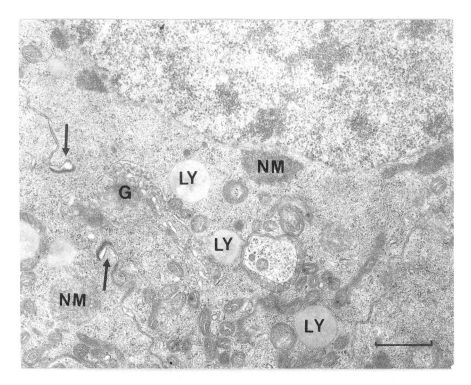

Figure 25. Higher magnification of nurse cell cytoplasm. Cell organelles pass through the cytoplasmic bridge bordered by a dense bridge rim (arrows). G, Golgi apparatus; LY, lipid yolk droplet; NM, 'nuage' material. Bar: 1 µm. (From Criel, 1989)

diately precedes ovulation. *Artemia* oocyte maturation thus seems correlated to moult, and germinal vesicle breakdown may be related to increased ecdysteroid titres that occur before moult (Walgraeve *et al.*, 1988). After ovulation, oocytes move into the oviducts that expand to create so-called lateral pouches. Metaphasic spindles move from tangential to perpendicular positions, and a second meiotic block is maintained while oocytes are in the extended oviducts.

3.2. SPERMATOGENESIS

Spermatogenesis comprises the sequence of events by which spermatogonia are transformed into spermatozoa. Spermatogenesis in *Artemia* males is not restricted to testis, and sperm cells have yet to achieve their definitive characteristics upon entering the testicular lumen (Wingstrand, 1978; Criel, 1980, 1991). Therefore, testis structure will be described when spermatogenesis is considered, as will spermatozoa in the *vas deferens* and in the median part of the ovisac, the latter functioning as a *receptaculum seminis*. For convenience, description of spermatogenesis is divided into three principal phases. In the first phase, termed spermatocytogenesis, spermatogonia proliferate by

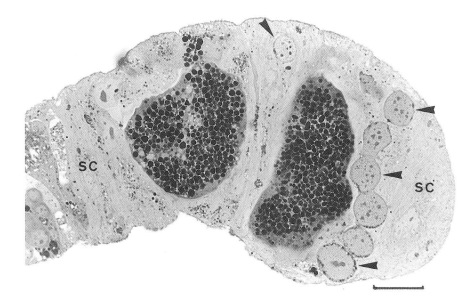

Figure 26. Phase-contrast micrograph of 2 µm section through oocytes at the end of vitellogenesis. The nurse cells (arrowhead) are released by the oocyte and phagocytosed by somatic cells (SC). Bar: 20 µm.

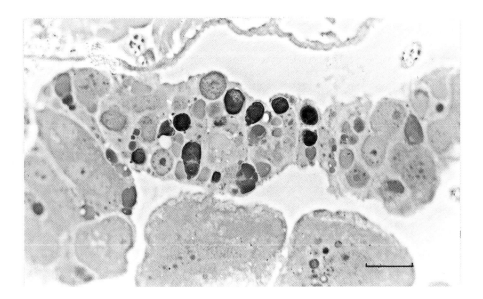

Figure 27. Phase-contrast micrograph from the central part of an ovary shortly after ovulation: phagocytosed nurse cells can still be recognised in the somatic cells. Bar: 20 µm. (From Criel, 1989)

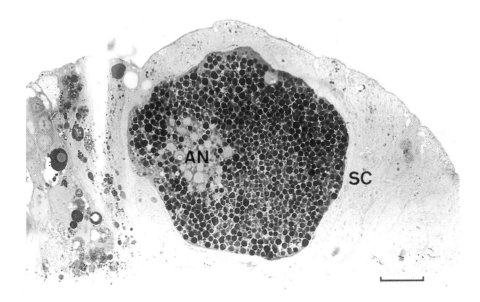

Figure 28. Phase contrast micrograph of a 2 µm section through an ovary in early postmoult. In the oocyte surrounded by clear somatic cells (SC) an accumulation of accessory nuclei (AN) is seen. Bar: 20 µm.

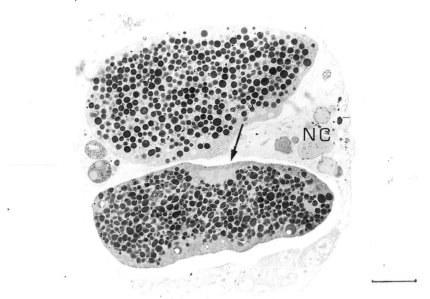

Figure 29. Phase contrast micrograph of a 2 µm section through an ovary in early postmoult. Metaphase chromosomes are seen in the equator of the meiotic spindle (arrow) in an oocyte which has become loose from its surrounding cells. NC, rejected nurse cells. Bar: 20 µm.

mitotic divisions, giving rise to successive generations of spermatogonia and eventually to primary spermatocytes. The second phase is meiosis, where spermatocyte chromosome number is halved, yielding spermatids. In the third phase, termed spermiogenesis, spermatid transformation leads to spermatozoa.

Some ultrastructural aspects of spermatogenesis have been described (Brown, 1970a) and the testis was shown to consist of two major categories of ontogenetically distinct cells, namely supporting somatic cells and spermatogenetic germ cells. The uniform supporting cells are tentatively called 'Sertoli' cells because of similarity with vertebrate Sertoli cells. Spermatogenic cells, on the other hand, result from the continuous differentiation of male germ cells, and are comprised of spermatogonia, primary spermatocytes, secondary spermatocytes, spermatids and spermatozoa. Bridges connecting these cells throughout spermatogenesis are equivalent to those found in the ovary.

Spermatogonia are located at the periphery of testis tubules, separated from the tubule wall by a thin rim of 'Sertoli' cells, and, in most cases, only a few spermatogonia group at a single location. Typical spermatogonia are small (8 to 10 µm) polygonal cells, with a large round to slightly oval nucleus of 7 to 8 µm, surrounded by a small amount of cytoplasm. The cytoplasm, packed with ribosomes, contains dispersed mitochondria and very little rough endoplasmic reticulum. Clumps of heterochromatin are scattered throughout the nucleoplasm, some adhering to the nuclear envelope. A prominent nucleolus surrounded by chromatin lies almost centrally. Occasionally, more than one nucleolus is found. Spermatogonia contain one centriole pair located in a nucleus indentation.

The germ cells are linked by occluded intercellular bridges in pro-metaphase and metaphase, with spindle axes often at right angles to bridges. The bridges are 0.8 to 1.5 µm wide, and blocked by approximately ten bowed transverse lamellae, perhaps composed of endoplasmic reticulum (Dym and Fawcett, 1971; Durfort et al. 1980; Schindelmeiser et al. 1985; Weber and Russel, 1987). Closed bridges are found between cells of the same and of different mitotic stages (Figure 30). Occluded intercellular bridges are rare in ovaries (Anteunis et al. 1966), but frequent in the testes, and they have been described previously in Artemia (Brown, 1970a). Transverse septa in the bridges may bring about a transient isolation of dividing spermatogonia and spermatocytes, preventing cell death from spreading to linked cells (Dym and Fawcett, 1971). On the other hand, septa could stabilise pre-existing bridges as cells undergo mitotic and meiotic divisions (Weber and Russel, 1987). The origin and fate of obstructing membranes are not clear in Artemia.

Chromosomes become tightly packed in pro-metaphase, forging a single large chromosomal mass. The nuclear envelope breaks down into small vesicles surrounding the spindle in late pro-metaphase, and during metaphase, the chromosomal mass, mitotic spindle and centrioles are the most conspicuous cell structures. Two centrioles, oriented at right angles to each other, reside at spindle poles (Figure 31), and a limited number of microtubules constitute each

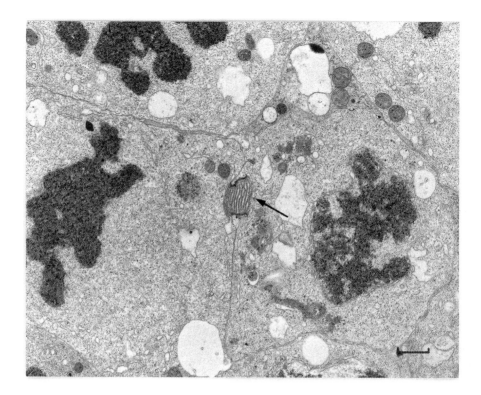

Figure 30. Closed intercellular bridge (arrow) between a metaphase and a telophase cell. Bar: 1 μm.

mitotic spindle. *Artemia* has holokinetic chromosomes (Stefani, 1963a, b; Stefani and Cadeddu, 1967; Abatzopoulos *et al.* 1986; Barigozzi, 1989). It is the only crustacean known to have this chromosome type. During anaphase, holokinetic chromosome arms remain parallel to the spindle equator as they move poleward, contrasting monokinetic chromosomes where arms lag behind kinetochores. This difference is due to location of the centromere, the chromosomal locus that nucleates kinetochore formation. Monokinetic chromosomes usually have one centromere, but holokinetic chromosomes have non-localised centromeres. Kinetochore material may either be arranged diffusely along the entire chromosome length, or distributed as multiple discrete packets. Vesicles originating as products of previous membrane disruption accumulate at both poles and reform the nuclear envelope during telophase.

Up to eight early primary spermatocytes are linked by cytoplasmic bridges containing fusomal-like material composed of microtubules, vesicles and a fine granular material that disintegrates in pachytene and disappears by the time meiotic divisions start (Figure 32). Although considered to be a mitotic spindle remnant, we have never observed fusome formation, nor have we seen mitosis in fusome linked cells. Primary spermatocytes initially resemble spermatogonia

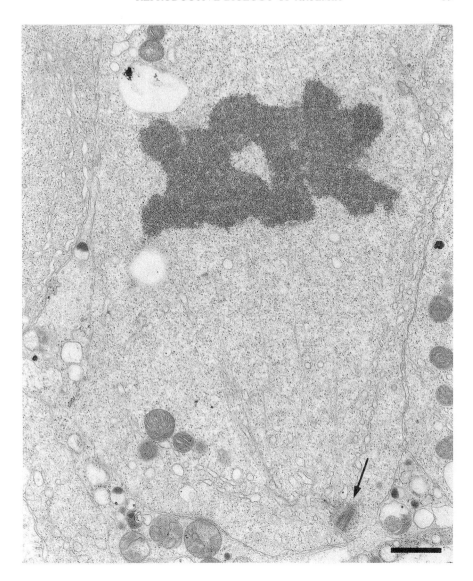

Figure 31. Details of the mitotic spindle: short cisternae of rough endoplasmic reticulum, and remnants of the nuclear envelope run on the outer side of the spindle, vacuoles and electron dense mitochondria are located outside the spindle area (arrow, one of the centrioles). Bar: 1 µm.

from which they arise, but as cytoplasm accumulates, they become distinctly larger and the nucleocytoplasmic ratio decreases. Free ribosomes and polyribosomes are prominent in primary spermatocyte cytoplasm, endoplasmic reticulum increases, limited Golgi develops, and mitochondria cluster. Long cisternae, perhaps derived from annulate membranes, envelop clustered mitochondria. Chromatin blocks enlarge, clearing the nucleoplasm, and the

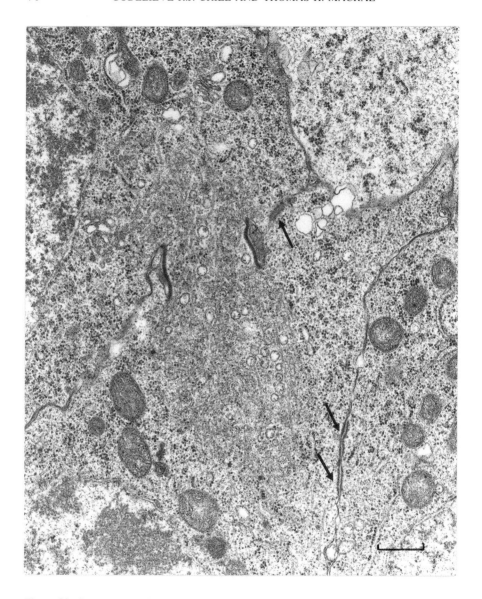

Figure 32. Open cytoplasmic bridge containing fusomal material linking preleptotene or 'resting primary spermatocytes'. Arrows: *maculae adhaerentes* between adjacent spermatocytes. Bar: 0.5 μm.

nucleolus lies close to the nuclear envelope. Centrioles are located in a more or less shallow indentation of the nuclear envelope and centrosomes duplicate in preparation for meiosis (Figure 33).

The meiotic zygotene stage features initiation of synaptonemal complex assembly, ensuring precise matching of homologous chromosomes. Most often,

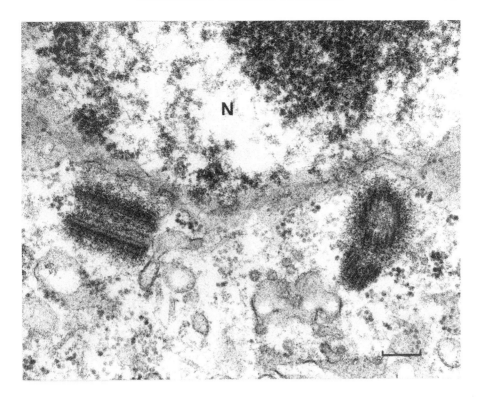

Figure 33. Duplication of the centrosome in preparation for meiosis. (N), nucleus. Bar: 200 nm.

all synaptonemal complexes attach at the same nuclear envelope region and a 'bouquet' arrangement arises. The most distinct appearance of synaptonemal complexes is attained in pachytene, after which they are arranged as bipartite structures attached to the nuclear membrane. The central region of synaptonemal complexes is striated and about 50 nm wide, while lateral elements are amorphous (Figure 34). Condensed chromatin is apparent along the synaptonemal complex, firmly attached to the outer surface of lateral elements. Occasionally, recombination nodules (Figure 35), structures correlated with pairing, are observed. Their distribution is similar to that seen in recombination events (Dresser, 1987), and they may contain enzymes required for recombination. The nucleolus is circular in pachytene, positioned close to the nuclear envelope, but not adhering to it. The nuclear envelope becomes irregular in pachytene, although perinuclear cisternae maintain regular width. Additionally, the two adjacent centriolar pairs surrounded by electron dense pericentriolar material (Figure 36) are near the nuclear envelope.

Pachytene-diplotene transition is marked by the initiation of synaptonemal complex degradation, accompanied by progressive but transient chromatin decondensation. Chromatin recondenses in diakinesis, resulting in compact

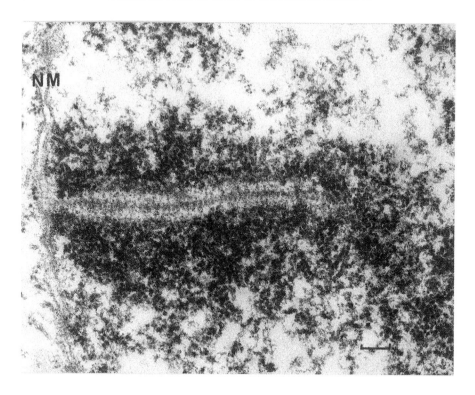

Figure 34. Synaptonemal complex attached to the nuclear membrane (NM) showing its bipartite structure: a striated central region and amorphous lateral elements. Condensed heterochromatin is apparent along the synaptonemal complex. Bar: 200 nm.

metaphase I bivalents. The centriole pairs migrate toward the opposite poles in late diakinesis; fusomal material within intercellular bridges disintegrates and disappears as meiotic divisions start. After pachytene, cytoplasm fills open intercellular bridges during interphase, with partitioning complexes again developing in mitosis and meiosis seeming to close pre-existing bridges. Complex membranous systems develop near disintegrating fusomes, with cisternae and straight short microtubule-like structures surrounding mitochondria (Figure 37). A correlation is suggested between fusome disintegration and development of the membranous system. The fusome may contain tubulin remnants from the mitotic spindle, some of which is used in construction of microtubules found between membranous system lamellae. No later stages of first meiotic divisions are identified.

Secondary spermatocytes are round to cuboid cells, about 7 to 9 μm wide and dispersed in groups of 2 to 4 cells. Their round nuclei lack nucleoli, but contain scattered clumps of chromatin. The outer membrane of nuclear envelope is often irregular, yielding a perinuclear cisterna of varying dimensions. Centrioles lie in shallow indentations of nuclear membranes at variable

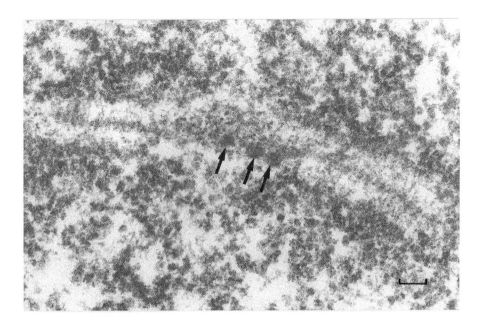

Figure 35. Dense spherical accumulations of material associated with the central region thought to be recombination nodules (arrows). Bar: 100nm.

angles to each other. The most prominent organelle of secondary spermatocytes is the complex membranous system which concentrates around nuclei. This abundant membranous system was not noticed by Brown (1970a). Microtubules remain associated with membrane complexes (Figure 38), but mitochondria have almost completely dissociated, appearing in clusters intermingled with coarse granular material resembling very electron dense ribosomes (Figure 39). A system of microtubule-associated smooth tubules develops on the plasma membrane facing other spermatocytes (Figure 40). Tubules may be seen bridging this space if it is narrow.

Only two groups of four cells presumed to be in second meiotic prophase have been encountered in all secondary spermatocytes examined and one group of four cells was in metaphase II. Membranous system elements which develop fully in secondary spermatocytes, and positioning of cells near the testis lumen, strengthen this conclusion because these characteristics allow identification of cell stages. At the end of prophase in secondary spermatocytes, chromosomes produce dense patches in clear nucleoplasm, nuclear membrane indents and microtubules, radiating from the centrosome (Figure 41), enter nuclei by passing through their pores. The centrosomal centrioles are oriented at right angles and surrounded by dense pericentriolar material (Figure 42). Occluding septae reappear in cytoplasmic bridges. Contrasting the metaphases described previously, chromosomes do not seem to aggregate. Because chromosome number is reduced, this finding favours the hypothesis that an artifactual aggre-

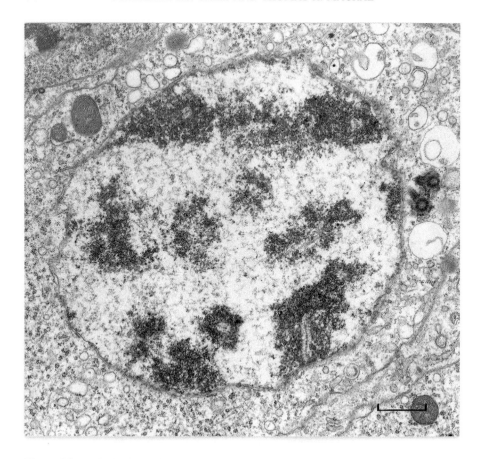

Figure 36. Reduplicating pair of centrioles located near the nuclear envelope surrounded by an electron dense pericentriolar material in a pachytene primary spermatocyte. Bar: 1 μm.

gation of chromosomes is provoked by their high number in first metaphase (Rieder *et al.* 1990). *Artemia* exhibit a trace of meiotic metaphase, anaphase and telophase during spermatogenesis.

Though a common structural plan is found for all spermatids, variation due mainly to membrane system development and distribution does exist. Spermatids are round, with a diameter of 7 to 8 μm and a large, centrally positioned nucleus encased by a cytoplasmic rim. Chromatin is more or less condensed and a small lamellated nuclear body is the most distinctive characteristic of the spermatid nucleus (Figure 43). It is still seen in the fertilising spermatozoa. The maximum nuclear body diameter is 0.6 μm, and it consists of concentric electron dense layers surrounding a fibrillar centre. Because the composition of *Artemia* nuclear bodies was not investigated, their function is uncertain, but their study in rat plasmocytes suggests a role in protein synthesis (Simar, 1969). The structural resemblance of rat nuclear bodies with those of *Artemia* is striking, nonetheless a function in sperm cell protein syn-

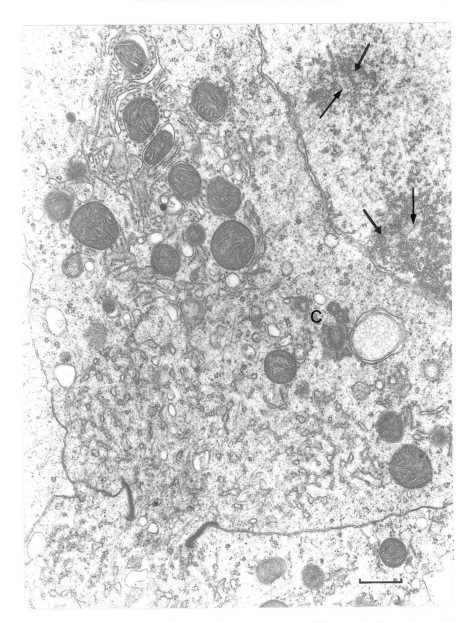

Figure 37. Zygotene or early pachytene primary spermatocyte: disintegrating fusomal material and developing complex membranous system surrounding the mitochondria. Microtubules longitudinally and transversely sectioned separate the cisternae at regular intervals. (C, centrioles; arrows, incipient synaptonemal complexes). Bar: 0.5 μm.

thesis is unlikely. Additionally, their persistence in sperm after fertilisation indicates possible function in genetic information transfer.

Spermatid centrioles lie at variable angles to each other, sometimes arranged

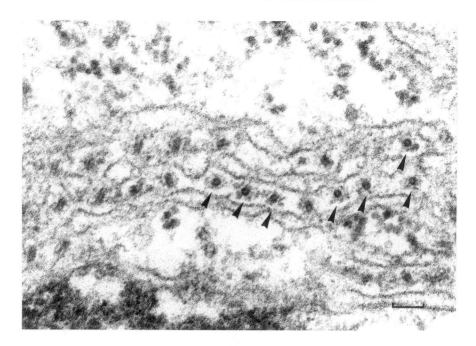

Figure 38. High magnification of transversely sectioned microtubules (arrowheads) from the membranous system. They are surrounded with filamentous material. Bar: 100 nm.

in parallel fashion in a shallow indentation of the nuclear membrane, and occasionally a small basal plate is seen at the end of some longitudinally sectioned centrioles (Figure 43). The persistence of two centrioles in spermatids and mature spermatozoa is a characteristic *Artemia* shares with other euphyllopods (Wingstrand, 1978) and with several mature malacostracan spermatozoa (Chevaillier and Maillet, 1965; Reger, 1966; Pochon-Masson, 1968a, b, 1969; Cotelli *et al.* 1975, 1976; Fain-Maurel *et al.* 1975; Cotelli and Lora Lamia Donin, 1983; Jespersen, 1983). There is, however, no agreement on the generalised presence of centrioles in other nonflagellate crustacean spermatozoa: no centrioles in mature spermatozoa of the coconut crab, *Birgus latro* (Tudge and Jamieson, 1991). Both centrioles remain in the copepods *Naobranchia cygniformis* (Manier *et al.* 1977) and *Ergasilus* (Coste *et al.* 1982), while in other copepods, *Labidocera* (Blades-Eckelbarger and Youngbluth, 1982) *Tisbe* and *Myticola* (Coste *et al.* 1982), centrioles disappear during spermiogenesis.

The cytoplasmic membranous system almost completely disappears during spermiogenesis, leaving a few lamellae surrounding nuclei and extending into the cytoplasm, a situation described in other crustaceans (Fain-Maurel *et al.* 1975; Reger *et al.* 1984). In many spermatids, microtubules persist between lamellae (Figure 43), but they disappear in other spermatids and the lamellae adhere to nuclear membranes (Figure 44). It is thought that the

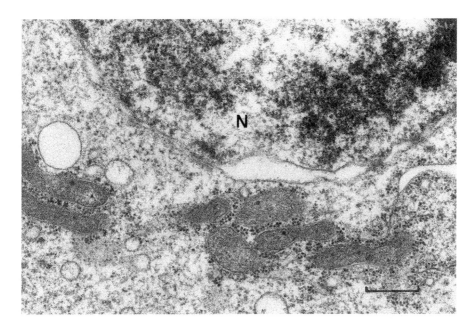

Figure 39. Enlargement of a cluster of mitochondria which have separated from the membranous system and lie intermingled with coarse granular material resembling very electron dense ribosomes (N, nucleus). Bar: 0.5 µm.

spermatids just described are most mature. The smooth tubular system adjacent to the plasma membrane persists in many spermatids, often forming a dark layer under the plasma membrane.

Spermatid mitochondria are associated with electron-dense, ribosome-like granules, perhaps derived from fibrogranular or 'nuage' material occasionally associated with nuclear pores in primary spermatocytes. The fibrogranular material does not further differentiate during *Artemia* spermiogenesis, and the significance of fibrogranular material and mitochondria association is unclear. Mitochondrial changes were not observed during spermiogenesis. Additionally, spermatid cytoplasm contains free ribosomes, microtubules, occasional lipid droplets, several irregular clear vacuoles and assorted vesicles, among which are some with a very electron-dense homogeneous granule about 200 nm in diameter persistent in mature and even fertilising sperm cells (Figure 46). To our knowledge, there is no equivalent of the membrane-bound, dense granules previously reported, but they must be considered because they enter *Artemia* oocytes upon fertilisation.

Germ cells found in the *vas deferens* (Figure 45) differ from those observed in the testis, and they are thought to be spermatozoa. Our study supports the conclusion that *Artemia* spermatozoa are 'unique and not comparable to any other crustacean nor other animal spermatozoa' (Brown, 1966, 1970a, b; Wingstrand, 1978). Wingstrand (1978) defines the branchiopod-type of sper-

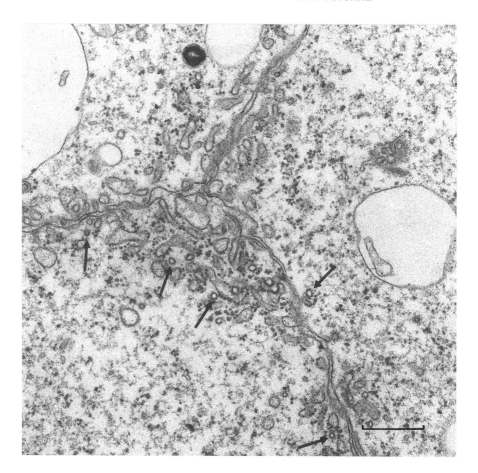

Figure 40. Enlargement of the smooth tubular system on the plasma membrane of adjacent spermatocytes. Arrowheads: microtubules in cross section. Bar: 0.5 μm.

matozoon as lacking a flagellum and acrosome, having a well-defined nucleus with a typical double envelope, and unmodified mitochondria. He states 'the branchiopod spermatozoa lack all the characteristic features usually associated with the concept 'spermatozoon' except that they are used to fertilize the eggs'. According to Fränzen (1970), branchiopod spermatozoa are highly modified, and for invertebrates such sperm cells are characteristic of internal fertilisation (Hodgson, 1986). On the other hand, it has been proposed that *Artemia* spermatozoa are morphologically unspecialized because they have changed little from the spermatid stage when compared with other crustacean spermatozoa (Brown, 1970a). Just as there is considerable variation from one *Artemia* spermatid to another, we also observed marked differences between spermatozoa in a single section through the *vas deferens*. Differences mostly concern general cell density and compactness, possibly due to unequal fixation.

Spermatozoa are spherical structures with about the same diameter as sper-

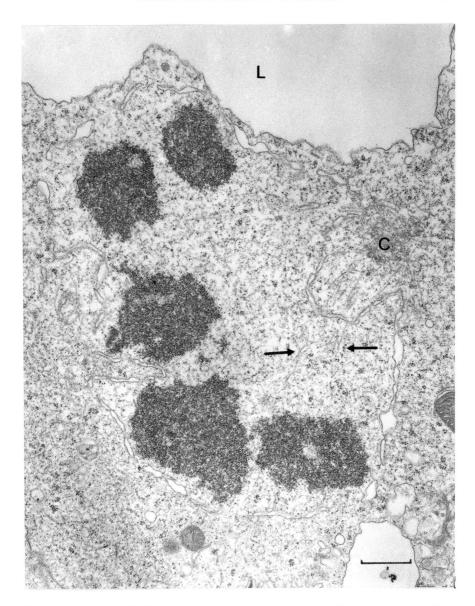

Figure 41. Low magnification of a nucleus at the end of prophase II: the chromosomes form dense patches in a clear nucleoplasm. The nucleus is indented and microtubules radiating from the centrosome enter the nucleus (arrows). C: tangential section through pericentriolar material, L, lumen of the testis. Bar: 1μm.

matids. They most often show small protuberances and are covered with a thin extracellular coat. The centrally located nucleus, at 3 to 4 μm in diameter, is somewhat smaller than the spermatid nucleus. As in the majority of crustaceans the chromatin in *Artemia* sperm cells is almost uncondensed,

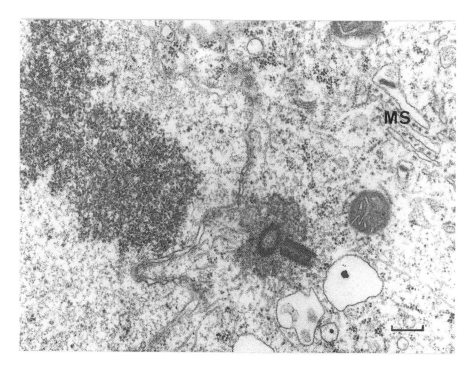

Figure 42. Centrosome with two centrioles oriented at right angles and surrounded by pericentriolar material (MS, part of membranous system). Bar: 0.5 μm.

contrasting the tightly packed chromatin found in 'typical' spermatozoa (Hinsch, 1980). The ultrastructure of this uncondensed chromatin varies among species (McKnight and Hinsch, 1986), and nuclear condensation varies from sperm cell to sperm cell in *Artemia*. Patches of chromatin and a nuclear body are found in every nucleus. Irregular granules fill some dilations of the perinuclear cisterna (Figure 46) and they persist in the fertilising sperm cell. The nuclear envelope is maintained in *Artemia* sperm cell, which is typical for all branchiopods (Wingstrand, 1978) but unlike most decapods (Arsenault, 1984). Two neighbouring centrioles are located near close to the nuclear membrane. As in spermatids, development of the membranous system varies, with some long, tightly packed cisternae running through the cytoplasm and at least partly surrounding the nucleus, often giving the nuclear membrane a pentalaminar appearance (Figure 47). These cisternae apparently arise from the membranous system, but the function of lamellar profiles is obscure. Circular profiles of concentric membranes, probably of similar origin, are found in most sperm cells. Some small flattened vesicles, probably issued from the tubular system, are located close to the plasma membrane (Figure 47). Golgi complexes are mostly absent but clusters of small vesicles and multivesicular bodies are found. Other cytoplasmic constituents show negligible change, with mitochondria not displaying structural modifications other than associa-

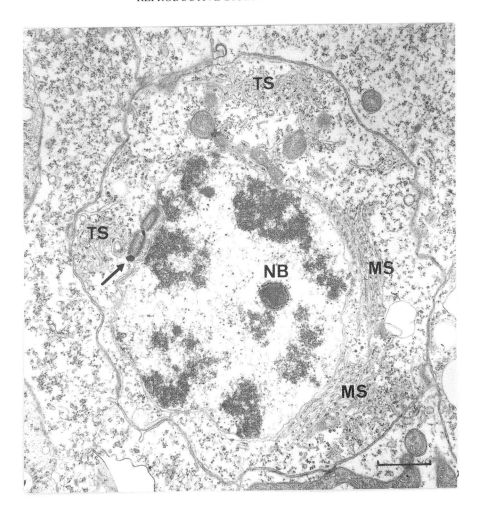

Figure 43. Young spermatid in the germinal epithelium of the testis. NB, nuclear body, characteristic for spermatids; MS, membranous system; TS, tubular system; arrow, basal plate attached at the end of the longitudinally sectioned parallel centrioles. Bar: 1 µm.

tion with ribosome-like granules. In agreement with Brown (1966, 1970a, b), we did not find a clearly defined acrosome; this is common in branchiopods (Delavault and Bérard, 1974; Wingstrand, 1978; Zaffagnini, 1987). Branchiopods are not the only group with spermatozoa lacking an acrosome, and they are often absent when spermatozoa encounter naked eggs (Clark and Griffin, 1988) or eggs with poorly developed envelopes (Villa and Tripepi, 1983; Hodgson *et al.* 1988). In this context, *Artemia* eggs have no chorion presenting a barrier to spermatozoon entry. However, the cluster of small vesicles observed in *Artemia* sperm cells may represent a multiple acrosome system, as occurs in some Cnidaria (Lunger, 1971; Hinsch and Clark, 1973; West, 1980).

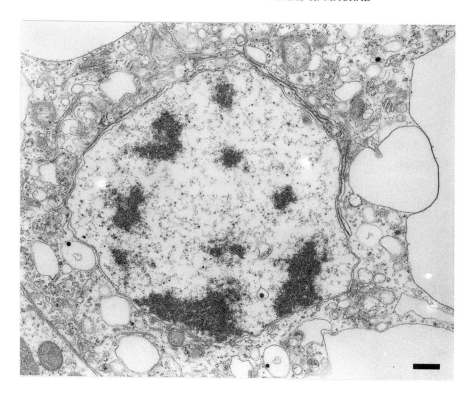

Figure 44. A more mature spermatid in the germinal layer of the testis: microtubules have disappeared from the membranous system, the lamellae adhere to the nuclear membrane. Bar: 0.5 µm.

Fertilisation in *Artemia* occurs within an hour after copulation (Benesch, 1969), and before this event spermatozoa reside in the median area of the ovisac, which functions temporarily as a *receptaculum seminis* (Figure 48) (Criel, 1980). Ultrastructural investigation of spermatozoa *in utero* did not reveal significant differences from those found in the *vas deferens*, but spectacular changes occur after opening of the sphincter that separates lateral pouches from the ovisac.

3.3. FERTILISATION

Fertilisation has been described as a unique occurrence, constituting an interaction between totally dissimilar cell types leading to a cascade of events and resulting in a new individual (Cran and Esper, 1990). A fundamental series of changes upon cell-cell contact allows species recognition, passage of the spermatozoon across the oocyte investment, and penetration into maternal cytoplasm. Oocytes initiate embryonic development in response to the stimulus provided by fertilising sperm. One activational phenomenon is resumption

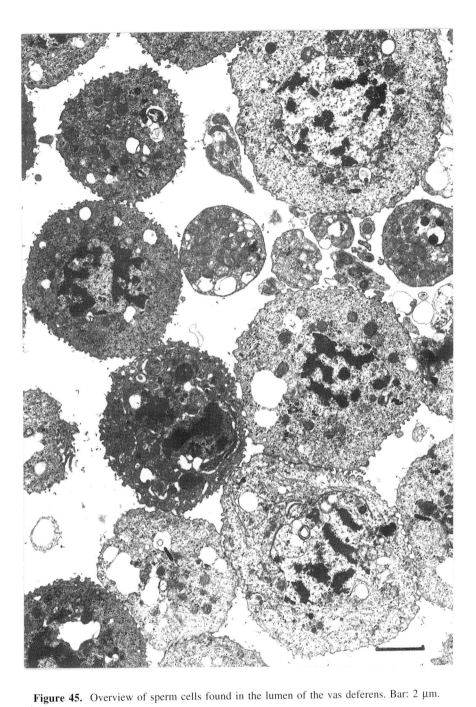

Figure 45. Overview of sperm cells found in the lumen of the vas deferens. Bar: 2 μm.

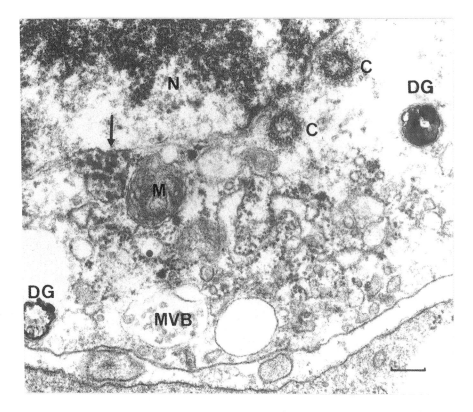

Figure 46. Detail of a sperm cell showing a delatation of the perinuclear cisterna containing irregular granules (arrow), a multivesicular body (MVB), a very electron dense membrane bound granule (DG), and two centrioles (C). (M: mitochondrion, N: nucleus). Bar: 200 nm.

of meiotic maturation, another is formation of the fertilisation membrane. These diverse aspects of fertilisation are dealt with later.

3.3.1. Sperm Cell Incorporation and Development of the Male Pronucleus
Artemia sperm cells emit pseudopodia upon release from the male genital tract into the surrounding medium (Brown, 1966). Similar changes occur in spermatozoa when lateral pouches open and eggs are released into the ovisac to be fertilised (Criel, 1980, 1989). Pseudopodia are best illustrated by scanning electron microscopy (Figure 49), appearing as numerous threadlike extensions that emerge from the sperm cell body. These changes are not categorised as an acrosomal reaction because *Artemia* sperm cell lack a typical acrosome.

Even though *Artemia* sperm cell development seems quite different from that of the decapod sperm cell, analogies can be drawn with regard to final maturation. Emission of extensions from *Artemia* sperm cells is as impressive as both the 'explosion' of reptantian sperm cells and the unfolding of radial arms in some natantian sperm cells. There is no consensus as to the

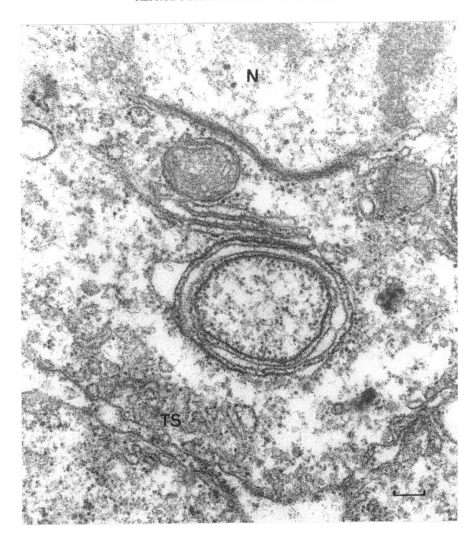

Figure 47. Detail of a sperm cell showing a cisterna tightly packed to the nuclear membrane and a circular profile of concentric membranes. Some smooth tubules lie close to the plasma membrane (TS) (N: nucleus). Bar: 200 nm.

role of spikes or radial arms during fertilisation (Pochon-Masson, 1965; McKnight and Hinsch, 1986). Moreover, spermatozoa in the bottom of the female *receptaculum*, remote from eggs, also undergo these modifications, indicating the changes are induced by alteration of uterine fluid upon opening of the oviducts, rather than through egg contact. Since these changes were also observed upon release of spermatozoa from the *vas deferens* into sea water (Brown, 1966) it is unknown if the morphological conversion results from a particular component of oviductal fluid, or merely from an environmental

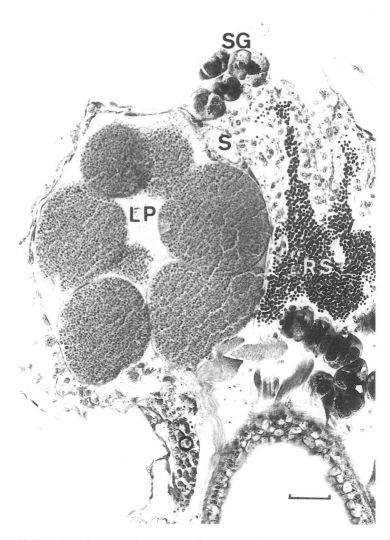

Figure 48. Frontal section through the ovisac of an animal with the eggs in the lateral pouches (LP) five minutes after copulation. (RS, receptaculum seminis; O, ovary; S, shutter separating the pouches from the ovisac; SG: shell gland). Bar: 50 μm.

change in either osmolarity, pH, or ionic composition. In this context, decapod sperm cells reaction is elicited by hypotonic conditions (Hinsch, 1971; Goudeau, 1982). As revealed by transmission electron microscopy, some sperm cell extensions (Figure 50) appear flexible, exhibiting only ribosomes, but most seem rigid and contain microfilaments. Their proximal part often possesses a central tubule, shown by serial sectioning to be an outer nuclear membrane extension (Figure 51). At this mature stage, spermatozoa are no longer round

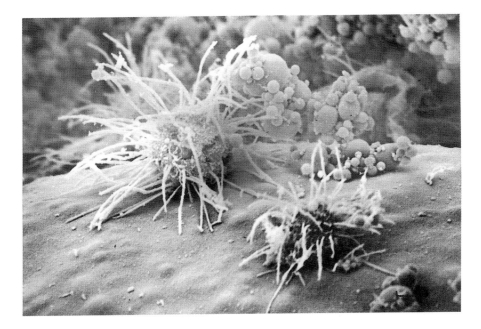

Figure 49. Scanning electron micrograph of a fertilising sperm cell. Bar: 2 μm.

or oval but triangular, still lacking polarity. Their nuclei and cytoplasm change little and centriole orientation is variable.

Spermatozoon incorporation takes place at an angle of 45° to the polar body, and at the penetration site a small surface protrusion containing cytoplasmic components but no microfilaments is formed, only to disappear within minutes (Figure 52). Because no remnants of sperm cell plasma membrane are observed inside the eggs, entry is probably by conventional membrane fusion. All sperm cell organelles, including centrioles, mitochondria, multivesicular bodies, membrane-bound electron dense vesicles, and some fibrous material derived from extensions are recovered inside the oocyte. It is thought that large swollen mitochondria surrounding nuclei are derived from sperm cell, and small dark mitochondria are from eggs. Clear vacuoles surround the incorporated sperm, but it is not clear if they originate from the spermatozoon or egg. They could be related to the intracellular digestion of sperm cell organelles which disappear by the time of male *pronucleus* formation.

The nuclear body remains in the nucleus during early fertilisation events, as do dense granules between nuclear lamellae and attached remnants of membranous systems (Figure 53). As for all eggs that are inseminated when arrested in meiotic maturation, sperm cell nuclei 'wait' in the cortical region for maternal chromatin to complete meiotic maturation. Moreover, the sperm cell nucleus must transform into a male *pronucleus* before association with the female *pronucleus* and participation in mitosis. The morphological events of male pronuclear development are generally divided into three stages, namely

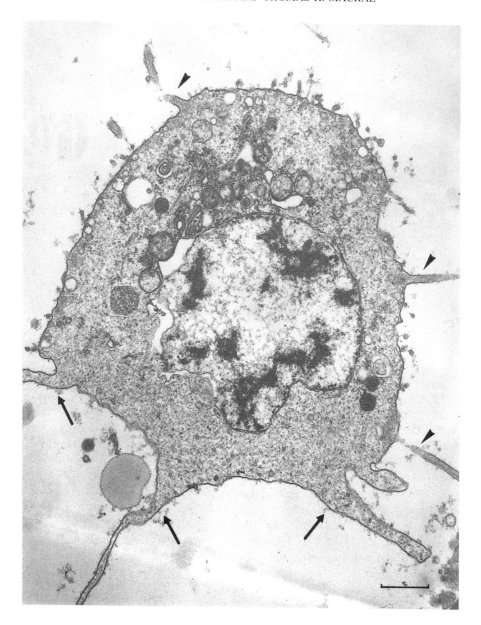

Figure 50. Electron micrograph of a fertilising sperm cell showing how some extensions are filled with cytoplasm (arrows) while others contain microfilaments (arrowheads). Bar: 1 μm.

breakdown of the incorporated spermatozoon nuclear envelope, dispersion of sperm cell nuclear chromatin, and *pronucleus* nuclear envelope reconstitution (Longo, 1973). In *Artemia*, fenestrations appear in sperm cell nuclear membranes almost immediately after penetration, but complete breakdown

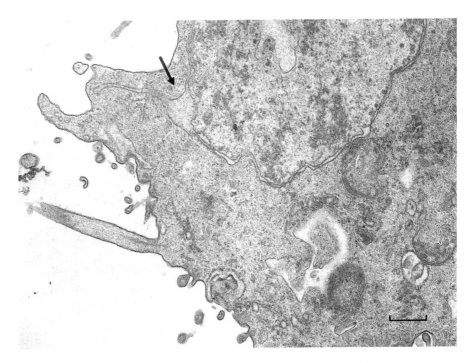

Figure 51. Detail of a fertilising sperm cell demonstrating an extension of the outer membrane of the nuclear envelope (arrow). Bar: 0.5 µm.

of nuclear envelopes is never seen, probably because these openings allow interaction with egg cytoplasm. Chromatin reorganisation and dispersion proceed irregularly.

There is a wealth of information indicating that transformation of sperm cell nuclei into male *pronuclei* depends upon factors that appear with ovum development (Longo, 1984; Lasalle and Testart, 1991; Longo *et al.* 1991) and that sperm cell nuclear enlargement is coupled to meiotic processing of maternal chromatin (Luttmer and Longo, 1987, 1988; Wright and Longo, 1988). The nuclear body eventually disappears, but the blebbing of outer nuclear lamella which liberates granules and membranes into the ooplasm goes on (Figure 54). The role of granule-containing vesicles pinched off from outer leaflets of sperm cell nuclear membranes is not clear. Upon formation of second polar bodies, but prior to full chromatin dispersion, developing male *pronuclei* start to migrate. Formation of sperm cell asters and evolution of sperm cell centrioles have not been observed. Sperm cell mitochondria are lost prior to onset of male *pronucleus* migration, and their influence on metabolic activation, embryogenesis and cytoplasmic inheritance is limited.

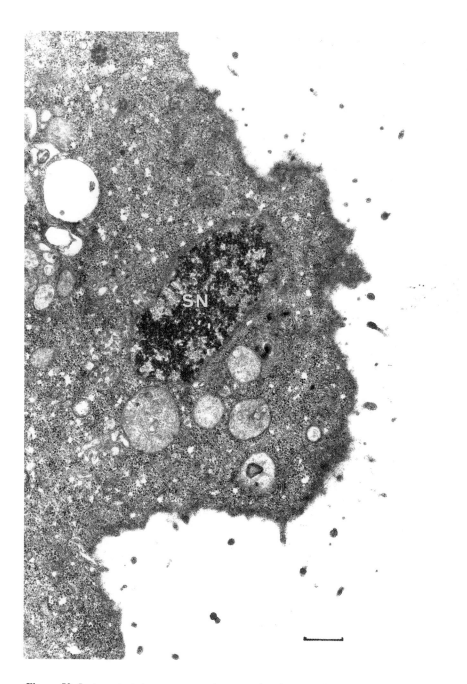

Figure 52. Just penetrated spermatozoon in a cytoplasmic protrusion (SN: sperm cell nucleus). Bar: 0.5 µm.

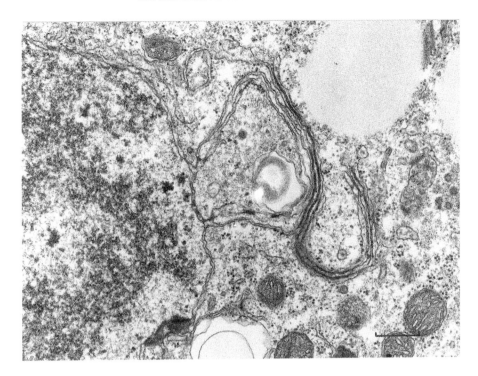

Figure 53. High magnification of a sperm cell nucleus shortly after penetration: a part of the membranous system is still attached to the nuclear envelope. Clear swollen sperm cell mitochondria are in contrast with the dense maternal ones. Bar: 0.5 µm.

3.3.2. Oocyte Maturation, Polar Body Formation and Female Pronucleus Development

Polar body formation is an extreme case of unequal oocyte cleavage, yielding a tiny cell, the polar body, and a large cell, the egg. This precise mechanism is designed to half egg chromosome number without disturbing developmental potential (Wilson, 1925 cited by Shimizu, 1990). No other aspect of anostracan cytology had aroused such an interest by the turn of the previous century. Anostracans were chosen for these studies because egg development is easily observed in living animals throughout the transparent cuticle. Since parthenogenesis often occurs in this group this allowed a comparison of amphigonic and parthenogenetic development. In amphigonic *Artemia* oocytes are blocked in metaphase I until release into the ovisac (Brauer, 1892; Artom, 1907), but fertilisation is required for normal egg maturation (personal observation). Oviductal secretion may play a role in the metaphase I block because *Artemia* oviductal cells are secretory (Criel, 1980).

Whereas *Artemia* metaphase I ovary spindles are situated tangentially, they lie either tangentially, radially or in an intermediate position in eggs prelevated from lateral pouches. On the other hand, eggs prelevated from the ovisac imme-

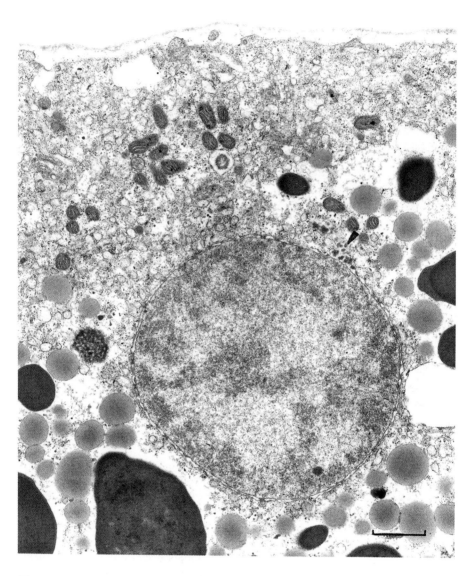

Figure 54. Male *pronucleus* in an egg in telophase II: the chromatin is almost decondensed. Many nuclear pores perforate the nuclear envelope. Arrowhead points to release of granular material by blebbing of the outer nuclear membrane. Bar: 1 µm.

diately after descent from oviducts have radially oriented spindles. Tangentially oriented spindles are somewhat fusiform while radially oriented spindles are shorter and barrel shaped. Metaphase proceeds to anaphase after insemination and meiotic spindles are anastral in amphigonic *Artemia* (Artom, 1907; Gross, 1935), whereas astral spindles are described in parthenogenetic *Artemia*

(Brauer, 1893, 1894; Petrunkewitz, 1902). The anastral condition is thought to be caused by the absence of centrosomes, structures that typically consist of two morphologically distinct entities, a pair of centrioles and pericentriolar material. The principle microtubule organising centre (MTOC) of most eukaryotic cells is pericentriolar material, and γ-tubulin, an important MTOC component, has a central role in microtubule nucleation. Neither centrioles nor pericentriolar material are found in meiotic spindles of amphigonic *Artemia*, and a typical functional centrosome is absent from mature eggs in other animals, although spindles occur. These observations suggest a fundamental difference between female and male meiosis with respect to centriole function (Kato *et al.* 1990). Within this context, however, centrosomes may not disappear completely from *Artemia* oocytes, because the mechanism of artificial activation upon egg release from oviducts would become difficult to explain. Thus, it is quite common that spermatocyte and oocyte centrosomes are reduced during gametogenesis and restored upon fertilisation, a process perhaps dependent on sperm (Schatten *et al.* 1989; Schatten, 1994). Even though centrosomes are not observed in meiotic spindles of *Artemia* oocytes, active spindle organization must occur. Studies other than those of a purely morphological nature are needed to answer this question, and one promising approach is to localise/identify centrioles and pericentriolar material with suitable antibodies.

Fautrez-Firlefyn and Roels (1968) provided the first electron micrographs of metaphase chromosomes in *Artemia* oocytes and this study has been extended. In contrast to meiotic divisions during spermatogenesis, maturing oocyte chromosomes maintain individuality when arranged as metaphase tetrads (Figure 55). Tetrads represent the sealed bivalents of homologous chromosomes (Barigozzi, 1944) and are characteristic of diploid bisexual and some parthenogenetic strains, whereas in other parthenogenetic strains diads seem to form. *Artemia* tetrads consist of four symmetrical, electron dense sections separated in cross section by fine, less electron dense, granular material. Material linking tetrads is thought to be remnants of synaptonemal complexes in *Bombyx* (Rasmussen and Holm, 1982), and it may be equivalent to granular material in *Artemia* tetrads. Tetrad sections are fenestrated and kinetochores span the entire chromosomal length facing the spindle pole. Microtubules abound in female meiotic spindles. Ribosomes reside within spindles and mitochondria gather around radial spindles.

Oocytes prelevated from the ovisac within minutes post-fertilisation are in anaphase. Their spindles are shorter and broader (Figure 56), no trace is left of either tetrads or kinetochores, and chromosomal material has moved toward poles, the latter dependent on attached microtubules. A slightly electron dense material remains between chromosomes at spindle equators and may contribute to midbody formation, the latter appearing as spindles elevate above egg surfaces. Polar body formation occurs upon protrusion of the cytoplasmic mass containing half of the anaphase spindle and formation of a cleavage furrow at the base of the protuberance. As anaphase terminates, membrane vesicles accumulate in spindle poles adjacent to chromosomes. The vesicles

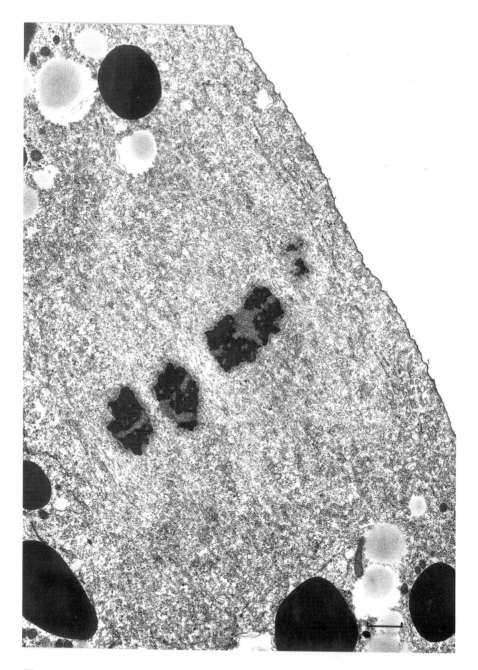

Figure 55. Tangential metaphase I spindle in an egg prelevated from the lateral pouches. A row of vesicles surrounds the spindle. Bar: 1 μm.

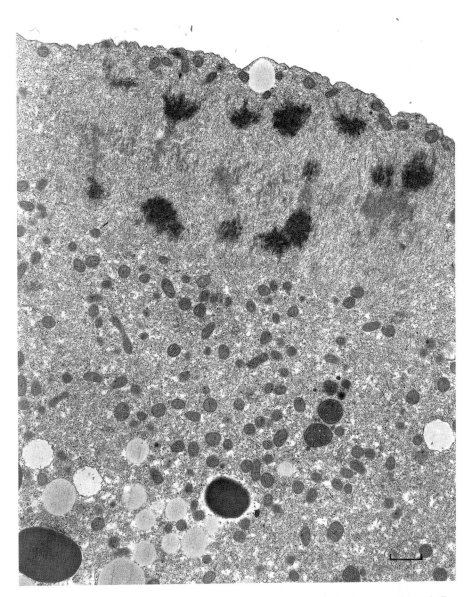

Figure 56. Anaphase I: the cement sealing the tetrads seems to remain in the centre of the spindle. Bar: 1 μm.

fuse, giving rise to the double-layered nuclear envelope, concomitant with reformation of nuclear pore complexes. Chromatin decondenses slightly when the nuclear envelope is complete. Preparing for polar body extrusion the oolemma increases in electron density due to accumulation of fine, dense material on the inside and a fuzzy coat on the exterior. Because polar body abstriction may be mediated by microfilaments (Le Guen *et al*. 1989; Shimizu,

1990), the dense material accumulating under the cell surface is probably actin and/or actin-associated proteins. The protruding polar body (Figure 57), surrounded by villi containing a core of microfilaments, contains ribosomes, mitochondria and occasionally a yolk body.

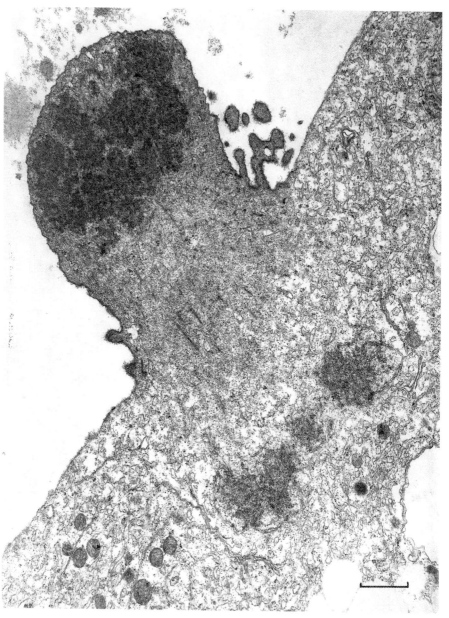

Figure 57. Protruding first polar body: microvilli surround its base; vesicles surround the telophase chromosomes. Bar: 1 µm.

The female nucleus enters prophase II immediately after polar body release. The prophase spindle of the second meiotic division is markedly smaller than the first meiotic spindle and forms in a tangential position, as does the spindle of metaphase I. The chromatids are fenestrated and almost encircled by kinetochores (Figure 58). Fine, electron-dense granules are associated with chromosome peripheries and appear as tiny clusters between microtubules (Figure 59). Anaphase spindles turn radially, protrude above the egg surface up to midbody level and are surrounded by electron dense granules of unknown origin. Karyomeres develop as in telophase I, eventually fusing in the polar body and egg to form kidney-shaped nuclei. Chromatin decondensation starts in female *pronuclei* before spindle remnants vanish (Figure 60), and round, completely clear *pronuclei* arise quickly under the egg surface. Second polar body extrusion proceeds by elevating the fertilisation membrane, followed immediately by female *pronucleus* migration to the egg centre.

3.3.3. Migration and Interaction of Pronuclei

Male and female *pronuclei* migrate to a zygotic region where they associate and establish the embryonic genome (Longo, 1976). *Pronuclei* movement is accompanied by ooplasm reorganisation and yolk-free cytoplasmic areas that gradually enlarge appear at random throughout eggs after fertilisation. Cytochrome oxidase staining illustrates mitochondrial profusion in these areas, even at the light microscopy level (Figure 61). Ultrathin sections show that mitochondria lie at the periphery of these yolk-free areas, the centres of which are filled with ribosomes, smooth and rough endoplasmic reticulum, Golgi and microtubules (Figure 62). The migrating *pronuclei* gradually enlarge and become surrounded with cytoplasm of the same composition as in yolk free areas. Perinuclear cytoplasm may associate more prominently with one of the two zygotic *pronuclei* but the difference is never great. Occasionally, a small aster radiates from a *pronucleus*. Cytochrome oxidase staining demonstrates mitochondria alignment, an arrangement shown ultrastructurally to depend upon elongated, smooth membrane cisternae running parallel to microtubules. Perinuclear plasms fuse when *pronuclei* approach the egg centre, and they lie in the narrow centre of an hour glass shaped, central ooplasmic mass, where yolk platelets are almost entirely excluded (Figure 63). Other cytoplasmic areas, previously scattered at random, accumulate in the zygote centre. A specialised cytoplasmic region develops between *pronuclei*, and mitochondria are almost completely excluded from the area. Bundled, radially-oriented, smooth endoplasmic reticulum cisternae extend along microtubules, radiating from both ends of the central ooplasmic mass. These structures penetrate clusters of protein yolk and lipid bodies, with lipid droplets being the only yolk containing bodies aligning in aster rays. The sperm cell aster, constructed in conjunction with the associated microtubules of the incorporated spermatozoon, may be required for male pronucleus movement to the female *pronucleus* (Longo, 1973; Longo, 1976). However, the poorly developed *Artemia* sperm cell asters are unlikely to generate pronuclear migration.

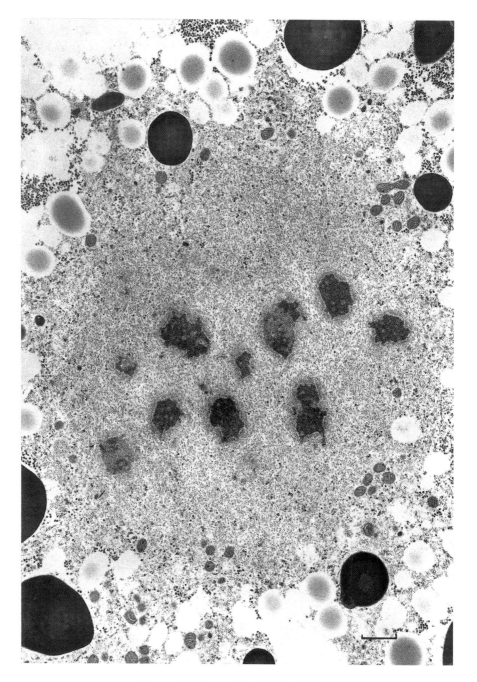

Figure 58. Prometaphase II. Bar: 1.5 μm.

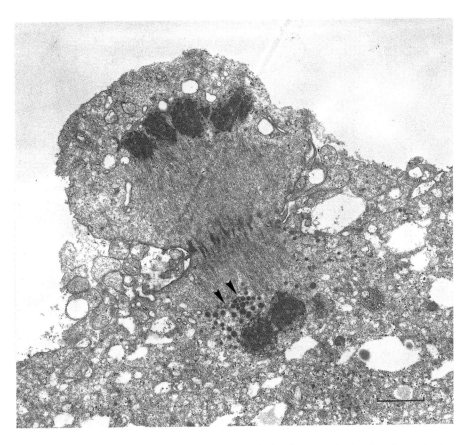

Figure 59. Protruding second polar body: a constriction forms at the level of the midbody. Vesicles surround the telophase chromosomes (arrowheads). Bar: 1 μm.

Migrating *pronuclei* are surrounded by nuclear envelopes perforated by many pores, and their nucleoplasm is clear except for small granules. *Pronuclei* can be seen very closely apposed to each other (Figure 64), but membrane fusion is not seen and chromosomes start to condense before fusion of nuclear membranes. Microtubules and smooth membranes fill the narrow space between closely apposed *pronuclei* (Figure 65). This indicates that fusion of *pronuclei* does not occur in *Artemia* although there is disagreement between earlier results. For example, Artom (1907) and Fautrez-Firlefyn (1951) describe pronuclear fusion in *Artemia*, while Brauer (1892) and Fries (1909) do not.

The yolk nucleus too migrates to the egg centre (Figure 64), but in contrast to migrating *pronuclei* there is no association with structurally specialised cytoplasm during migration. The yolk nuclei eventually settles near one of the *pronuclei*, suggesting a function other than induction of endogenous yolk synthesis (Fautrez and Fautrez-Firlefyn, 1964).

Figure 60. Just formed female *pronucleus*: nuclear pores (arrowheads) are formed simultaneously with the nuclear envelope. Arrow: remnants of the spindle microtubules. Bar: 0.5 µm.

3.3.4. Development of the Fertilisation Membrane

Most eggs have one or more extracellular coats that shape the vitelline envelope and play an important role in fertilisation. Vitelline envelopes are elaborated during oocyte growth and mediate egg-spermatozoon recognition. Structural modification of the vitelline envelope is initiated by the sperm cell and subsequently carried out by the egg as a consequence of egg-spermatozoon interaction. Cortical granules are released, producing a physical barrier to polyspermy transforming the vitelline envelope into a fertilisation envelope protecting the embryo during early development.

No vitelline envelope is observed with the light microscope before fertilisation in branchiopods (Claus, 1886; Brauer, 1892; Mawson and Yonge, 1938;

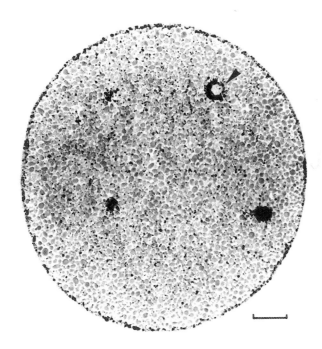

Figure 61. Light micrograph showing the distribution of the mitochondria in a two micron section of an egg at the onset of the migration of the *pronuclei*: mitochondria gather around a migrating *pronucleus* and in several other areas. Bar: 20 µm.

Kupka, 1940; Linder, 1960; Fautrez and Fautrez-Firlefyn, 1971; Zaffagnini, 1987). When branchiopod eggs arrive in lateral pouches after ovulation their membranes are covered with only a fuzzy coat, approximately 25 nm in width. A relatively broad layer of cortical cytoplasm separates the egg surface from deeper cytoplasm that contains mainly yolk. Smooth vesicles filled with intracisternal granules, mitochondria and free ribosomes are the main cortical components (Figure 66), with rough endoplasmic reticulum cisternae seen occasionally. These smooth vesicles appear very early in vitellogenesis when they are pinched off from rough endoplasmic reticulum cisternae. The formation of the fertilisation membrane begins after egg descent into the ovisac, a process triggered in bisexual strains by copulation or artificial stimulation, such as brief heating of unmated females. Development of the fertilisation membrane in branchiopods has been described ultrastructurally for *Tanymastix* (Garreau de Loubresse, 1974), *Artemia* (De Maeyer-Criel *et al.* 1977), and *Daphnia* (Zaffagnini, 1987).

After egg descent into the ovisac, Golgi complexes appear in the cortical layer. The Golgi is inconspicuous at first (Figure 67), but has developed considerably by the end of anaphase I. The smooth vesicles filled with intracisternal granules are conveyed to the Golgi complexes where they are trans-

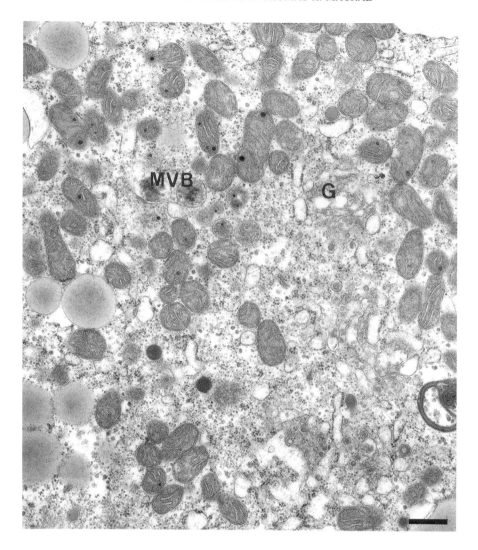

Figure 62. Electron micrograph of a cytoplasmic area surrounded with mitochondria, between the yolk platelets: Golgi complexes (G), short RER cisternae, smooth vesicles and ribosomes are seen in the centre. (MVB, multivesicular body). Bar: 0.5 µm.

formed into dark secretory granules. When the second meiotic division begins these dark secretory granules migrate to the oocyte surface, and are released (Figure 68). This transforms the fuzzy egg coat into a fertilisation membrane which loosens from the egg surface in anaphase II. Peroxidase is localized in some intracisternal granules, in the Golgi and in the dense granules formed by this organelle (Roels, 1970). Possibly, peroxidase functions in fertilisation membrane hardening (Somers and Shapiro, 1989; Larabell and Chandler, 1991).

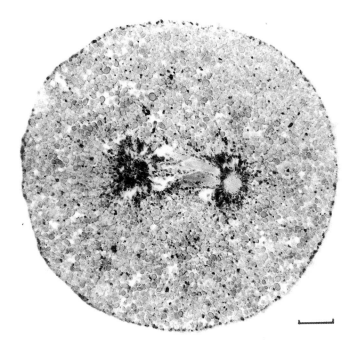

Figure 63. Light micrograph showing the distribution of the mitochondria: a clear zone has become apparent between both *pronuclei*. Yolk granules are excluded from this region. The scarce mitochondria align in the aster-rays. Bar: 20 µm.

The fertilisation membrane completely separates from the egg and is temporarily and locally elevated after formation of the second polar body, while small villi project into the perivitelline space (Figure 69). Several coated pits and vesicles are seen near egg surfaces, but how this temporary elevation of the fertilisation membrane occurs is unknown.

3.4. EGG ENVELOPE DEVELOPMENT

Generally, *Artemia* are thought to develop either oviparously or ovoviviparously. However, not all encysted embryos produced by oviparous animals enter diapause, sometimes nauplii emerge from some cysts without dehydration or other treatment (Jensen, 1918; Mathias, 1937; Lochhead and Lochhead, 1940; Dutrieu, 1960a, b; Morris and Afzelius, 1967; Benesch, 1969; Anderson *et al.* 1970). These cysts are surrounded by a much thinner shell than those that enter diapause (Lochhead and Lochhead, 1940) (Figure 70), consisting of a trilaminar fertilisation membrane with a thin chorion possessing irregular patches of dense material. The thin shell is covered superficially with a fibrous mat, but an embryonic cuticle is never formed (personal observation).

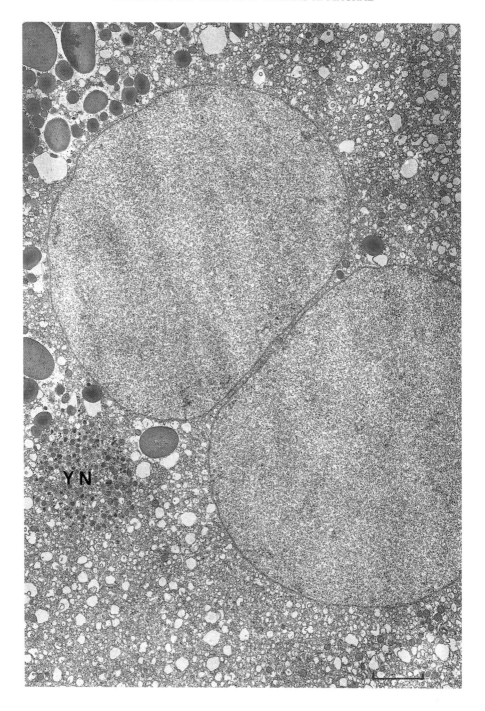

Figure 64. Encounter of the *pronuclei*: independently from the *pronuclei* the yolk nucleus (YN) too has migrated to the centre of the egg. Bar: 2 μm.

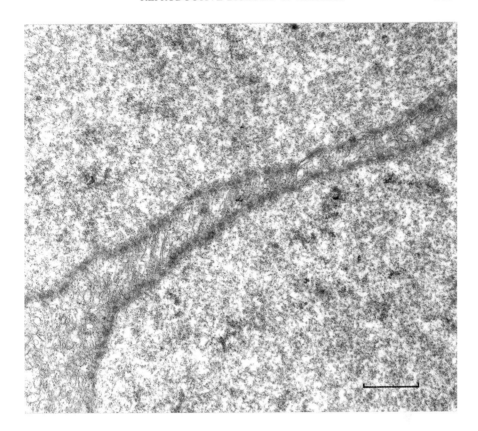

Figure 65. *Pronuclei* in very close contact: microtubules and smooth vesicles run between them. Bar: 1 µm.

The cryptobiotic cyst shell has two important layers in addition to the hypochlorite-soluble, double-layered outer chorion secreted by shell gland, and the hypochlorite-resistant embryonic cuticle formed by blastoderm cells (Morris and Afzelius, 1967). These are the outer cuticular membrane, a structure resistant to hypochlorite, impermeable to lead and separating chorion from embryonic cuticle, and the inner cuticular membrane, which delineates the embryo from the fibrous layer of the embryonic cuticle. Three main layers in the cyst envelope were also demonstrated by scanning electron microscopy (Mazzini, 1978).

Formation of the *Artemia* cyst shell is similar to that of *Tanymastix lacunae* (Garreau de Loubresse, 1974), although differences occur (Anderson *et al.* 1970; Fautrez and Fautrez-Firlefyn, 1971). For example, the *Artemia* fertilisation membranes are not split by insertion of shell gland secretion, but transform into trilaminar membranes covered with a fibrillar matrix in which dense secretion products accumulate. Additionally, the outer cuticular membrane achieves final structure after creation of the embryonic cuticle

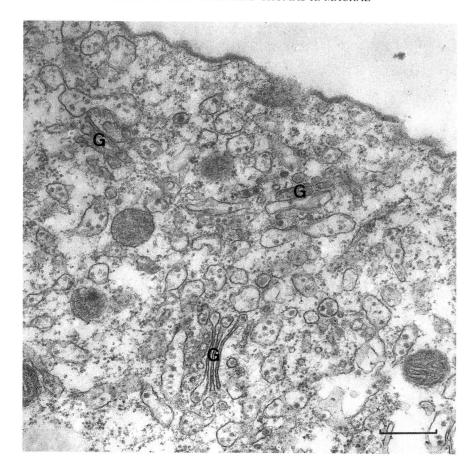

Figure 66. Periphery of a just fertilised egg: Golgi complexes (G) appear in the cortical layer. Bar: 0.5 µm.

fibrous layer. The outer cuticular membrane is missing from embryos excised and cultured *ex utero*, indicating that maternal factors are required for its establishment (De Chaffoy et al. 1978). The inner cuticular membrane appears immediately after the release of the cysts from the ovisac.

Early embryos of ovoviviparous animals are only surrounded by a trilaminar fertilisation membrane and thin fibrous matrix (Figure 71). Personal observations support the idea that the hatching membrane emerges as a moulted larval cuticle (Morris and Afzelius, 1967).

3.5. CHANGES IN THE RELEASED CRYPTOBIOTIC CYST

Changes in cysts after release suggest that dormancy is established gradually (Jardel, 1986), a process divided into changes preparative to dehydration and for dehydration itself. Within ninety-six hours of release, ultrastructural

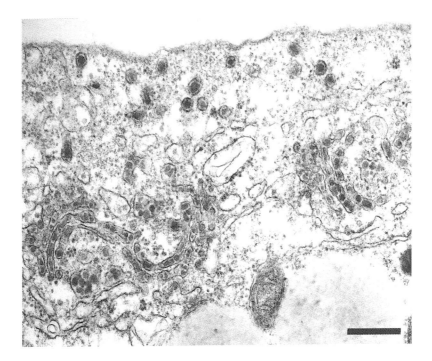

Figure 67. Cortex of an oocyte early in meiosis II: the Golgi complexes actively produce dense granules which move to the egg surface. Bar: 0.5 µm.

modifications occur in cysts, and a peak in cytochrome-C-oxidase activity suggests transformations are energy dependent. The rough endoplasmic reticulum is remodelled into narrow ribosome-free profiles called PERMs (paired endoplasmic reticulum membranes). PERMs gradually surround yolk platelets and lipid yolk droplets, possibly protecting against dehydration damage. They could be the membranes described around yolk platelets (Vallejo *et al.* 1979). Golgi complexes are transformed into an apparently non-functional state, gap junctions are reduced and septate junctions arise. Clegg *et al.* (1996), and Jackson and Clegg (1996) also found metabolic changes for days after release. Neither dehydration (Jardel, 1986) nor anoxia (Clegg *et al.* 2000) causes severe disruption of cellular organisation, depletion of cell components, or loss of typical trilaminar membrane structure.

4. Embryonic Development

The data presented here originate from Benesch (1969) unless cited otherwise. Contrasting earlier conclusions (Brauer, 1893, 1894; Fautrez-Firlefyn, 1951), *Artemia* cleavage is total, with an equal distribution of yolk among blastomeres, as revealed by light microscopy. 'Fusomes' are observed between

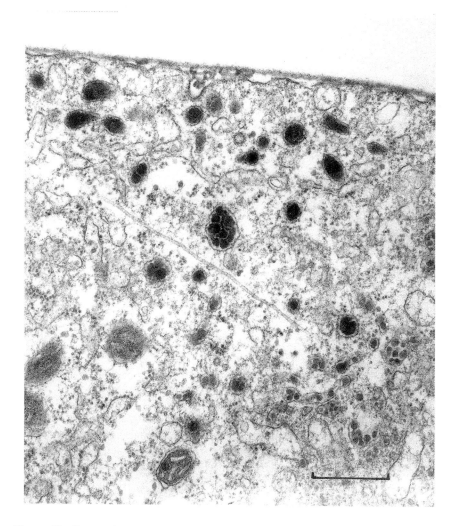

Figure 68. Cortex of an oocyte in meiosis II: exocytosis of the dense granules provokes a thickening of the surface coat which gradually gets loose from the egg surface. Bar: 0.5 μm.

some blastomeres (Jardel, 1986), but they are similar to cytoplasmic bridges observed in *Artemia* ovaries. The 'zones nécrotiques' (Bes *et al.* 1983) may represent the same cytoplasmic bridges after poor preservation for microscopy. Cleavage is most similar to this process in *Cladocera* (Fränsemeier, 1939). First and second cleavages are meridional. Total equal and radial cleavages proceed until the 512-cell blastula stage. The blastocoel enlarges from the 4 to the 256-cell stage, blastomere length decreases and embryo size increases. Nuclei lie in the blastomere centre until the 64-cell stage and peripherally after the 128-cell stage. Blastomeres grow and the blastocoel diminishes between cell stages 256 and 512. It is not possible to follow individual cell lineages,

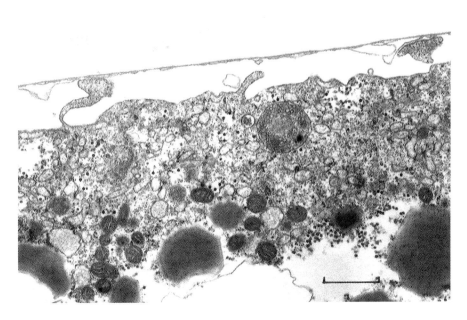

Figure 69. Elevated fertilisation membrane of an egg just after the release of the second polar body. Bar: 1 μm.

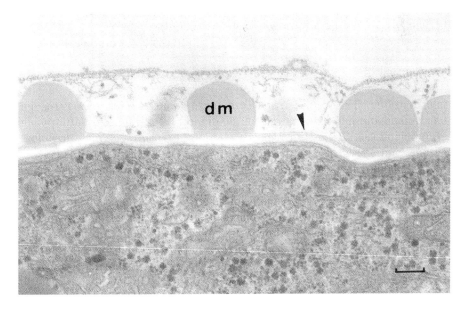

Figure 70. Transmission electron micrograph of the thin shell of a blastula presumebly developing into a non-diapausing cyst. Irregular patches of dense material (dm) superficially covered by a fibrous mat; fertilisation membrane (arrowhead). Bar: 200 nm. (From Criel, 1991)

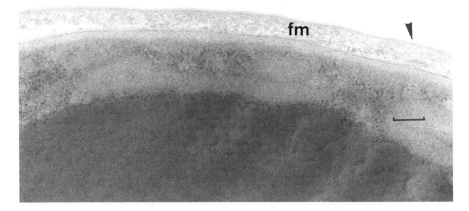

Figure 71. Transmission electron micrograph of the cuticle of a nauplius developing in the ovisac (fm, thin fibrous matrix; arrowhead, trilaminar fertilisation membrane. Bar: 200 nm. (From Criel, 1991)

although mitosis seems to occur preferentially in some regions. A hollow spherical blastula forms at the 512-cell stage.

Gastrulation starts late in *Artemia* and proceeds in two main phases termed G-I and G-II. Primordial germ cells and mesoderm migrate inward during G-I, and the embryo is symmetrical at the end of G-I with only a small remnant of blastocoel. Symmetry may be organised by the mesoderm, but whatever the case, twelve to sixteen primordial germ cells are distinguishable from other cells by their nuclei. A second invagination pole arises at G-II onset, and as a consequence, endoderm cells containing large nuclei enter from the posterior region where mesoderm is invaginating. Thus, presumptive mesoderm and endoderm move inward separately. Concurrently, a mass of mesodermal cells originating from a ring surrounding the primordial germ cells develops into naupliar segments, remaining separate from encroaching ectodermal cells. Metanaupliar mesoderm immigrates last in the initial immigration pole, arising behind the primordial germ cells. This interpretation (Benesch, 1969) seems better founded than earlier conclusions that metanaupliar mesoderm arises from ectoderm (Fränsemeier, 1939; Weisz, 1947). On the other hand, Anderson (1967, 1973) postulated that metanaupliar mesoderm arises from mesoteloblasts, as is the case for many non-Malacostraca.

When naupliar segmentation is evident in the mesoderm, ectodermal cells protrude in regions where subepidermal tendon cells appear, phenomena that may be interdependent. Many mitoses occur in mesodermal cells, but endodermal cells migrate without dividing and separate regions of naupliar mesoderm into somites. These somites remain attached to one another and form cross-arches.

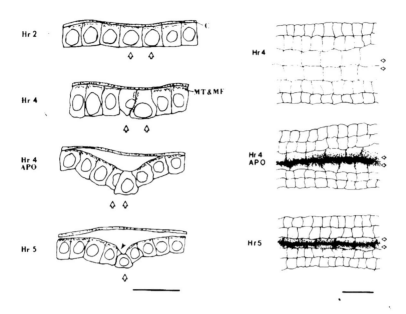

Figure 72. Diagrammatic view of cell shape changes in the ventrolateral epidermis. The transverse view of a 60 μm region of integument is shown on the left, and the surface view of the arthrodial membrane (AM) region and neighbouring thoracopod bud (ThB) cells (two from segment 1 and two from segment 2) is shown on the right. The bar at the bottom of each panel represents 20 μm. (Left) At hour 2, cells are similar in size, still attached to the cuticle (C), and demonstrate a polarized cytoskeleton with microtubules (MT) and microfilaments (MF) enriched in the apical region. The AM cells are indicated by arrows. By hour 4, ThB cells have undergone enhanced replication and are columnar. The AM cells are still attached to the cuticle and their basal regions have enlarged. The organization of microtubules and microfilaments has shifted from primarily apical to enriched along the lateral membranes of apposing AM cells. The basolateral regions of one AM cell may shift transversely with respect to the cell in the other AM file. Later in hour 4, AM cells and neighbouring ThB cells undergo apolysis (APO) and invagination begins. The neighbouring ThB cells now decrease in height and are not as densely packed. As invagination proceeds, contact of AM cells with ThB neighbours changes to a more apicolateral position. By hour 5, apolysis is complete and AM cells have invaginated to the greatest extent. The arrowhead indicates the position of the second AM cell. (Right) Surface view of the invaginating AM and neighbouring ThB cells. Depth of invagination is indicated by stippling. Anterior is to the left in the left panel and to the top in the right panel (from Freeman et al. 1992).

Creation of naupliar and metanaupliar mesoderms, which originate from the same invagination pole, differs. Naupliar segments differentiate simultaneously, whereas metanaupliar mesoderm fashions an unpaired mass of cells at the embryo's hind end. Mitosis occurs after hatching within the cell mass when metanaupliar proliferation starts, and somites arise one after another. Ectoderm gradually differentiates from blastoderm, with migration of nuclei in the ectoderm of antenna I and II and the mandible followed by elongation of intersegmental ectodermal cells, the latter representing presumptive tendon cells.

The first immigration pole fades away in the posterior region of the embryo. The hindmost mesodermal cells are probably gut constituents and ectoderm is the presumptive proliferation zone of metanaupliar segments. Migration of endodermal cells gradually results in invagination, wherein cells reach the dorsal ectoderm and push the mesoderm aside. No mitotic activity is seen in the endoderm, it decreases in mesoderm cells and initiates in ectoderm cells. Ectoderm, mesoderm and endoderm are clearly differentiated by the end of gastrulation. The proctodeum and stomodeum lay outside the embryo, with segmental ectoderm and mesoderm found in naupliar segments and preantennal mesoderm.

Seven stages in naupliar development, termed Na-0 to Na-6, have been observed before emergence (Benesch, 1969). Mitoses are seen throughout the entire process because Benesch examined ovoviviparously developing nauplii. Fränsemeier (1939), Nakanishi *et al.* (1963), and Olson and Clegg (1978) do not report mitoses during naupliar development because post-diapause cysts are studied.

The endoderm lumen has yet to develop in stage Na-0 and invagination of stomodeum ectoderm follows the endoderm. Mitosis is rare in the mesoderm and some muscle cells are differentiating. Presumptive proctodeum at the posterior end of the embryo, with ectodermal cells smaller than those of metanaupliar ectoderm, has not invaginated. The proctodeum mesodermal cells lie between metanaupliar mesoderm cells. Na-1 beginning is marked by the end of stomodeum invagination; endodermal cells order into an epithelium, and the stomodeum remains closed to the outside. Multiple mitoses cause endoderm proliferation, generating the midgut, and mitosis continues at a reduced level in the mesoderm. The ectoderm undergoes important changes wherein cells arrange into presumptive mandible, tritocerebral, and deuterocerebral ganglia, separated from one another by tendon cells, as well as into protocerebrum and deuterocerebrum. Optic ganglia and ganglia of the nauplius eye appear, as do paired pigment cells of the nauplius eye. Ganglion and retinula cells of the dorsal and ventral nauplius eye arise at this time.

Cells of the midgut proliferate during stage Na-2 and its junction with the stomodeum moves anteriorly. These organs are distinguished easily because nuclei of stomodeal cells lie close to the lumen, whereas those of the midgut are more peripheral. The future opening of the stomodeum shifts posteriorly, pushed by invaginated cerebral ganglia. Additionally, invagination of cerebral ganglia from one side and proliferation of midgut cells from the other force preantennal mesoderm ventrally and anteriorly from the stomodeum. Concurrently, dorsal and anterior mesoderm of the first antenna relocate posteriorly, contacting mesoderm of the second antenna. Developing blood cells arise in the medial dorsal area of first antenna mesoderm and lie between muscle cells of the second antenna. Second antenna and mandible mesoderm remain lateral to the gut, while metanaupliar mesoderm and primordial germ cells migrate posteriorly. Constrictor and dilator muscles also arise during Na-2 as elongated spindle-shaped cells in preantennal mesoderm. Constrictor cells lie transversely to the stomodeum and dilator cells are removed laterally, extending posteriorly into the second antenna segment. These mesodermal cell movements conceal whether constrictor and dilator muscles of the oesophagus and labrum arise from second antennal or preantennal mesoderm. Dorsal ectodermal cells enlarge and form the 'Atemplatte' (Benesch, 1969), the organ that later constitutes the nauplial salt gland (Conte et al. 1972). The 'Atemplatte' has a tripartite pattern generated by three internal projections, these acting as prospective attachment sites for antennal and mandibular muscles.

In stage Na-3, the growing midgut reaches invaginating proctodeum posteriorly and ectoderm anteriorly, in the region dorsal to the presumptive nauplius eye where lateral gut diverticles are established. The oesophagus narrows and moves anteriorly at the same time, followed by associated muscle cells. The proctodeum invaginates at the hind end, pushing constituent muscle cells and those of the gut laterally and dorsally. Muscle cells, connective tissue cells and the second antenna gland are discernible in the mesoderm. Pre-antennal mesoderm lies between the midgut, oesophagus and brain. Mesoderm of the first antenna is lateral to pre-antennal mesoderm, while being dorsal and lateral to the midgut. Mandible mesoderms touch dorsally, anteriorly and laterally, well separated from metanaupliar mesoderm that lies lateral to the gut and proctodeum invagination. Primordial germ cells are positioned ventrally and laterally between mandibular and metanaupliar mesoderm. Prospective muscle cells surround the midgut. The ectoderm further differentiates, resulting in *setae*-forming cells in presumptive appendages, and notches located behind them indicate where appendages will appear. Benesch (1969) traces all parts of the nervous and neurosecretory systems, but adult *Artemia* lack a neurosecretory system and only a few isolated cells of this type occur.

Proctodeum invagination terminates in stage Na-4, but does not open into the midgut yet. The stomodeum is in its final position. Endodermal cells are bigger than the epithelial-like ectodermal cells. Notches mark appendage locations and muscle fibrils appear in nauplial mesoderm. Only pre-antennal mesoderm has yet to reach its ultimate location in developing nauplii.

Stage Na-5 is characterised by embryo growth within surrounding membranes, compressing many organs and especially the midgut. Mesoderm includes antennal glands in addition to muscle, blood and fat-storing cells, the latter two filled with yolk platelets. At the transition of midgut and hindgut on either side of the proctodeum, two giant cells of unknown function, and previously termed ganglion cells (Fränsemeier, 1939), appear between the muscle cells (Benesch, 1969). Nauplial mesoderm delineates from metanaupliar mesoderm as stage Na-5 progresses, the labrum extends posteriorly and appendages grow, with three pairs of head appendages appearing in rudimentary form.

The thoraco-abdomen elongates without mitosis during stage Na-6, the gut opens anteriorly and the hindgut is closed. Differentiation of nauplii muscles is complete and antennal glands are functional. Metanaupliar mesoderm consists of a compact cell mass and primordial germ cells occur in maxillar regions. Nervous system fibres are visible, the peri-oesophageal nerve ring forms, and nerve fibres run from the protocerebrum to nauplius eye ganglion. Some caution is necessary in interpretation of these results because Benesch (1969) discerns innervation of the X-organ, a structure missing from *Artemia*. Compound eye rudiments are present as slight lateral proliferations just behind the antennule base (Weisz, 1947) and nauplius eyes have differentiated.

Hatching occurs in stage 0. *Artemia* exist as true nauplii immediately before and just after hatching, with undifferentiated pre-antennal and unsegmented metanaupliar mesoderm, and a functional neuromuscular system. Nauplii are pear-shaped, possessing antennular, antennal and mandibular segments, rudimentary appendages, and a more or less smooth telson fold. The gut has stretched in a way that its anterior end opens into the stomodeum, while it is still separated posteriorly from the proctodeum by a double epithelial cell layer (Anderson, 1967). Gut muscles are distinguishable from somatic muscles over the entire nauplius length. Labrum fat storing cells are located dorsally between the nauplius eye, brain and stomodeum, proceeding ventrally in front of gut diverticles. Primary blood cells round up and lose yolk while primordial germ cells reside near maxillary and first thoracic segments. Multilayered ectodermal cell groups form rudiments of protocerebral ganglion and other ganglia up to the mandibulary region. The ectoderm is a single layer, with elongated cylindrical cells in the neck organ and a zone of proliferation. Subepidermal tendon cells protrude between naupliar segments, while flattened ectodermal cells establish the integument. Presumptive brain parts are recognisable, yolk platelets exist in all cells, and the haemocoel is relatively large.

5. Development of *Artemia* Larvae

5.1. DESCRIPTION OF LARVAL STAGES

Artemia postembryonic development progresses from newly hatched nauplii containing three cephalic segments, through larval and juvenile stages, successively adding two maxillar, eleven thoracic, two genital, and six abdominal segments. Postmandibular segments are added in the unsegmented trunk after hatching from a growth zone lying immediately in front of the telson. Subdivision of *Artemia* larval development into stages, based mainly on changes in external morphology and/or histological surveys of internal events, has been proposed. Morphological changes have been related to larval moults (Heath, 1924; Reynier, 1959; Hentschel, 1968; Blake, 1979; Freeman, 1986, 1989; Schrehardt, 1987), but moult number estimates differ from nine to nineteen, preventing agreement among staging systems (Heath, 1924; Weisz, 1947; Reynier, 1959; Anderson, 1967; Hentschel, 1968; Freeman, 1986, 1989; Schrehardt, 1987). Larval moult number is reported to vary with rearing conditions (Weisz, 1947) while Benesch (1969) proposed that segment differentiation proceeds individually, unrelated to moulting, and more than one developmental step may occur during an intermoult. Weisz (1947) suggested a system whereby each of nineteen stages is defined by the number of delineated segments. Freeman (1986, 1989) discerns 19 stages, with appearance of each segment linked to a moult, thus giving nineteen instars. Benesch (1969) describes twenty-four larval stages without clear evidence. Many details leading to adult *Artemia franciscana* have been established by scanning electron microscopy indicating seventeen moult stages (Schrehardt, 1987). Although Schrehardt's descriptions are based on moults, as are Hentschel's (1968), he attempted to fit *Artemia* development into a general scheme of the type proposed by Kaestner (1967) for Euanostracans (in Schrehardt, 1987). One naupliar, four metanaupliar, seven postmetanaupliar and five postlarval stages are described, corresponding to stages described by Hentschel (1968), and numbered L-1 to L-17. Recent studies usually follow Schrehardt's scheme.

Differences in either the development or reduction of antenullae, antennae and larval mandibles are used to distinguish naupliar and metanaupliar stages. Meta- and postmetanaupliar stages are characterised by segmentation and development of maxillules, maxilles, thoracomeres (thoracic segments) and thoracopods. Segmentation, as well as different steps in appendage differentiation, proceed two by two during each intermoult period. Work by Schrehardt (1987) invalidates the proposal that the first six thoracic segments become functional simultaneously, coincident with loss of the specialised naupliar feeding apparatus at the ninth moult (Anderson, 1967). Postlarval stages are characterised by antennae transformation toward their definitive shape and function, and by development of genital structures. The seventeenth moult concludes postembryonic development and both sexes attain adult organisation. Schrehardt's staging system was used by Peterson and Rosowski (1994) to describe development of molar surfaces by scanning electron microscopy. Additionally, Schrehardt's system was employed when post-embryonic development of *Artemia persimilis*, including two non-feeding nauplius, three metanauplius, seven postmetanauplius, and five post-larval stages was investigated (Cohen *et al.* 1998).

Segmentation of a free-living organism with a transparent cuticle makes *Artemia* an interesting system for studying the influence of environmental factors on gene expression. For example, how deficiencies in diet disrupt *Artemia* pattern formation have been examined (Hernadorena, 1993), and Manzanares *et al.* (1993, 1996) characterised an *Artemia* engrailed gene that shares characteristics with these genes from higher insects. The progressive nature of segmentation and maturation, yielding an entire segmentation series in single animals, enabled the spatial expression pattern of the engrailed-type gene to be examined during development. Using an antibody raised against the *Artemia* engrailed protein, it was shown that cephalic and trunk segmentation are both independent and superimposed events. One segmentation process in the embryo's rostral zone generates cephalic segments, and it is superimposed on another mechanism at the caudal end responsible for trunk segmentation. Zones of engrailed expression in the trunk are added one at a time as segments are generated from the posterior growth region. No apparent correspondence exists between engrailed expression domains and cell division patterns in larval ectoderm.

5.2. INTERNAL DEVELOPMENT OF *ARTEMIA* LARVAE

The ectoderm differentiates into integument, exoskeleton, nervous system, oesophagus and hindgut, the latter two distinguished from midgut by longitudinal muscles, strong circular muscles, and a chitinous inner lining associated with the hindgut. Other ectoderm structures are paired *ductus ejaculatorius*, the unpaired ovisac, integumentary glands, ectodermal glands of the labrum, thoracopodal glands in thoracic segments, shell glands in female genital segments, and accessory glands in male genital segments (Benesch, 1969).

The larval integument secretes an external cuticle and an internal basement membrane (Freeman, 1986, 1989). Unlike most crustaceans, the larval cuticle of *Artemia* nauplii is much thinner than underlying epidermal cells. The cuticle has the same appearance in all integument regions and is divided into an outer epicuticle and a fibrous procuticle. The nauplius epidermis is a cell monolayer of variable width, with head and antennal areas thinner than developing thoracic and abdominal regions. Tendon cells, *setae*-forming cells and salt gland (neck organ) differentiate from integument. Epidermally-derived tendon cells are either modified for attachment to muscles or may act as an endoskeleton (Freeman, 1989), but whether they arise from arthrodial membrane or thoracopod bud cells is unclear. Tendon cells are described from gastrula stage-II onward (Benesch, 1969), and during segmentation they probably establish permanent connections between the arthrodial membrane and dorsal segment surface. The *seta*-forming cells differentiate from the epidermal layer before larval emergence and hatching (Freeman, 1989). Their fine structure has not been described, but light microscopic observations show tetraploid nuclei. Benesch (1969) draws the first setal cells at stage Na-4.

Growth and differentiation were described as distinct but related processes in *Artemia* ectodermal cells during segmentation of the first two instars (Freeman, 1986, 1989). Examination of microtubule and microfilament organisation discloses how cell shape change and division contribute to segmentation of the epidermis in *Artemia* (Freeman *et al.* 1992). Thorax segment morphogenesis involves a spatio-temporal pattern of cell replication in the ventral epidermis, and division between segments results from differential replication activities. Specifically, as a segment is produced, the epidermal cell density of the thoracopod bud increases relative to the arthrodial membrane because cells in the former have a greater replication rate. Regions of lower cell density become intersegmental arthrodial membranes, representing locations of thoracopod articulation, whereas areas with high mitotic activity achieve greater cell density, evaginate and become epidermal cells. Differences in the orientation pattern of division axes in segmenting and nonsegmenting regions occur, as does variation in relative growth. Although epidermal cells are organised hexagonally, their nuclei are arranged in rows or files, running in an anterio-posterior direction. The division axes orientation influences the direction of rows, the latter shown by use of bromodeoxyuridine to be a determining factor in differentiation.

Cells on the ventro-lateral surface of segments, the budding regions of future appendages, are active during thoracopod morphogenesis, whereas regions destined to become arthrodial membranes initially lack mitotic activity and consist of two transverse cell files. Nuclei are displaced laterally, perhaps partly due to the action of microtubules, such that nucleus-free zones in each cell file are contiguous and a region of cytoplasm separates segments. Arthrodial membrane cells change from cuboidal to squamous spindle-shape, while adjacent thoracopodal bud cells remain cuboidal. The density of thoracopodal bud cells increases 60% more than those in the arthrodial membrane, leading to invagination of the arthrodial membrane and evagination of the thoracopod epithelium. Additionally, moulting cycle events may play a role in evagination. The cells of the arthrodial membrane undergo a shape transformation prior to apolysis, during which microfilaments and microtubules are enriched in the lateral borders (Figure 72). These changes are completed after apolysis and the arthrodial membrane cells compose the innermost edge of the invaginating segment, eventually leading to epidermal segment formation. Apolysis proceeds in an antero-posterior direction such that loss of contact with the exoskeleton in anterior regions permits areas of greater cell density to curve outward slightly and adjacent thinner regions to deform inward. More posterior thorax domains are not susceptible to deformation because epidermal cell density remains roughly equivalent in all areas. Microtubules and microfilaments may play a role in evagination. Endites, exites and epipodites differentiate from the central appendage body as the thoracopod develops, with arthrodial membranes and tendon cells forming in each structure.

The midgut elongates as part of endoderm development (Benesch, 1969), with mitosis occurring over its whole length, but slightly enhanced in distal regions. The partition between midgut and hindgut is disrupted, the gut opens externally, and feeding begins. The diverticles differentiate further and the gut wall bulges as it attaches to the ectoderm by tendons. The cytoplasm clears as yolk in midgut cells disappears, and an apocrine secretion, together with formation of a peritrophic membrane, are observed.

Visceral mesoderm develops independently from somatic mesoderm, wherein elongated mesodermal cells, discernible from somatic mesodermal cells, surround the alimentary canal (Benesch, 1969). Fibrils initiate in the muscle cell cytoplasm once yolk is resorbed . Gut muscles are striated and mononucleated. Two giant mesodermal cells remain on each side of the hindgut. Somatic mesoderm is found in the naupliar, as well as metanaupliar region, producing multilayered arcs in the former. Mesoderm segmentation starts at gastrulation stage-II and occurs simultaneously over the entire region by close conjunction of mesoderm and tendon cells when many mitoses are visible. Segmentation ends when mesoderm cell movements cease, organs can be distinguished, and ectoderm ganglia are obvious.

Pre-antennal mesoderm gives rise to oesophagus muscle, labrum levator muscle, eyestalk adductors, transverse eye muscle, brain connective tissue, and labrum fat-storing cells in the naupliar region. In contrast, *antennula* mesoderm forms *antennula* muscle, brain connective tissue cells, and primary blood cells. Antenna mesoderm generates muscle, in addition to connective tissue cells for the mandibular ganglion and the mandibular gland or secondary blood forming organ. Pre-antennal segments differentiate somewhat slower than other naupliar segments, with labrum levator muscle and oesophagus muscles developing simultaneously with other naupliar muscles, but prior to eye muscles, labrum fat-storing cells and connective tissue of the brain. The disposition of naupliar region muscles can be compared to those of the *maxillae* and thorax because the former are segmental.

Metanaupliar region mesoderm cells differentiate and divide as broad paired lateral bands beneath the naupliar embryonic ectoderm, ultimately establishing cell groups that separate as somites. Initially, somites consist of one cell layer, but they become multilayered as cells divide. Production of mesodermal somites and the external appearance of segments occur simultaneously. The metanaupliar mesoderm accumulates sufficient material during embryonic segmentation to produce maxillar and first thoracic somites, and formation of the second thoracic somite is initiated (Benesch, 1969). The third somite appears after hatching. Maxillar and first thoracic somites separate first from the mesodermal budding zone, after which somites develop one by one. The somites fuse ventrally, open dorsally, and are comprised of two to three cell rows. Before somites separate, prospective tendon cells emerge laterally and ventrolaterally in intersegmental regions. Cell division in the mesoderm progressively enlarges previously formed somites and they fuse to budding mesoderm. Cell division in the ectoderm causes elongation and sets the following somite free. Mitotic waves proceed in the antero-posterior direction in each developing segment and the budding mesoderm. Whether ectoderm or mesoderm initiates segmentation is unknown, but more mitoses occur in the ectoderm. his may happen because ectodermal cells are smaller than mesodermal cells, thereby requiring more divisions to obtain equivalent elongation. Later, during somite growth and differentiation, mesodermal cells decrease in size and mitotic proportion changes. The segmentation furrow becomes visible externally when dorsoventral muscle cells posteriorly, which first differentiate anteriorly in the somite, attach to tendon cells (Figure 72). The somite is attached to proliferating mesoderm posteriorly, with loosening preceded by formation of epidermal tendon cells at the hind end and ectoderm straightening by cell division. Cytokinesis bulges the segmental ectoderm after insertion of dorsoventral muscle cells and mesoderm becomes multilayered, filling the gap that results. The ectoderm becomes multilayered and ganglia arise.

The segmental somites initiate differentiation upon thoracopod evagination, differentiating into muscle, connective tissue, heart and pericardial septum, fat-storing cells, coelom and segmental organs. Heat shock affects development of the metanaupliar somatic mesoderm but not the post naupliar visceral mesoderm (Herandorena and Marco, 1991), demonstrating in *Artemia*, as for *Drosophila*, that ectoderm and endoderm achieve proper development even during strong inhibition of mesoderm structure formation.

6. References

Abatzopoulos, T.J., Kastritsis, C.D. and Triantaphyllidis, C.D. (1986) A study of karyotypes and heterochromatic associations in *Artemia*, with special references to two N. Greek populations. *Genetica* **71**, 3–10.

Adiyodi, R.G. and Subramoniam, T. (1983) Arthropoda – Crustacea. In K.G. King and R.G. Adiyodi (eds.), *Reproductive Biology of Invertebrates*, Vol. 1, John Wiley & Sons Ltd., pp. 443–495.

Anderson, D.T. (1967) Larval development and segment formation in the branchiopod crustaceans *Limnadia stanleyana* King (Conchostraca) and *Artemia salina* (L.) (Anostraca). *Australian Journal of Zoology* **15**, 47–91.

Anderson, D.T. (1973) Embryology and phylogeny in annelids and arthropods. *International Series of Monographs in Pure and Applied Biology, Zoology Division* **50**, 263–364.

Anderson, E., Lochhead, J.H., Lochhead, M.S. and Huebner, E. (1970) The origin and structure of the tertiary envelope in thick-shelled eggs of the brine shrimp *Artemia*. *Journal of Ultrastucture Research* **32**, 497–525.

Anteunis, A., Fautrez-Firlefyn, N. and Fautrez, J. (1964) L'ultrastructure du noyau vitellin de l'oeuf d'*Artemia salina*. *Experimental Cell Research* **35**, 239–247.

Anteunis, A., Fautrez-Firlefyn, N., and Fautrez, J. (1966) La structure des ponts intercellulaires 'obturés' et 'ouverts' entre les oogonies et les oocytes dans l'ovaire d'*Artemia salina*. *Archives de Biologie* **77**, 645–664.

Anteunis, A., Fautrez-Firlefyn, N. and Fautrez, J. (1968) Ultrastructure du nucléole expulsé dans le cytoplasme de l'oocyte d'*Artemia salina*. *Comptes Rendus de l'Académie des Sciences (Paris)* **266**, 1862–1863.

Arrigo, A.-P. (1998) Small stress proteins: chaperones that act as regulators of intracellular redox state and programmed cell death. *Biological Chemistry* **379**, 19–26.

Arsenault, A.L. (1984) Changes in the nuclear envelope associated with spermatid differentiation in the shrimp, *Crangon septemspinosa*. *Journal of Ultrastructure Research* **86**, 294–308.

Artom, C. (1907) La maturazione, la fecondazione e i primi stadii di sviluppo dell'uovo dell'*Artemia salina*. *Lin. di Cagliari, Biologica (Rac. Sci. Biol.)* **1**, 495–515.

Barigozzi, C. (1942) I fenomeni cromosomici nelle cellule somatiche di *Artemia salina* Leach. *Chromosoma* **2**, 251–292.

Barigozzi, C. (1944) I fenomeni cromosomici delle cellule germinali in *Artemia salina* Leach. *Chromosoma* **2**, 549–575.

Barigozzi, C. (1989) Cytogenetics and speciation of the brine-shrimp *Artemia*. *Atti della Accademia Nazionale dei Lincei Memorie*, S. VIII **19**, fasc. 4, 57–96.

Benesch, R. (1969) Zur Ontogenie und Morphologie von *Artemia salina* L. *Zoologische Jahrbucher Abteilung für Anatomie und Ontogenie der Tiere* **86**, 307–458.

Bes, J.C., Caratero, C., Caratero, A., Gaubin, Y. and Planel, H. (1983) Aspects ultrastructuraux de l'embryon dormant *Artemia*. *Bulletin de l'Association des Anatomistes* **67**, 255–264.

Bier, K. (1965) Zur Funktion des Nahrzellen im meroistischen Insektenovar unter besonderer Berücksichtigung der Oogenese adephager Coleopteren. *Zoologische Jahrbucher Abteilung für Anatomie und Ontogenie der Tiere* **71**, 371–384.

Blades-Eckelbarger, P.I. and Youngbluth, M.J. (1982) The ultrastructure of spermatogenesis in *Labidocera aestiva* (Copepoda: Calanoida). *Journal of Morphology* **174**, 1–15.

Blake, R.W. (1979) The development of the brine shrimp *Artemia salina* (L.): a morphometric approach. *Zoological Journal of the Linnean Society* **65**, 255–260.

Brauer, A. (1892) Über das Ei von *Branchipus grubii* v. Dyb. von der Bildung bis zur Ablage. *Abhandlungen der königlichen Akademie der Wissenschaften zu Berlin (Phys. Abh. nicht zur Akad.gehör Gelehrter)* **II**, 1–66.

Brauer, A. (1893) Zur Kenntnis der Reifung des parthenogenetisch sich entwickelnden Eier von *Artemia salina*. *Zoologischer Anzeiger* **16**, 138–140.

Brauer, A. (1894) Zur Kenntnis der Reifung des parthenogenetisch sich entwickelnden Eier von *Artemia salina*. *Archiv für mikroskopische Anatomie* **43**, 162–222.

Brown, G.G. (1966) Ultrastructural studies on crustacean spermatozoa and fertilization, Ph.D. Thesis, University of Miami, Coral Gables, California.

Brown, G.G. (1970) Some ultrastructural aspects of spermatogenesis and sperm morphology in the brine shrimp *Artemia salina* Leach (Crustacea: Branchiopoda). *Proceedings of the Iowa Academy of Sciences* **76**, 473–485.

Brown, G.G. (1970b) Some comparative aspects of selected crustacean spermatozoa and crustacean phylogeny. In B. Baccetti (ed.), *Comparative Spermatology*, Academic Press N.Y., London, pp.183–203.

Busa, W.B. and Crowe, J.H. (1983) Intracellular pH regulates transitions between dormancy and development of brine shrimp (*Artemia salina*) embryos. *Science* **221**, 366–386.

Busa, W.B., Crowe, J.H. and Matson, G.B. (1982) Intracellular pH and the metabolic status of dormant and developing *Artemia* embryos. *Archives of Biochemistry and Biophysics* **216**, 711–718.

Carpenter, J.F. and Hand, S.C. (1986) Arrestment of carbohydrate metabolism during anaerobic dormancy and aerobic acidosis in *Artemia* embryos: determination of pH-sensitive control points. *Journal of Comparative Physiology B* **156**, 451–459.

Cassidy, J.D. and King, R.C. (1972) Ovarian development in *Habrobracon juglandis* (Ashmead) (Hymenoptera: Braconidae) I. The origin and differentiation of the oocyte-nurse cell complex. *Biological Bulletin* **143**, 483–505.

Cave, M.D. (1982) Morphological manifestation of ribosomal DNA amplification during insect oogenesis. In R.C. King and H. Akai (eds.), *Insect Ultrastructure*, Vol. 1, Plenum Press, New York, London, pp. 86–117.

Charniaux-Cotton, H. (1985) Vitellogenesis and its control in Malacostracan Crustacea. *American Zoologist* **25**, 197–206.

Chevaillier, P. and Maillet, P.L. (1965) Structure fine et constitution cytochimique du spermatozoïde de la langoustine *Nephrops norvegicus* L. (Crustacé Décapode). *Journal de Microscopie* **4**, 679–700.

Choi, W.C. and Nagl, W. (1976) Electron microscopic study of the differentiation and development of trophocytes and oocytes in *Gerris najas* (Heteroptera). *Cytobios* **17**, 47–62.

Clark, W.H.Jr. and Griffin, F.J. (1988) The morphology and physiology of the acrosome reaction in the sperm of the decapod *Sicyonia ingentis*. *Development Growth & Differentiation* **30**, 451–462.

Claus, C. (1886) Untersuchungen über die Organisation und Entwicklung von *Branchipus* und *Artemia* nebst vergleichenden Bemerkungen über andere Phyllopoden. *Arbeiten aus dem Zoologische Institute Wien* **6**, 267–370.

Clegg, J.S. (1964) The control of emergence and metabolism by external osmotic pressure and the role of free glycerol in developing cysts of *Artemia salina*. *Journal of Experimental Biology* **41**, 879–892.

Clegg, J.S. (1965) The origin of trehalose and its significance during the formation of encysted dormant embryos of *Artemia salina*. *Comparative Biochemistry and Physiology A* **14**, 135–143.

Clegg, J.S. (1997) Embryos of *Artemia franciscana* survive four years of continuous anoxia: the case for complete metabolic rate depression. *Journal of Experimental Biology* **200**, 467–475.

Clegg, J.S. and Golub, A.L. (1969) Protein synthesis in *Artemia salina* embryos. II. Resumption of RNA and protein synthesis upon cessation of dormancy in the encysted gastrula. *Developmental Biology* **19**, 178–200.

Clegg, J.S. and Conte, F.P. (1980) A review of the cellular and developmental biology of *Artemia*. In G. Persoone, P. Sorgeloos, O. Roels and E. Jaspers (eds.), *The Brine Shrimp Artemia*, Vol. 2, Universa Press, Wetteren, Belgium, pp. 11–54.

Clegg, J.S. and Jackson, S.A. (1992) Aerobic heat shock activates trehalose synthesis in embryos of *Artemia franciscana*. *FEBS Letters* **303**, 45–47.

Clegg, J.S. and Jackson, S.A. (1998) The metabolic status of quiescent and diapause embryos of *Artemia franciscana* (Kellogg). *Archiv für Hydrobiologie Special Issues Advances in Limnology* **52**, 425–439.

Clegg, J.S., Jackson, S.A. and Warner, A.H. (1994) Extensive intracellular translocations of a major protein accompany anoxia in embryos of *Artemia franciscana*. *Experimental Cell Research* **212**, 77–83.

Clegg, J.S., Jackson, S.A., Liang, P. and MacRae, T.H. (1995) Nuclear-cytoplasmic translocations of protein p26 during aerobic-anoxic transitions in embryos of *Artemia franciscana*. *Experimental Cell Research* **219**, 1–7.

Clegg, J.S., Drinkwater, L.E. and Sorgeloos, P. (1996) The metabolic status of diapause embryos of *Artemia franciscana* (SFB). *Physiological Zoology* **69**, 49–66.

Clegg, J.S., Willsie, J.K. and Jackson, S.A. (1999) Adaptive significance of a small heat shock/α-crystallin protein (p26) in encysted embryos of the brine shrimp, *Artemia franciscana*. *American Zoologist* **39**, 836–847.

Clegg, J.S., Jackson, S.A. and Popov, V.I. (2000) Long-term anoxia in encysted embryos of the crustacean, *Artemia franciscana*: viability, ultrastructure, and stress proteins. *Cell and Tissue Research* **301**, 433–446.

Cohen, R.G., Rodríguez Gil, S.G. and Vélez, C.G. (1998) The postembryonic development of *Artemia persimilis* Piccinelli & Prosdocimi. *Hydrobiologia* **391**, 63–80.

Conte, F.P., Hootman, S.R. and Harris, P.J. (1972) Neck organ of *Artemia salina* nauplii. A larval salt gland. *Journal of Comparative Physiology* **80**, 239–246.

Coste, F., Manier, J.-F. and Raibaut, A. (1982) Un type structural de spermatozoïde chez les Copépodes. *Crustaceana* **43**, 249–260.

Cotelli, F. and Lora Lamia Donin, C. (1983) Ultrastructure of the spermatozoon of *Squila mantis*. *Acta Zoologica (Stockh.)* **64**, 131–137.

Cotelli, F., Ferraguti, M. and Lora Lamia Donin, C. (1975) An unusual centriolar pattern in isopoda sperm cells. *Journal of Submicroscopic Cytology* **7**, 289–292.

Cotelli, F., Ferraguti, M., Lanzavecchia, G. and Lora Lamia Donin, C. (1976) The spermatozoon of Peracarida. I. The spermatozoon of terrestrial isopods. *Journal of Ultrastructure Research* **55**, 378–390.

Cran, D.G. and Esper, C.R. (1990) Cortical granules and the cortical reaction in mammals. *Journal of Reproduction and Fertility (Suppl.)* **42**, 177–188.

Criel, G. (1980) Morphology of the female genital apparatus of Artemia: a review. In G. Persoone, P. Sorgeloos, O. Roels and E. Jaspers (eds.), *The Brine Shrimp Artemia*, Vol. 1, Universa Press, Wetteren, Belgium, pp. 75–86.

Criel, G.R.J. (1989) Morphological study of the ovary of Artemia. In A.H. Warner, T.H. MacRae and J.C. Bagshaw (eds.), *Cell and Molecular Biology of Artemia Development*, Plenum Publishing Corporation, pp. 99–129.

Criel, G.R.J. (1991) Ontogeny of Artemia. In R.A. Browne, P. Sorgeloos, and C.N.A. Trotman (eds.), *Artemia Biology*, CRC Press, Boca Raton, Florida, pp. 155–185.

Crowe, J.H., Carpenter, J.F. and Crowe L.M. (1998) The role of vitrification in anhydrobiosis. *Annual Review of Physiology* **60**, 73–103.

Cuoc, C., Brunet, M., Arnaud, J. and Mazza, J. (1993) Differentiation of cytoplasmic organelles and storage of yolk during vitellogenesis in *Hemidiaptomus ingens* and *Mixodiaptomus kupelwieseri* (Copepoda, Calanoida). *Journal of Morphology* **217**, 87–103.

De Chaffoy, D., De Maeyer-Criel, G. and Kondo, M. (1978) On the permeability and formation of the embryonic cuticle during development in vivo and in vitro of *Artemia salina* embryos. *Differentiation* **12**, 99–109.

de Cuevas, M. and Spradling, A.C. (1998) Morphogenesis of the *Drosophila* fusome and its implication for oocyte specification. *Development* **125**, 2781–2789.

Delavault, R. and Bérard, J.J. (1974) Etude ultrastructurale de la spermatogénèse chez *Daphnia magna* S. (Entomostracés, Branchiopodes, Cladocères). *Comptes Rendus de l'Académie des Sciences (Paris)* **278**, 1589–1592.

De Maeyer-Criel, G., Fautrez-Firlefyn, N. and Fautrez, J. (1977) Formation de la membrane de fécondation dans l'oeuf d'*Artemia salina*. *Wilhelm Roux's Archives of Developmental Biology* **183**, 223–231.

Dohmen, M.R. (1985) Nuage material, the origin of dense-core vesicles in oocytes of *Nassarius reticulatus* (Mollusca Gastropoda). *International Journal of Invertebrate Reproduction* **8**, 117–125.

Dresser, M.E. (1987) The synaptonemal complex and meiosis: an immunochemical approach, in P.B. Moens (ed.), *Meiosis*, Academic Press Inc., pp. 245–274.

Drinkwater, L.E. and Clegg, J.S. (1991) Experimental biology of cyst diapause. In R.A. Browne, P. Sorgeloos and C.N.A. Trotman (eds.), *Artemia Biology*, CRC Press, Boca Raton, Florida, pp. 93–117.

Durfort, M., Lopez Camps, J., Bargallo, R., Bozzo, M.G. and Fontarnau, R. (1980) Présence de ponts intercellulaires entre les cellules germinales de *Mytilicola intestinalis* Steuer (Crustacea, Copepoda). *La Cellule* **73**, 205–213.

Dutrieu, J. (1960a) Observations biochimiques et physiologiques sur le développement d'*Artemia salina* Leach. *Archives de Zoologie Expérimentale et Générale* **99**, 1–134.

Dutrieu, J. (1960b) Quelques observations biochimiques et physiologiques sur le développement d'*Artemia salina* Leach. *Rendiconti degli Istituti Scientifici della Universitá di Camerino* **1**, 196–224.

Dym, M. and Fawcett, D.W. (1971) Further observations on the numbers of spermatogonia, spermatocytes, and spermatids connected by intercellular bridges in the mammalian testis. *Biology of Reproduction* **4**, 195–215.

Fain-Maurel, M.A., Reger, J.F. and Cassier, P. (1975) Le gamète mâle des Schizopodes, et ses analogies avec celui des autres Péracarides. II Particularité de la différentiation cellulaire au cours de la gametogenèse. *Journal of Ultrastructure Research* **51**, 281–292.

Fautrez, J. and Fautrez-Firlefyn, N. (1964) Sur la présence et la persistance d'un noyau vitellin atypique dans l'oeuf d'*Artemia salina*. *Developmental Biology* **9**, 81–91.

Fautrez, J. and Fautrez-Firlefyn, N. (1971) Contribution a l'étude des glandes coquillières et des coques de l'oeuf d'*Artemia salina*. *Archives de Biologie* **82**, 41–83.

Fautrez-Firlefyn, N. (1951) Etude cytochimique des acides nucléiques au cours de la gamétogénèse et des premiers stades du développement embryonnaire chez *Artemia salina* L. *Archives de Biologie* **62**, 391–438.

Fautrez-Firlefyn, N. and Roels, F. (1968) Liaison entre fibres fusoriales et chromosomes au cours de la meiose. *Comptes Rendus de l'Académie des Sciences (Paris)* **267**, 1521.

Finamore, F.J. and Clegg, J.S. (1969) Biochemical aspects of morphogenesis in the brine shrimp, *Artemia salina*. In G.M. Padilla, G.L. Whitson and I.L. Cameron (eds.), *The Cell Cycle*, Academic Press, New York, pp. 249–278.

Fränsenmeier, L. (1939) Zur Frage der Herkunft des metanauplialen Mesoderms und der Segmentbildung bei *Artemia salina* Leach. *Zeitschrift für Wissenschaftliche Zoologie* **152**, 439–472.

Fränzen, A. (1970) Phylogenetic aspects of the morphology of spermatozoa and spermiogenesis. In B. Baccetti (ed.), *Comparative Spermatology*, Academic Press, New York & London, pp. 29–46.

Freeman, J.A. (1986) Epidermal cell proliferation during thoracic development in larvae of *Artemia*. *Journal of Crustacean Biology* **6**, 37–48.

Freeman, J.A. (1989) The integument of *Artemia* during early development. In T.H. Mac Rae, J.C. Bagshaw and A.H. Warner (eds.), *Biochemistry and Cell Biology of Artemia*, CRC Press, Boca Raton, Florida, pp. 233–256.

Freeman, J.A., Cheshire, L.B. and MacRae, T.H. (1992) Epithelial morphogenesis in developing *Artemia*: the role of cell replication, cell shape change, and the cytoskeleton. *Developmental Biology* **152**, 279–292.

Fries, W. (1909) Die Entwicklung der Chromosomen im Ei von Branchipus Grub. und der parthenogenetischen Generationen von *Artemia salina*. *Archiv für Zellforschung* **4**, 44–80.

Garreau de Loubresse, N. (1974) Etude chronologique de la mise en place des enveloppes de l'oeuf d'un crustacé phyllopode: *Tanymastyx lacunae*. *Journal de Microscopie* **20**, 21–38.

Go, E.C., Pandey, A.S. and MacRae, T.H. (1990) Effect of inorganic mercury on the emergence and hatching of the brine shrimp, *Artemia franciscana*. *Marine Biology* **107**, 93–102.

Golub, A. and Clegg, J.S. (1968) Protein synthesis in *Artemia salina* embryos. I. Studies on polyribosomes. *Developmental Biology* **17**, 644–656.

Goudeau, M. (1982) Fertilization in a crab: I. Early events in the ovary, and cytological aspects of the acrosome reaction and gamete contacts. *Tissue & Cell* **14**, 97–111.

Gross, F. (1935) Die Reifungs- und Furchungsteilungen von *Artemia salina* im Zusammenhang mit dem Problem des Kernteilungsmechanismus. *Zeitschrift für Zellforschung* **23**, 522–565.

Hand, S.C. and Gnaiger, E. (1988) Anaerobic dormancy quantified in *Artemia* embryos: a calorimetric test of the control mechanism. *Science* **239**, 1425–1427.

Heath, H. (1924) The external development of certain phyllopods. *Journal of Morphology* **38**, 453–483.

Hentschel, E. (1968) Die postembryonalen Entwicklungsstadien von *Artemia salina* Leach bei verschiedenen Temperaturen (Anostraca, Crustacea). *Zoologischer Anzeiger* **180**, 372–384.

Hernandorena, A. (1993) Guanylate requirement for patterning the postcephalic body region of the brine shrimp *Artemia*. *Roux's Archives of Developmental Biology* **203**, 74–82.

Hernandorena, A. and Marco, R. (1991) Heat induced developmental uncoupling of mesoderm from ectoderm and endoderm germ layer derivatives during *Artemia* postembryonic segmentation. *Roux's Archives of Developmental Biology* **200**, 300–305.

Hinsch, G.W. (1971) Penetration of the oocyte envelope by spermatozoa in the spider crab. *Journal of Ultrastructure Research* **35**, 86–97.

Hinsch, G.W. (1980) Spermiogenesis in *Coenobita clypeatus*, I. Sperm structure. *International Journal of Invertebrate Reproduction* **2**, 189–198.

Hinsch, G.W. and Clark, W.H.Jr. (1973) Comparative fine structure of Cnidaria spermatozoa. *Biology of Reproduction* **8**, 62–73.

Hodgson, A.N. (1986) Invertebrate spermatozoa – structure and spermatogenesis. *Archives of Andrology* **17**, 105–114.

Hodgson, A.N., Baxter, J.M., Sturrock, M.G. and Bernard, R.T.F. (1988) Comparative spermatology of 11 species of polyplacophora (Mollusca) from the suborders Lepidopleurina, Chitonina and Acanthochitonina. *Proceedings of the Royal Society of London* B **235**, 161–177.

Iwasaki, T. (1970) Incorporation of 3H-thymidine during oogenesis in *Artemia salina*. *Annotationes Zoologicae Japonenses* **43**, 132–141.

Jackson, S.A. and Clegg, J.S. (1996) Ontogeny of low molecular weight stress protein p26 during early development of the brine shrimp, *Artemia franciscana*. *Development Growth & Differentiation* **38**, 153–160.

Jardel, J.P.L. (1986) Ultrastructural modifications associated with anhydrobiosis in the brine shrimp, *Artemia* (L.), Ph.D. Thesis, Monash University, Australia.

Jensen, A.C. (1918) Some observations on *Artemia gracilis*, the brine shrimp of Great Salt Lake. *Biological Bulletin* **34**, 18.

Jespersen, A. (1983) Spermiogenesis in *Anaspides tasmaniae* (Thomson) (Crustacea, Malacostraca, Syncarida). *Acta Zoologica (Stockh.)* **64**, 39–46.

Kato, K.H., Washitani-Nemoto, S., Hino, A. and Nemoto, S. (1990) Ultrastructural studies on the behavior of centrioles during meiosis of starfish oocytes. *Development Growth & Differentiation* **32**, 41–49.

Kessel, R.G. (1989) The cytoarchitecture of the cortical ooplasm of *Styela* oocytes with special reference to annulate lamellae-endoplasmic reticulum relationships. *Journal of Submicroscopic Cytology and Pathology* **21**, 509–519.

King, R.C., Cassidy, J.D. and Rousset, A. (1982) The formation of clones of interconnected cells during gametogenesis in insects. In R.C. King and H. Akai (eds.), *Insect Ultrastructure*, Vol. 1, Plenum Press, New York, London, pp. 3–31.

Kubrakiewicz, J., Asamski, R.T. and Bilinski, S.M. (1991) Ultrastuctural studies on accessory nuclei in developing oocytes of the crustacean, Siphonophanes grubei. *Tissue & Cell* **23**, 903–307.

Kupka, E. (1940) Untersuchungen über die Schalenbildung und Schalenstruktur bei den Eiern von *Branchipus schaefferi* (Fischer). *Zoologischer Anzeiger* **132**, 130–139.

Langdon, C.M., Bagshaw, J.C. and MacRae, T.H. (1990) Tubulin isoforms in the brine shrimp, *Artemia*: primary gene products and their posttranslational modification. *European Journal of Cell Biology* **52**, 17–26.

Langdon, C.M., Freeman, J.A. and MacRae, T.H. (1991a) Post-translationally modified tubulins in *Artemia*: prelarval development in the absence of detyrosinated tubulin. *Developmental Biology* **148**, 147–155.

Langdon, C.M., Rafiee, P. and MacRae, T.H. (1991b) Synthesis of tubulin during early postgastrula development of *Artemia*: isotubulin generation and translational regulation. *Developmental Biology* **148**, 138–146.

Larabell, C. and Chandler, D.E. (1991) Fertilization-induced changes in the vitelline envelope of echinoderm and amphibian eggs: Self-assembly of an extracellular matrix. *Journal of Electron Microscopy Technique* **17**, 294–318.

Lassalle, B. and Testart, J. (1991) Sequential transformations of human sperm nucleus in human egg. *Journal of Reproduction and Fertility* **91**, 393–402.

Le Guen, P., Crozet, N., Huneau, D. and Gall, L. (1989) Distribution and role of microfilaments during early events of sheep fertilization. *Gamete Research* **22**, 411–425.

Liang, P. and MacRae, T.H. (1999) The synthesis of a small heat shock/a-crystallin protein in *Artemia* and its relationship to stress tolerance during development. *Developmental Biology* **207**, 445–456.

Liang, P., Amons, R., Clegg, J.S. and MacRae, T.H. (1997a) Molecular characterization of a small heat shock/a-crystallin protein in encysted *Artemia* embryos. *Journal of Biological Chemistry* **272**, 19051–19058.

Liang, P., Amons, R., MacRae, T.H. and Clegg, J.S. (1997b) Purification, structure and in vitro molecular-chaperone activity of *Artemia* p26, a small heat-shock/a-crystallin protein. *European Journal of Biochemistry* **243**, 225–232.

Lin, H., Lin Y. and Spradling, A.C. (1994) The *Drosophila* fusome, a germline-specific organelle, contains membrane skeletal proteins and functions in cyst formation. *Development* **120**, 947–956.

Linder, H.J. (1960) Studies on the fresh water fairy shrimp *Chirocephalopsis bundyi* (Forbes) II. Histochemistry of egg-shell formation. *Journal of Morphology* **107**, 259–280.

Lison, L. and Fautrez-Firlefyn, N. (1950) Deoxyribonucleic acid content of ovary cells in *Artemia salina*. *Nature* **166**, 610–611.

Lochhead, J.H. and Lochhead, M.S. (1940) The egg shells of the brine shrimp, *Artemia*. *Anatomical Record (suppl.)* **78**, 75–76.

Lochhead, J.H. and Lochhead, M.S. (1967) The development of oocytes in the brine shrimp, *Artemia*. *Biological Bulletin* **133**, 453–454.

Longo, F.J. (1973) Fertilization: A comparative ultrastructural review. *Biology of Reproduction* **9**, 149–215.

Longo, F.J. (1976) Sperm aster in rabbit zygotes. *Journal of Cell Biology* **69**, 539–547.

Longo, F.J. (1984) Transformation of sperm nuclei incorporated into sea urchin (*Arbacia punctulata*) embryos at different stages of the cell cycle. *Developmental Biology* **103**, 168–181.

Longo, F.J., Cook, S. and Mathews, L. (1991) Pronuclear formation in starfish eggs inseminated at different stages of meiotic maturation: Correlation of sperm nuclear transformations and activity of the maternal chromatin. *Developmental Biology* **147**, 62–72.

Lunger, P.D. (1971) Early stages of spermatozoan development in the colonial hydroid *Campanularia flexuosa*. *Zeitschrift für Zellforschung und mikroskopische Anatomie* **116**, 37–51.

Luttmer, S.J. and Longo, F.J. (1987) Rates of male pronuclear enlargement in sea urchin zygotes. *Journal of Experimental Zoology* **243**, 289–298.

Luttmer, S.J. and Longo, F.J. (1988) Sperm nuclear transformations consist of enlargement and condensation coordinate with stages of meiotic maturation in fertilized *Spisula solidissima* oocytes. *Developmental Biology* **128**, 86–96.

MacRae, T.H. (1997) Tubulin post-translational modifications. Enzymes and their mechanisms of action. *European Journal of Biochemistry* **244**, 265–278.

MacRae, T.H. and Ludueña, R.F. (1984) Developmental and comparative aspects of brine shrimp tubulin. *Biochemical Journal* **219**, 137–148.

MacRae, T.H. and Pandey, A.S. (1991) Effects of metals on early life stages of the brine shrimp, *Artemia*: a developmental toxicity assay. *Archives of Environmental Contamination and Toxicology* **20**, 247–252.

MacRae, T.H. and Liang, P. (1998) Molecular characterization of p26, a cyst-specific, small heat shock/(-crystallin protein from *Artemia franciscana*. *Archiv für Hydrobiologie Special Issues Advances in Limnology* **52**, 393–409.

Manier, J.-F., Raibaut, A., Rousset, V. and Coste, F. (1977) L'appareil génital m,le et la spermiogénèse du Copépode parasite *Naobranchia cyniformis* Hess, 1863. *Annales des Sciences Naturelles Zoologie* **19**, 439–548.

Manzanares, M., Marco, M. and Garesse, R. (1993) Genomic organization and developmental pattern of expression of the engrailed gene from the brine shrimp *Artemia*. *Development* **118**, 1209–1219.

Manzanares, M., Williams, T.A., Marco, R. and Garesse, R. (1996) Segmentation in the crustacean *Artemia*: Engrailed staining studied with an antibody raised against the *Artemia* protein. *Roux's Archives of Developmental Biology* **205**, 424–431.

Marco, R. and Vallejo, C.G. (1976) Mitochondrial biogenesis during *Artemia salina* development: Storage of precursors in yolk platelet. *Journal of Cell Biology* **70**, Abst. No. 963.

Marco, R., Garesse, R. and Vallejo, C.G. (1980) Mitochondrial unmasking and yolk platelets metabolization during early development in *Artemia*. In G. Persoone, P. Sorgeloos, O. Roels and E. Jaspers (eds.), *The Brine Shrimp Artemia*, Vol. 2, Universa Press, Wetteren, Belgium, pp. 481–490.

Marco, R., Garesse, R. and Vallejo, C.G. (1981) Storage of mitochondria in the yolk platelets of *Artemia* dormant gastrulae. *Cellular and Molecular Biology* **27**, 515–522.

Marco, R., Batuecas, B. and Garesse, R. (1983) The isolation and characterization of the yolk cytoplasmic DNA and the storage of mitochondria in the yolk granules of *Artemia* and *Drosophila* early embryos. In H.E.A. Schenk and W. Schwemmler (eds.), *Endocytobiology II. Intracellular Space as Oligogenetic Ecosystem*, Walter de Gruyter and Company, Berlin and New York, p. 824.

Mathias, P. (1937) Biologie des Crustacés Phyllopodes. *Actualités scientifiques et Industrielles* **447**, 1–107.

Mawson, M.L. and Yonge, C.M. (1938) The origin nature of the egg membranes in *Chirocephalus diaphanus*. *Quarterly Journal of Microscopical Science* **80**, 533–565.

Mazzini, M. (1978) Scanning electron microscope/Morphology and amino-acid/Analysis of the egg-shell of encysted brine shrimp, *Artemia salina* Leach (Crustacea Anostraca). *Monitore Zoologico Italiano* **12**, 243–252.

McKnight, C.E. and Hinsch, G.W. (1986) Sperm maturation and ultrastructure in *Scyllarus chacei*. *Tissue & Cell* **18**, 257–266.

Mehlen, P., Coronas, V., Ljubic-Thibal, V., Ducasse, C., Granger, L., Jourdan, F. and Arrigo, A.-P. (1999) Small stress protein Hsp27 accumulation during dopamine-mediated differentiation of rat olfactory neurons counteracts apoptosis. *Cell Death and Differentiation* **6**, 227–233.

Meusy, J.J. and Charniaux-Cotton, H. (1984) Endocrine control of vitellogenesis in Malacostraca crustaceans. In W. Engels, W.H. Clark, A. Fischer, P.J.W. Olive and D.F. Went (eds.), *Advances in Invertebrate Reproduction*, Elsevier, North Holland.

Morris J.E. (1971) Hydration, its reversibilty, and the beginning of development in the brine shrimp, *Artemia salina*. *Comparative Biochemistry and Physiology* **39**, 843–857.

Morris, J.E. and Afzelius, B.A. (1967) The structure of the shell and other membranes in encysted *Artemia salina* embryos during cryptobiosis and development. *Journal of Ultrastructure Research* **20**, 244–259.

Nakanishi, Y.H., Iwasaki, T., Okigaki, T. and Kato, H. (1962) Cytological studies of *Artemia salina*. I. Embryonic development without cell multiplication after the blastula stage in encysted dry eggs. *Annotationes Zoologicae Japonenses* **35**, 223–228.

Nakanishi, Y.H., Okigaki, T., Kato, H. and Iwasaki, T. (1963) Cytological studies of *Artemia salina*. II. Deoxyribonucleic acid (DNA) content and the chromosomes in encysted dry eggs and nauplii. *Proceedings of the Japan Academy* **139**, 306–309.

Olson, C.S. and Clegg, J.S. (1976) Nuclear numbers in encysted dormant embryos of different *Artemia salina* populations. *Experientia* **32**, 864–865.

Olson, C.S. and Clegg, J.S. (1978) Cell division during the development of *Artemia salina*. *Wilhelm Roux's Archives of Developmental Biology* **184**, 1–13.

Pandey, A.S. and MacRae, T.H. (1991) Toxicity of organic mercury compounds to the developing brine shrimp, *Artemia*. *Ecotoxicology and Environmental Safety* **21**, 68–79.

Peterson, J.A. and Rosowski, J.R. (1994) Scanning electron micrscope study of molar-surface development of *Artemia franciscana* Kellogg (Anostraca). *Journal of Crustacean Biology* **14**, 97–112.

Petrunkewitsch, A. (1902) Die Reifung der parthenogenetischen Eier von *Artemia salina*. *Anatomischer Anzeiger* **21**, 256–263.

Pochon-Masson, J. (1965) L'ultrastructure des épines du spermatozoïde chez les décapodes (Macroures, Anomoures, Brachyoures). *Comptes Rendus de l'Académie des Sciences* **260**, 3762–3764.

Pochon-Masson, J. (1968a) L'ultrastructure des spermatozoides vesiculaires chez les Crustaces Décapodes avant et au cours de leur dévagination experimentale I. Brachyoures et Anomoures. *Annales des Sciences Naturelles Zoologie* **10**, 1–100.

Pochon-Masson, J. (1968b) L'ultrastructure des spermatozoides vésiculaires chez les crustacés décapodes avant et au cours de leur dévagination experimentale II. Macroures. Discussion et conclusion. *Annales des Sciences Naturelles Zoologie* **10**, 367–454.

Pochon-Masson, J. (1969) Infrastructure du spermatozoide de *Palaemon elegans* (De Man), (Crustacé Décapode). *Archives de Zoologie Expérimentale et Générale* **110**, 363–372.

Rafiee, P., MacKinlay, S.A. and MacRae, T.H. (1986a) Taxol-induced assembly and characterization of microtubule proteins from developing brine shrimp (*Artemia*). *Biochemistry and Cell Biology* **64**, 238–249.

Rafiee, P., Matthews, C.O., Bagshaw, J.C. and MacRae, T.H. (1986b) Reversible arrest of *Artemia* development by cadmium. *Canadian Journal of Zoology* **64**, 1633–1641.

Rasmussen, S.W. and Holm, P.B. (1982) The meiotic prophase in Bombyx mori. In R.C. King, and H. Akai (eds.), *Insect Ultrastructure*, Plenum Press, New York, pp. 61–85.

Reger, J.F. (1966) A comparative study in the fine structure of developing spermatozoa in the isopod, *Oniscus asellus*, and the amphipod, *Orchestoidea* sp. *Zeitschrift für Zellforschung und mikroskopische Anatomie* **75**, 579–590.

Reger, J.F., Escaig, F., Pochon-Masson, J. and Fitzgerald, M.E.C. (1984) Observations on crab, *Carcinus maenas*, spermatozoa following rapid-freeze and conventional fixation techniques. *Journal of Ultrastructure Research* **89**, 12–22.

Reynier, M. (1959) Recherches sur le développement et la reproduction d'*Artemia salina*. *Bulletin de la Société des Sciences de Nancy* **18**, 155–182.

Rieder, C.L., Bowser, S.S., Cole, R., Rupp, G., Peterson, A. and Alexander, S.P. (1990) Diffuse kinetochores and holokinetic anaphase chromatin movement during mitosis in the hemipteran *Agallia constricta* (Leafhopper) cell line AC-20. *Cell Motility and the Cytoskeleton* **15**, 245–249.

Roels, F. (1968) Mise en évidence au microscope électronique de la phosphatase acide dans l'oeuf d'*Artemia salina*. *Annales de la Société Royale Zoologique de Belgique* **98**, 266–267.

Roels, F. (1970) Localisation d'activités peroxidasiques dans l'oeuf d'*Artemia salina* à l'aide de 3,3'-diaminobenzidine et de pyrogallol. *Archives de Biologie* **81**, 229–274.

Roels, F. and Wisse, E. (1973) Distinction cytochimique entre catalases et peroxidases. *Comptes Rendus de l'Académie des Sciences* **276**, 391–393.

Rosowski, J.R., Belk, D., Gouthro, M.A. and Lee, K.W. (1997) Ultrastructure of the cyst shell and underlying membranes of the brine shrimp, *Artemia franciscana* Kellogg (Anostraca) during postencystic development, emergence, and hatching. *Journal of Shellfish Research* **16**, 233–249.

Sabelli Scanabissi, F. and Trentini, M. (1979) Ultrastructural observations on oogenesis in *Triops cancriformis* (Crustacea, Notostraca). *Cell and Tissue Research* **201**, 361–168.

Santos, J.L. (1999) The relationship between synapsis and recombination: two different views. *Heredity* **82**, 1–6.

Schatten, G. (1994) The centrosome and its mode of inheritance: the reduction of the centrosome during gametogenesis and its restoration during fertilization. *Developmental Biology* **165**, 299–335.

Schatten, H., Thompson-Coffe, C., Coffe, G., Simerly, C. and Schatten, G. (1989) Centrosomes, centrioles, and posttranslationally modified α-tubulins during fertilization. In H. Schatten and G. Schatten (eds.), *The Molecular Biology of Fertilization*, Academic Press, pp. 189–210.

Schindelmeiser, J., Greven, H. and Bergmann, M. (1985) Meiosis and spermiogenesis in the testis of *Salamandra salamandra* (L) (Amphibia, Urodela). *Anatomischer Anzeiger* **158**, 169–180.

Schrehardt, A. (1987) Scanning electron-microscope study of the post-embryonic development of *Artemia*. In P. Sorgeloos, D.A. Bengtson, W. Decleir and E. Jaspers (eds.), *Artemia Research and its Applications*, Vol. 1, Universa Press, Wetteren, Belgium, pp. 5–32.

Shimizu, T. (1990) Polar body formation in Tubifex eggs. *Annals of the New York Academy of Sciences* **582**, 260–272.

Simar, L.J. (1969) Ultrastructure et constitution des corps nucleaires dans les plasmocytes. *Zeitschrift für Zellforschung und mikroskopische Anatomie* **99**, 235–251.

Somers, C.E. and Shapiro, B.M. (1989) Insights into the molecular mechanisms involved in sea urchin fertilization envelope assembly. *Development Growth & Differentiation* **31**, 1–7.

Stefani, R. (1963a) Il centromero non localizzato in *Artemia salina* Leach. *Atti della Accademia Nazionale dei Lincei Rendiconti-Classe di Scienze Fisiche, Matematiche e Naturali* **35**, 375–378.

Stefani, R. (1963b) La digametia femminile in *Artemia salina* Leach e la constitutzione del corredo cromosomico nei biotipi diploide anfigonico e diploide partenogenetico. *Caryologia* **16**, 625–636.

Stefani, R. and Cadeddu, G. (1967) L'attivita centromerica in *Artemia salina* Leach. *Rend. Sem. Fac. Scienze Univ. Cagliari* **37**, 287–291.

Tate, W.P. and Marshall, C.J. (1991) Post-dormancy transcription and translation in the brine shrimp. In R.A. Browne, P. Sorgeloos, and C.N.A. Trotman (eds.), *Artemia Biology*, CRC Press, Boca Raton, Florida, pp. 21–36.

Trotman, C.N.A. (1991) Normality and abnormality in early development. In R.A. Browne, P. Sorgeloos, and C.N.A. Trotman (eds.), *Artemia Biology*, CRC Press, Boca Raton, Florida, pp. 75–92.

Trotman, C.N.A., Mansfield, B.C. and Tate, W.P. (1980) Inhibition of emergence, hatching, and protein biosynthesis in embryonic *Artemia salina*. *Developmental Biology* **80**, 167–174.

Trotman, C.N.A., Gieseg, S.P., Pirie, R.S. and Tate, W.P. (1987) Abnormal development in *Artemia*: Defective emergence of the prenauplius with bicarbonate deficiency. *Journal of Experimental Zoology* **243**, 225–232.

Tudge, C.C. and Jamieson, B.G.M. (1991) Ultrastructure of the mature spermatozoon of the coconut crab *Birgus latro* (Coenobitidae: Paguroidea: Decapoda). *Marine Biology* **108**, 395–402.

Vallejo, C.G. and Marco, R. (1976a) Unmasking of mitochondrial precursors stored in the yolk platelets of *Artemia salina* dormant gastrulae. In T. Bucher, W. Neupert, W. Sebald and S. Werner (eds.), *Genetics and Biogenesis of Chloroplasts and Mitochondria*, Elsevier/North-Holland Biomedical Press, Amsterdam, pp. 847–850.

Vallejo, C.G. and Marco, R. (1976b) Yolk platelets as storage sites of mitochondria during

Artemia salina development: In 10th International Congress of Biochemistry, Hamburg, F.R. Germany (Abst.), p. 536.
Vallejo, C.G. and Perona, R. (1982) Microscopical study of the mechanisms of yolk degradation in *Artemia* and the concomitant release of mitochondria and other stored informational material. *Biologie Cellulaire* **45**, 97.
Vallejo, C.G., Gunther Sillero, G. and Marco, R. (1979) Mitochondrial maturation during *Artemia salina* embryogenesis. General description of the process. *Cellular and Molecular Biology* **25**, 113–124.
Vallejo, C.G., Marco, R., De Chaffoy, D. and Kondo, M. (1980) Yolk metabolization and mitochondrial release in *Artemia*. In W. Schwemmler and H.E.A. Schenk (eds.), *Endocytobiology: A Synthesis of Recent Research*, Proceedings of the International Colloquium on Endosymbiosis and Cell Research, Vol. 1, Walter de Gruyter and Company, New York, pp. 747–762.
Van Beek, E., Van Brussel, M., Criel, G. and De Loof, A. (1987) A possible extra-ovarian site for synthesis of lipovitellin during vitellogenesis in *Artemia* sp. (Crustacea; Anostraca). *International Journal of Invertebrate Reproduction and Development* **12**, 227–240.
Villa, L. and Tripepi, S. (1983) An electron microscope study of spermatogenesis and spermatozoa of *Ascidia malaca*, *Ascidiella aspersa* and *Phallusia mamillata* (Ascidiacea, Tunicata). *Acta Embryologiae et Morphologiae Experimentalis* **4**, 157–168.
Wagstaff, M.J.D., Collaço-Moraes, Y., Smith, J., de Belleroche, J.S., Coffin, R.S. and Latchman, D.S. (1999) Protection of neuronal cells from apoptosis by Hsp27 delivered with a Herpes simplex virus-based vector. *Journal of Biological Chemistry* **274**, 5061–5069.
Wahba, A.J. and Woodley, C.L. (1984) Molecular aspects of development in the brine shrimp *Artemia*. In W.E. Cohn and K. Moldave (eds.), *Progress in Nucleic Acid Research and Molecular Biology*, Vol. 31, Academic Press Inc., New York, pp. 221–265.
Walgraeve, H.R., Criel, G.R., Sorgeloos, P. and De Leenheer, A.P. (1988) Determination of ecdysteroids during moult cycle of adult *Artemia*. *Journal of Insect Physiology* **34**, 597–602.
Warner, A.H., Chu, P.P.Y., Shaw, M.F. and Criel, G. (2002) Yolk platelets in artemia embryos: are they really storage sites of immature mitochondria? *Comparative Biochemistry and Physiology Part B* **132**, 491–503.
Weber, J.E. and Russell, L.D. (1987) A study of intercellular bridges during spermatogenesis in the rat. *American Journal of Anatomy* **180**, 1–24.
Weisz, P.B. (1947) The histological pattern of metameric development in *Artemia salina*. *Journal of Morphology* **81**, 45–95.
Weitzman, M.C. (1966) Oogenesis in the tropical land crab, *Gecarcinus lateralis* (Fréminville). *Zeitschrift für Zellforschung und mikroskopische Anatomie* **75**, 109–119.
West, D.L. (1980) Spermiogenesis in the anthozoan, *Aiptasia pallida*. *Tissue & Cell* **12**, 243–253.
Wheeler, R., Yudin, A.I. and Clark, W.H.Jr. (1979) Hatching events in the cysts of *Artemia salina*. *Aquaculture* **18**, 59–77.
Wingstrand, K.G. (1978) Comparative spermatology of the Crustacea Entomostraca I/Subclass Branchiopoda. *Biologiske Skrifter* **22**, 23–66.
Wright, S.J. and Longo, F.J. (1988) Sperm nuclear enlargement in fertilized hamster eggs is related to meiotic maturation of the maternal chromatin. *Journal of Experimental Zoology* **247**, 155–165.
Xiang, H. and MacRae, T.H. (1995) Production and utilization of detyrosinated tubulin in developing *Artemia* larvae: evidence for a tubulin-reactive carboxypeptidase. *Biochemistry and Cell Biology* **73**, 673–685.
Zaffagnini, F. (1987) Reproduction in *Daphnia*. *Memorie dell'Istituto de Idrobiologia 'Dott. Marco De Marchi'* **45**, 245–284.
Zerbib, C. (1980) Ultrastructural observation of oogenesis in the crustacea amphipoda *Orchestia gammarellus* (Pallas). *Tissue & Cell* **12**, 47–62.
Zheng, L.M., Zychlinsky, A., Liu, C.C., Ojcius, D.M. and Young, J.D.E. (1991) Extracellular ATP as a trigger for apoptosis or programmed cell death. *Journal of Cell Biology* **112**, 279–288.

CHAPTER III

PHYSIOLOGICAL AND BIOCHEMICAL ASPECTS OF *ARTEMIA* ECOLOGY

JAMES S. CLEGG
Molecular and Cellular Biology and Bodega Marine Laboratory
University of California (Davis)
Bodega Bay, CA 94923
USA

CLIVE N.A. TROTMAN
Department of Biochemistry
University of Otago, Box 56
Dunedin, New Zealand

1. Introduction

This chapter considers life cycle-dependent biochemical and physiological adaptations critical to the survival of *Artemia* in nature. Thus, the encysted gastrula embryo ('cyst') is arguably the most resistant of all animal life history stages to extremes of environmental stress, while the motile stages are among the best osmoregulators in the animal kingdom. These and other adaptations are emphasized not only because they are essential to the success of brine shrimp, enabling them to cope with the many challenges presented by their harsh environment, but because this animal is a useful model system for the study of a variety of general biological processes. We hope our coverage will encourage others to study this remarkable organism.

A number of books (Persoone *et al.* 1980; Decleir *et al.* 1987; MacRae *et al.* 1989; Warner *et al.* 1989; Browne *et al.* 1991) and reviews (Clegg, 1986a,b; Hand and Hardewig, 1996; Guppy and Withers, 1999) have been published that treat these aspects to varying degrees. As a result, the reference list is not exhaustive, particularly for many of the earlier publications and for biochemical or physiological studies that are not easily related to the ecology of *Artemia*, the major focus of this chapter (completed in November 2000). We apologize to authors of those papers, and to those whose work we overlooked inadvertently.

Finally, we note that almost all of the research to be considered here involves populations of a single species, *Artemia franciscana*, mainly from the South San Francisco Bay of California and the Great Salt Lake, Utah, USA. Unless

---

Th. J. Abatzopoulos et al. (eds.), Artemia: Basic and Applied Biology, 129–170.
© 2002 Kluwer Academic Publishers. Printed in the Netherlands.

stated otherwise, the coverage refers to this species. Whether generalizations developed here apply to other *Artemia* species remains to be seen.

2. The Two Paths of Development

Whether bisexual or parthenogenetic, development of a given clutch of eggs follows one of two paths: ovoviviparous, resulting in nauplius larvae, or oviparous resulting in encysted gastrula embryos (cysts) that enter diapause, a state of obligate dormancy, released into their, usually hypersaline, environment. Diapause is terminated by environmental cues which differ in detail for the various species and even in geographically separated populations of the same species (Lavens and Sorgeloos, 1987; Drinkwater and Clegg, 1991). Embryos released from diapause resume development under permissive conditions of oxygen availability, water content and temperature. The embryo's inner embryonic cuticle is apparently impermeable to non-volatile solutes (De Chaffoy *et al.* 1978; Clegg and Conte, 1980) and, as we shall see, this relative independence from the environment has substantial adaptive significance in terms of inorganic ion homeostasis. Larvae produced from cysts appear to be morphologically the same as those produced ovoviviparously; however, significant biochemical differences exist (Liang and MacRae, 1999). It is certainly true that newly hatched larvae have adaptive repertoires very different from the encysted embryos, which is not surprising since the environmental challenges they face are so different.

These two paths involve adaptive features related in an obvious way to their ecological setting. Although exceptions exist, ovoviviparity is commonly taken when conditions are favourable (such as low or moderate salinity, high oxygen tension, and abundance of food) whereas diapause cysts often result when these and other conditions are not as favourable. This scenario is a common one in organisms that experience wide fluctuations in environmental parameters and are capable of producing either active or dormant stages.

For the sections that follow it may be useful to refer to the structure of encysted embryos, one of which is illustrated in Figure 1. This cyst, representative of many examined, had been dried to terminate diapause, then fully re-hydrated (at 2–4 °C to suppress metabolism) before being prepared for electron microscopy by conventional methods (Clegg *et al.* 2000a). To our knowledge the only detailed electron microscopic description of diapause embryos is the doctoral dissertation of Jardel (1986), although Criel (1991a) describes a few of their ultrastructural features. The early study by Morris (1968) on electron microscopy of dried (activated) cysts revealed what one might expect for cells that are able to desiccate reversibly: although dense and highly compact, ultrastructural integrity was retained and most cellular components were easily recognized, although some with modification. Later we consider some possible mechanisms underlying the retention of ultrastructural integrity in the desiccated embryo. Other ultrastructural observations

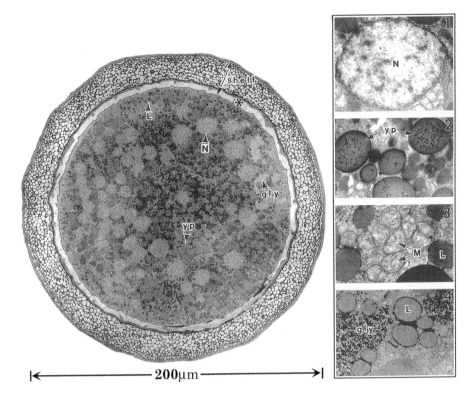

Figure 1. Ultrastructure of a post-diapause (activated) encysted gastrula embryo of *Artemia franciscana*. The panels to the right illustrate major organelles at higher magnification. Abbreviations are: *L*, liqid droplets; *N*, nuclei; *yp*, yolk platelets; *gly*, glycogen; *m*, mitichondria; *, inner embryonic cuticle. Modified from Clegg *et al.* (1999)

on activated cysts have been published (Morris and Afzelius, 1967; Hultin and Morris, 1968; Criel, 1991a; Clegg and Jackson, 1997) including some that had experienced anoxia for different periods of time (Hofmann and Hand, 1990b; Clegg *et al.* 2000a). On balance, hydrated cyst ultrastructure seems typical of yolky crustacean embryos, suggesting that the basis of their tolerance to desiccation is not revealed at this level of resolution, but must be sought at the sub-cellular level.

A distinctive feature of these embryos is their ability to endure virtually complete desiccation; indeed, desiccation can be an important diapause-terminating condition in at least some brine shrimps (Drinkwater and Crowe, 1987; Lavens and Sorgeloos, 1987; Drinkwater and Clegg, 1991).

3. The Desiccated Cyst

The ability of these embryos to resist a wide variety of severe environmental stresses is nothing less than spectacular. For example, they can be desiccated to the point where no liquid water remains, based on high vacuum/surface chemistry methodology (Clegg, 1978; Clegg et al. 1978), nuclear magnetic resonance spectroscopy (see Kasturi et al. 1990), quasi-elastic neutron scattering (Trantham et al. 1984) and other techniques (Clegg, 1986a; Clegg and Drost-Hansen, 1991). Do these laboratory studies have ecological relevance? Cysts in nature undergo severe dehydration, osmotically or by air exposure, and can survive cycles of hydration/desiccation, all without warning. These capabilities are built into diapause cysts during their development, preparing them for stresses they will encounter soon, or years later. A significant question arises: how do these embryos survive the complete removal of cellular water, a condition that quickly destroys the vast majority of animal cells? We know something about the answer.

3.1. THE WATER REPLACEMENT HYPOTHESIS (WRH) AND VITRIFICATION

Detailed studies over more than two decades by John and Lois Crowe, and their many students and colleagues, have revealed the importance of the disaccharide trehalose in desiccation tolerance (for access to their many papers see Crowe et al. 1992, 1996, 1998a, b). This sugar is present in massive amounts in *Artemia* cysts, making up about 15% of their dried weight (Dutrieu, 1960; Clegg and Conte, 1980). There is good *in vitro* evidence from the Crowe's laboratory that trehalose protects membranes and proteins against the destructive effects of dehydration, chiefly through its ability to serve as a substitute for structural water associated with their surfaces. These and other results have provided a basis for the water replacement hypothesis (see Crowe and Clegg, 1973; Clegg, 1986b; Crowe et al. 1998b) and offer a good explanation for at least part of the mechanisms underlying desiccation tolerance *in vivo*. It is certainly the case that many animal systems capable of reversible desiccation accumulate substantial concentrations of trehalose as well as other polyhydroxy compounds (see Crowe and Clegg, 1973; Yancey et al. 1983; Crowe et al. 1992, 1998b; Clegg, 2001).

In an adaptive context, it appears that the ability to desiccate reversibly was accomplished, in this case, through evolutionary acquisition of 'water substitutes' instead of evolving membranes and macromolecules unique in terms of their interactions with water (if indeed that could even be accomplished). The Crowe laboratory has recently taken advantage of this insight into the biological role of trehalose by demonstrating that it can also be used effectively for practical applications. For example, trehalose protects liposomes against dehydration damage (Crowe et al. 1998a,b) and prevents chill-damage to human pancreatic islets, allowing them to have a much longer 'shelf-life'

(Beattie et al. 1997). That discovery has dramatic clinical and medical implications, and also provides an excellent example of how high quality basic science can result in major benefits to humans. The Crowes and colleagues are currently documenting the ability of trehalose to protect human blood platelets against chill-damage (personal communication), another discovery of major clinical importance.

Water-replacement is not the only contribution that trehalose provides for tolerance to desiccation – recent work has indicated that vitrification (amorphous sugar glass formation) might also play a major role, and that trehalose is particularly good in that respect (reviewed by Sun and Leopold, 1997; Crowe et al. 1996, 1998a). It seems likely that the preservation of ultrastructure in dried encysted embryos is due at least in part to the embedment of cell structures in the vitrified matrix of these embryos. The vitreous state can also account for the chemical inertness of organisms capable of anhydrobiosis since the viscosities of these glasses are truly enormous. Suppression of adventitious chemical reactions must be critical to cells not capable of repair, and vitrification would be advantageous in that respect as well.

From this scenario one might expect that the accumulation of trehalose would be restricted to the desiccation-tolerant encysted embryo, and that is indeed the case. Trehalose synthesis begins at about 2 days after fertilization, but only in diapause-destined embryos (Clegg, 1965; Clegg et al. 1999). This sugar continues to accumulate as the complete diapause embryo is formed (about 5 days post-fertilization) and a further slight increase takes place during the first few days after release from the female (Clegg et al. 1999), then stops as embryonic metabolism becomes undetectable (Clegg et al. 1996; Clegg and Jackson, 1998). Trehalose is not present in detectable amounts in any other stage of the life cycle.

Another compatible solute, glycerol, is also accumulated in substantial amounts in diapause embryos, amounting to 2 to 5% of the dry weight (Clegg, 1962). Using the same approach as for trehalose, the Crowe laboratory found that glycerol was a poor protector against desiccation damage and could actually be detrimental (Crowe et al. 1984). That puzzling result remains to be understood. However, glycerol might be involved with another physiological process related to *Artemia* ecology, as we shall see.

Trehalose is also a major substrate for the energy metabolism that supports the conversion of the encysted gastrula into a hatched swimming nauplius (Clegg and Conte, 1980; Slegers, 1991). Trehalose that is not used for that process serves as a substrate for the synthesis of glycerol (the amount depending on the external osmotic pressure – Clegg, 1964) and glycogen, the dominant stored carbohydrate in all other *Artemia* life history stages. More will be said later about trehalose metabolism.

3.2. ANTIOXIDANTS AND RELEVANT ENZYMES

It has long been appreciated that dried biological systems are particularly sensitive to attack by molecular oxygen (Keilin, 1959; Lion et al. 1961; Crowe and Clegg, 1973; Clegg, 2001). Consequently, one might expect that dried cysts would contain high levels of the low molecular weight antioxidants and the enzymes involved with antioxidation mechanisms. It is usually the case that algae provide a major food source for *Artemia*, and these autotrophs contain abundant concentrations of carotenoids, particularly when growing under hypersaline conditions. These fat-soluble compounds are incorporated into the oocytes of *Artemia* and, therefore, become a major component of the lipids of encysted embryos. Carotenoids, in general, are considered to be excellent antioxidants, playing important roles in the detoxification processes that neutralize free radicals, and the literature on these compounds is large (see Packer, 1993). Nelis et al. (1989) have written the definitive review on carotenoids in life cycle stages of *Artemia*. Their review, highly recommended to supplement our brief coverage, shows that canthaxanthin is the major carotenoid of cysts, being synthesized from dietary β-carotene. It appears that *cis*-canthaxanthin is the dominant isomeric form and Nelis et al. (1989) discuss at length the possible significance of *cis-trans* conversions. Of considerable interest is the conjugation of carotenoids to proteins which they suggest might involve stabilization of the protein, while also protecting the carotenoid from photo-oxidation. The full significance of carotenoproteins remains to be realized, as is the extent to which carotenoids serve as effective antioxidants in cysts. These are fertile areas for further investigation.

Surprisingly little work has been done on other antioxidant systems in encysted *Artemia* embryos. Shanmugasundaram et al. (1996a) reported only low concentrations of α-tocopherol (vitamin E) in cysts of parthenogenetic *Artemia* collected from salt pans near Madras, India. In another study they measured catalase and glutathione reductase activities, and concentrations of reduced glutathione and ascorbic acid (Shanmugasundaram et al. 1996b). Although it was concluded that the levels of these antioxidant components changed in the dry cysts over time, it is not clear how that was estimated. In any event, none of these antioxidants was present in high concentration. Rudneva (1999) used cysts collected from Sakskoe Lake (Crimea, Ukraine), probably also a parthenogenetic strain (Triantaphyllidis et al. 1998) in a more detailed and comparative study. Interestingly, cysts contained similar or even lower levels of vitamins A and C, and glutathione compared to eggs of mussels (*Mytilus galloprovincialis*) and to eggs and larval stages of three fish species. In addition, these cysts did not contain elevated levels of lipoxygenase, superoxide dismutase, catalase, peroxidase or glutathione reductase activities compared to eggs and larvae of the other species. To explain these rather curious results Rudneva (1999) suggested that the thick shell of *Artemia* cysts might protect the embryo from oxidative damage. It is not clear whether the basis for this suggestion involves the screening of ultraviolet radiation by the shell, or prevention of the passage of molecular oxygen across it. We are not aware of any evidence indicating that dried cysts are impermeable to molecular oxygen. It is conceivable that shells contain substances that act as oxygen sinks, a possibility we explore later (section 11.4). Finally, it seems that damaging oxidation reactions would be suppressed greatly by the vitrification that probably occurs in severely desiccated cysts, so this might reduce the need for the various anti-oxidant mechanisms mentioned above.

Cysts contain substantial amounts of vitamin C in the form of stable ascorbic acid-2-sulfate (Mead and Finamore, 1969). It has been suggested that this compound serves as a stored form of the vitamin or, possibly, as a source of sulphate for the developing embryo. Additional information on this interesting compound can be found in Bond et al. (1972), Nelis et al. (1994) and Merchie et al. (1995).

One source of cellular oxidation damage comes from the interaction of ionic iron (Fe^{2+}, Fe^{3+}) with a wide variety of compounds to produce damaging products (Fenton reaction). We assume that ionic iron is present in Artemia cysts, but do not know the amounts or the mechanism of sequestration. In general, ionic iron is bound by various proteins, particularly ferritin (Henle and Linn, 1997; Winzerling and Law, 1997). Of some interest, therefore, is the presence in cysts of large amounts of a protein called 'artemin' (De Graaf et al. 1990; De Herdt et al. 1979). Although this multi-subunit protein of Mr 500,000 shows weak sequence similarity to ferritin it contains no iron after purification (De Graaf et al. 1990). Nevertheless, artemin remains a potential ionic iron-binding protein. It is interesting that this protein shows an ontogenetic pattern very similar to another dominant and very important cyst protein (p26) that we consider in section 4.3. Both proteins are present in about the same large amounts, both appear as the diapause embryo forms, and then both disappear in the early larval stages. Potential relationships between artemin and p26 present interesting opportunities for study.

4. The Hydrated and Activated Cyst

4.1. ANOXIA

Hydrated post-diapause cysts require only adequate temperature and molecular oxygen to resume metabolism and development. Recently it has come to be appreciated that the abilities of cysts to survive continuous anoxia rival their tolerance for reversible desiccation. Life history stages of most animals die within hours if deprived of molecular oxygen (Bryant, 1991; Hochachka et al. 1993) and even well-adapted forms such as sessile intertidal invertebrates rarely survive more than a month of anoxia, even at low temperatures (Hochachka and Guppy, 1987; Storey, 1998; Storey and Storey, 1990; Guppy et al. 1994; Guppy and Withers, 1999). In contrast, activated (hydrated) Artemia cysts survived an entire year of continuous anoxia without significant reduction in hatching, and only a modest reduction after the second year (Clegg, 1994). About 60% survived four years of anoxia (Clegg, 1997) and almost 10% survived more than six years (Clegg et al. 1999); indeed, almost seven years of continuous anoxia were required to reduce hatching to zero (unpublished results). Are these laboratory studies on anoxia relevant to the natural situation? Cysts blown on shore are often found in piles or windrows, embedded in decaying debris. It seems certain that cysts buried beneath the surface of this material encounter severe hypoxia or anoxia. In addition, cysts that sink and become buried in the sediments will definitely become anoxic in these usually hydrogen sulphide-rich locations. Remember also that cysts are strictly endogenous when it comes to energy sources, so that even a slow rate of metabolism under anoxic conditions would eventually deplete substrates required to complete embryonic development when permissive conditions return.

What, then, is the metabolic status of the hydrated anoxic embryo? Anoxic survival in well-adapted animals is achieved through reductions in the metabolic rate, or 'metabolic rate depression' (MRD), commonly falling to 1 to 10% of the aerobic rate (see Guppy and Withers, 1999). The case has been made (Hontoria et al. 1993; Clegg, 1997; Clegg et al. 1999), despite opinions to the contrary (see Hand, 1995; Guppy and Withers, 1999) that *Artemia* cysts bring their metabolism to a reversible standstill during the first day or so of anoxia. At least that seems to occur for the conventional pathways of energy metabolism and macromolecular turnover, but the presence of minor pathways that have escaped detection obviously cannot be ruled out. Not only does this lack of a measurable metabolism represent the ultimate in MRD, it raises some puzzling questions about anoxic cyst survival to be considered later. First, however, we inquire into the mechanisms involved in anoxia-induced metabolic rate depression, and its reversal upon re-oxygenation: what controls these transitions?

4.2. THE pH_i SWITCH

Pioneering research by Busa et al. (1982) and Busa and Crowe (1983), followed by an extensive series of studies by them and their colleagues (Busa and Nuccitelli, 1984; Busa, 1986; Carpenter and Hand, 1986; Hand and Carpenter, 1986; Drinkwater and Crowe, 1987; Crowe et al. 1987) implicated intracellular pH changes (ΔpH_i) as a major factor regulating aerobic-anoxic transitions in metabolism. More recently a series of studies by Hand and colleagues has supported and extended considerably the original findings (Hofmann and Hand, 1990a,b, 1992, 1994; Hand, 1993, 1997, 1998; Anchordoguy and Hand, 1994, 1995; Kwast et al. 1995; Kwast and Hand, 1996 a,b; Hand and Hardewig, 1996; Hardewig et al. 1996; Eads and Hand, 1999; van Breukelen and Hand, 2000). These detailed contributions have provided great insight into the metabolic transitions associated with varying concentrations of molecular oxygen. Anoxia causes a rapid reduction in pH_i from 7.9 or higher in aerobic embryos, eventually falling to about 6.5 after a few days of anoxia, to our knowledge the largest ΔpH_i associated with dormancy in any system. Acidic pH_i is maintained over at least 5 months of anoxia (Clegg *et al.* 1995) and, we assume, remains there until molecular oxygen became available. A great deal of research from Hand's laboratory (references above) indicates that anoxia-induced intracellular acidification inhibits RNA and protein synthesis and degradation, and several key enzymes in the pathway of trehalose utilization, shutting down metabolism and development. Likewise, the restoration of oxygen results in an increase in pH_i (Kwast *et al.* 1995) presumably reactivating all these processes and turning on metabolism and development once again. These and other relationships between oxygen, pH and metabolism have been summarized in a recent review by Hand (1997). The rate of post-anoxic metabolism is an inverse function of the duration of the previous anoxic period (Clegg, 1993) and can be greatly depressed together with the rate of development that is being driven by it (see Clegg, 1997).

There exists some uncertainty about the pH_i of cysts in diapause. Drinkwater and Crowe (1987) found that diapause cysts had a pH_i similar to that of activated ones, suggesting that pH changes were not involved in the induction of diapause. However, subsequent study showed that cysts newly released from females actively metabolize for several days and, therefore, are not in metabolic diapause (Clegg *et al.* 1996, 1999). Thus, the key question concerns the timing of the ^{31}P-NMR studies of Drinkwater and Crowe (1987). Although this information was not given in the paper, it appears that these measurements were done within a day or so after cysts were released (personal communication, Laurie Drinkwater). In that case, pH_i remains a viable candidate for a role in the regulation of diapause-induction. Our expectation is that pH_i in diapause cysts will decrease as their metabolic rate slows following release from females, but additional NMR work is needed to evaluate the suggested causal connection between metabolic rate and pH_i in embryos destined for diapause.

That uncertainty aside, the pH-switch is a major regulator of metabolism during aerobic-anoxic transitions. It provides an interesting and well-documented hypothesis, and represents not only a major contribution to our understanding of metabolic control in *Artemia* embryos, but also to similar transitions in other systems (see Busa and Nuccitelli, 1984; Busa, 1986; Hochachka *et al.* 1993). This is yet another example of how useful these encysted embryos can be for the study of biological problems, in this case the control of cellular metabolism by pH_i.

4.3. STRESS PROTEINS

As mentioned, results on anoxic cysts raise interesting questions about how their cells maintain integrity in the apparent absence of free energy flow and biosynthesis. We use the term 'apparent' recognizing that it is not possible to prove experimentally the absence of a metabolic rate at physiological temperatures (see Clegg, 1997). Two processes that require large amounts of free energy in animal cells are those involved with ionic/osmotic homeostasis, and the turnover of macromolecules. Encysted embryos have escaped the problem of ion transport across the shell by tolerating desiccation and becoming impermeable to inorganic ions. However, anoxic cysts face potentially severe problems from hydrolases and macromolecular denaturation and aggregation, notably since these molecules cannot be replaced by synthesis throughout the period of anoxia (see Clegg et al. 1999). The possibility that evolution would select for proteins that are thermally stable seems extremely unlikely because protein function requires flexibility, and that causes instability at physiological temperatures (Somero, 1995).

Anchordoguy and Hand (1994, 1995) and van Breukelen and Hand (2000) found that the ubiquitin pathway of protein degradation was greatly suppressed during anoxia, while other evidence indicated the absence of proteolytic activity of any kind (Warner, 1989; Clegg, 1997; Warner et al. 1997). Our view is that all proteolytic activity must be completely inhibited in anoxic embryos, otherwise it is difficult to understand how they could possibly survive such prolonged bouts. To our knowledge no study has been made of nuclease activity in anoxic cysts, but circumstantial evidence suggests that these enzymes must also be shut down.

Several years ago we set out to explore the integrity of globular proteins in anoxic cysts. We and our collaborators found massive amounts of what turned out to be a small heat shock protein, named p26, whose amino acid sequence (Liang et al. 1997a, b) showed that it belongs to the α-crystallin family (de Jong et al. 1998; MacRae, 2000). This protein is a particle of M_r 500,000–700,000 (15nm diameter), composed of sub-units of M_r 26,000, and has been studied in detail (Clegg et al. 1994, 1995; Jackson and Clegg, 1996; Liang et al. 1997a, b; Liang and MacRae, 1999; Clegg et al. 1999; Criel and MacRae, Chapter 2 of this volume). An earlier study by Grosfeld and Littauer (1976) detected two proteins in cysts that could be p26 and artemin, based on relative abundance and electrophoretic migration, but that observation was not followed by further analysis.

A major feature of p26 is its ability to act as a molecular chaperone, preventing denatured proteins from aggregating and, to some extent, renaturing those that have begun to unfold. In vitro studies revealed that p26 functions without a source of free energy, such as ATP or GTP, an important feature in view of the apparent lack of a conventional energy metabolism in anoxic cysts. Like trehalose, p26 is restricted to the encysted embryo, with small amounts present in the instar-I nauplius, but none in any other life cycle stage, in spite of repeated attempts to induce its synthesis (Liang and MacRae, 1999; Clegg et al. 1999). It first appears at 3 days after fertilization, but only in embryos destined to enter diapause, eventually reaching concentrations of 10–15% of the diapause embryo's total non-yolk protein (Clegg et al. 1994; Jackson and Clegg, 1996; Liang and MacRae, 1999). To our knowledge this is the highest constitutive level for any stress (heat shock) protein in non-transfected cells. A key observation was that about half of the total p26 in diapause embryos is located in the nucleus where it presumably protects nuclear components (Jackson and Clegg, 1996), although other functions are certainly possible (see Clegg *et al.* 1999). Upon termination of diapause most of the nuclear p26 is translocated to the cytoplasm as the activated embryo resumes metabolism and development. However, should the embryo encounter appreciable environmental stress (such as prolonged anoxia or heat shock) p26 translocates back into nuclei. These translocations are strongly pH-dependent (Clegg *et al.* 1995) in a manner fully consistent with the pH switch. Much remains to be learned about the roles and significance of this remarkable protein, but its importance in the adaptive repertoire of *Artemia* has been established. Further discussion of p26 in *Artemia* is given by Criel and MacRae in Chapter II of this volume.

Other stress proteins are found in encysted embryos, the major ones being those of the Hsp70 and 90 families (Miller and McLennan, 1988a, b; Clegg *et al.* 1999). The extent to which these proteins are involved with cyst stress resistance is not known, although they do not appear to undergo stress-induced translocation to nuclei (Clegg *et al.* 1999) which is unusual since other cell types carry out stress-induced nuclear translocations of both of these stress protein families (Morimoto *et al.* 1994).

Research on molecular chaperones and stress proteins has been intense during the last decade or so, and the massive literature continues to grow (see Feder and Hofmann, 1999). While the vast majority of this work has been directed toward understanding the molecular and cellular functions of these proteins (for reviews: Hartl, 1996; Beissinger and Buchner, 1998; Richardson *et al.* 1998; Ellis, 1999; Ellis and Hartl, 1999; van den IJssel *et al.* 1999; Whitley *et al.* 1999) their study in an ecological and evolutionary context has only recently begun. The reader is referred to excellent reviews by Feder (1999) and Feder and Hofmann (1999) for a thorough account of stress proteins and the stress response in the latter context (also see Coleman *et al.* 1995). *Artemia* provides excellent opportunities for research of this kind.

5. Thermal Resistance of Cysts, Larvae and Adults

We now turn to the response of cysts to temperature extremes. We are not aware of any published work on measurements of cyst temperatures under natural conditions, but it can be assumed safely that cysts encounter thermal stress (high and low), notably when they are washed onto shore, existing in either dried or hydrated condition, and exposed to the warmth of sunlight or to freezing temperatures.

Cysts dried to low water contents, say about 10% water by weight, are extremely tolerant of temperature extremes, which is not surprising since dried systems are, in general, more stable than hydrated ones. Dried cysts survive exposure to 2.2 °K (Skoultchi and Morowitz, 1964), and have been reported to survive an hour or so at 103 °C with little reduction in hatching (Hinton, 1968). Other studies have been considered by Clegg and Conte (1980). It seems certain that trehalose is involved with this impressive thermal tolerance (see Yancey et al. 1983; Somero, 1986; Winzor et al. 1992; Hottiger et al. 1994; Crowe et al. 1998b; Wera et al. 1999). As expected, the situation is much different in hydrated cysts. For one example, cysts with water contents below about 0.6 g H_2O/g dry weight survived prolonged exposure to – 24 °C, whereas those with higher water contents were killed by this treatment (Crowe et al. 1981). Studies on the responses of cysts of differing water content to high temperatures should yield interesting results.

Miller and McLennan (1988a) showed that hydrated cysts were reasonably tolerant to high temperatures: the LT_{50} of cysts from the Great Salt Lake, Utah, was 49 °C (using 1 hour exposures), fairly impressive for animal cells. They also demonstrated (Miller and McLennan, 1988b) the presence of several isoforms of the Hsp-70 and 90 families in cysts, some of which were induced by heat shock, as well as a small Hsp, reported to be Mr 31,000, which might be the previously mentioned artemin which has at least some features of stress proteins (Clegg and Jackson, 1998). Surprisingly, Miller and McLennan (1988a) did not observe the induction of thermotolerance in cysts (details were not given) although that was readily achieved using instar-I nauplii. Lack of induced thermotolerance in cysts has been reported by others (Liang and MacRae, 1999; Clegg et al. 1999) using a variety of time and temperature exposures. Recognizing that the induction of thermotolerance is essentially universal (Nover, 1991) it was suggested (see Clegg et al. 1999) that the encysted embryo is pre-loaded with all the ingredients required to survive the various stresses it will encounter, often without warning and perhaps years later. It is also possible that the cost attached to the repeated synthesis of large amounts of stress proteins might confer difficulty for a system that is endogenous and cannot obtain organic substrates from the environment. These two possibilities are not mutually exclusive.

When hydrated cysts are exposed to temperatures in the range of 42 to 50 °C, the nuclear translocation of p26 takes place (Abatzopoulos et al. 1994; Clegg et al. 1999), but the extent and speed of translocation depend heavily on the time/temperature protocol. Studies on the Hsp-70 and Hsp-90 families indicated that heat shock did not result in their translocation to nuclei of encysted embryos (Clegg et al. 1999) an unusual result in view of their nuclear translocation in the cells of other organisms (Morimoto et al. 1994, and references given previously on molecular chaperones). We are not aware of any data published on pH_i of cysts after heat shock, but predict it will be acidic and similar to that of anoxic and diapause cysts. In fact, it is becoming increasingly likely that cysts undergoing diapause, heat shock and anoxia utilize very similar regulatory and protective mechanisms at the cellular and molecular levels.

In contrast to cysts and larvae little is known about the heat shock response of adults. As expected, adults are much less tolerant to high temperatures than other life history stages, and are only moderately resistant when compared to other adult aquatic crustaceans. An LT_{50} near 38 °C (30 minute exposures) was obtained for *Artemia* adults that had reached sexual maturity within a week of study (Frankenberg *et al.* 2000). Unlike cysts, adults mounted a strong response to heat shock involving induced thermotolerance lasting about a week, and upregulated proteins of the Hsp-70 family. As mentioned, adults lack measurable amounts of p26, glycerol and trehalose, and it seems likely that these deficiencies are related to their poorer thermal performance, although there are likely to be other reasons as well.

In 1996 cysts of *A. franciscana* from Francisco Bay (SFB) were used to inoculate much warmer growth ponds in the Mekong Delta region of Vietnam (V). Cysts produced from the resulting adults were harvested and used to inoculate the ponds in 1997, and so on thereafter in sequential fashion. V-cysts produced by the first generation of adults in 1996 were shown to be substantially more heat-resistant than those produced in SFB (Clegg *et al.* 2001). Furthermore, adults produced in the laboratory from these V-cysts were also more tolerant of high temperatures than those cultured from SFB cysts (Clegg *et al.* 2000b). These and other results from this ongoing 'Vietnam Experiment' suggest that genetic selection for thermal stability is extremely rapid, being incorporated into the developmental program in a single generation. Related to these results is the detailed study by Browne and Wanigasekera (2000) on adults of five *Artemia* species subjected to various temperature-salinity combinations in the laboratory. They found significant interactions between these two parameters in terms of survival and several reproductive traits, and compared these with the temperature and salinity conditions of the habitats of these species. The authors conclude that *A. franciscana* has the greatest temperature tolerance range of any *Artemia* species and a high degree of phenotypic plasticity, both of which fit our results (Clegg *et al.* 2000b, 2001). Because of the diversity of *Artemia* habitats around the world this organism provides unique opportunities for the study of phenotypic plasticity and the evolution of thermal and salinity tolerance. Browne and Wanigasekera (2000) consider virtually all of the literature on the salinity and temperature tolerance of *Artemia*.

An interesting relationship between a major stress protein/chaperone (Hsp104) and trehalose has been revealed by genetic and heat shock experiments on the yeast, *Saccharomyces cerevisiae* (Elliott *et al.* 1996; Iwahashi *et al.* 1998; Singer and Lindquist, 1998). In this case, trehalose minimizes protein aggregation during heat shock, as expected (Hottiger, 1994; Crowe *et al.* 1998b; Wera *et al.* 1999). However, soon after the cells are returned to physiological temperatures, trehalose is rapidly metabolized, the reason being that high concentrations of trehalose interfere with Hsp104 chaperoning. Relationships between trehalose and p26 in *Artemia* cysts should prove to be interesting. Our expectation is that physiological concentrations of trehalose will enhance the chaperone function of p26 *in vitro*.

We have focused on adaptations in encysted embryos that are associated with three common and major environmental stressors: lack of water and molecular oxygen, and thermal extremes. Cysts also exhibit impressive resistance to other hazards that are beyond the scope of the present paper: for example, UV and ionizing radiations, soaking in a variety of organic solvents and, when dry, surviving the conditions of outer space (see articles in Persoone *et al.* 1980; Decleir *et al.* 1987; MacRae *et al.* 1989; Warner *et al.* 1989). These abilities justify the opinion that *Artemia* cysts are the 'extremophiles' of the animal kingdom.

6. Selected Biochemical Features of Encysted Embryo Development

Given permissive environmental conditions the metabolism and development of encysted embryos (Figure 2) are rapidly re-initiated. Respiration, RNA and protein synthesis begin within minutes, (Clegg and Conte, 1980) supporting the conclusion that encysted embryos are pre-loaded with all the components needed for these activities. An important recent find in this regard is the activation of ribosomal S6 kinase in less than 15 minutes following the hydration of dried cysts (Malarkey *et al.* 1998). That kinase is believed to play a key role in the up-regulation of mRNA translation in many cell types, and that appears to be the case for the rapid onset of protein synthesis in encysted embryos after hydration. In another important study, Brandsma *et al.* (1997) suggested that the major rate-limiting step in protein synthesis (initiation and elongation) in activated cysts is the rate at which aminoacyl-tRNAs are replenished. It appears that most tRNA molecules are not attached to their appropriate amino acids in quiescent embryos.

Figure 2 describes some highlights of encysted embryo development. After a period of post-diapause development (PDD), on the order of 12–24 hours depending on temperature and salinity, the embryo first emerges from the shell (E-1), continues developing for 2–4 hours (E-2) followed by hatching of the first stage nauplius (N-I). An adaptation of major importance is the absence

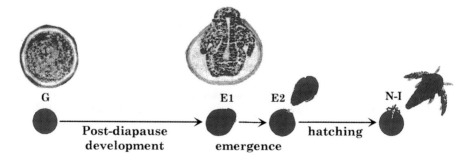

Figure 2. Development of a post-diapause (activated) encysted gastrula (G). Emerged embryos (E1 and E2) give rise to the swimming nauplius larva (N-I) after post-diapause development. Modified from Clegg and Conte (1980).

of DNA synthesis and cell division throughout PDD, first demonstrated by Nakanishi et al. (1962, 1963) then confirmed and extended by others (Finamore and Clegg, 1969; Olson and Clegg, 1978). Why should *Artemia* have acquired such an unusual way to develop, to undergo morphogenesis without cell division? We believe this is yet another adaptive mechanism that confers stability and minimizes damage from the various environmental threats that can arrive at any time. Cells in the act of making DNA and undergoing cell division are particularly vulnerable, so it seems that these gastrulae have found a way to develop into larvae without the participation of either process. It is also possible that mechanisms that protect nuclear DNA prevent its replication – in fact, that seems very likely.

Early work on changes in numerous enzyme activities and metabolite concentrations during PDD has been summarized (Clegg and Conte, 1980). The more recent review by Slegers (1991) is an excellent account of *Artemia* developmental biochemistry and is highly recommended. In what follows we draw selectively from these reviews, but also consider research appearing since 1990.

6.1. YOLK PLATELETS

Much research has been done on these abundant organelles (Figure 1) showing that they are not inert storage structures but play important roles during development to the nauplius and beyond. The research groups in Madrid have contributed substantially to our understanding of yolk platelet structure and function, as well as other aspects of the developmental biochemistry and genetics of *Artemia*. We can consider only briefly their contributions, one of which involved the first demonstration of mitochondria-like structures in yolk platelets of the encysted gastrula (see Vallejo et al. 1979) an initially controversial observation that has since been confirmed by Hofmann and Hand (1990b) but still questioned (Criel and MacRae, Chapters I and II, this volume). Careful analysis of mitochondrial differentiation during development (Vallejo et al. 1996) revealed that maturation of pre-existing mitochondria, rather than mitochondrial proliferation, accounts for the marked increase in respiratory activity that has often been demonstrated in developing encysted embryos (Slegers, 1991). These results agreed with studies showing increases in levels of mitochondrial cytochrome oxidase (Hofmann and Hand, 1990b) and timing of the digestion of yolk platelets (Utterback and Hand, 1987; Perona et al. 1988). An adaptive interpretation of these findings would be that a substantial fraction of mitochondria (or their precursors) is protected by sequestration in yolk platelets, maintaining their integrity in the face of stressful challenges. That protection is important since mitochondria can not be made *de novo*, but must be derived from pre-existing organelles. Slegers (1991) and Warner (1992) describe the enzymes and metabolites associated with yolk platelets, sites of significant guanosine nucleotide metabolism. There are two phases of yolk utilization – an initial pH-dependent one, followed by enzymatic attack (Slegers, 1991). Once again, the pH switch seems to play an important metabolic role since acidic pH inhibits yolk platelet utilization, while that is enhanced by alkaline pH. Other information on yolk platelet structure and function can be found in the books and reviews cited in the introduction to this chapter.

6.2. TREHALOSE AND TREHALASE

Most enzyme activities in the encysted gastrula do not change significantly during the first 4–6 hours of post-diapause development, and many remain about the same until the time of emergence and hatching (Clegg and Conte, 1980; Slegers, 1991). The presence of these enzymes at more or less constant levels reflects the ability of cysts to resist stress while retaining the capacity to resume metabolism and development rapidly, eventually resulting in the formation of nauplius larvae (Figure 2). Energy metabolism is based chiefly on trehalose utilization, presumably via the glycolytic pathway and Krebs' cycle/electron transport chain. Thus, control of the initial step in trehalose catabolism is critical, apparently being catalysed by the hydrolytic enzyme trehalase, reviewed thoroughly by Slegers (1991) whose account will be briefly summarized here (also see Vallejo, 1989). Acidic and neutral trehalases have been described, based on the pH-dependence of their activity *in vitro*. The former appears to be largely associated with intracellular components that sediment at low speeds (1,000 g) in extracts. Interestingly, the pH optimum of the acidic form is close to the pH_i of anoxic cysts in which, nevertheless, trehalase activity is completely suppressed, perhaps through its association with cell particulates. The neutral trehalases are soluble and may exist in two interconvertible forms having different Mr. Once again, pH seems to be a key regulator since alkaline pH, that of the activated metabolising embryo, results in the more active form of the enzyme. However, Vallejo (1989) found that the particulate acidic trehalase, when solubilized with detergent, can be activated *in vitro* at neutral pH simply by heating in the physiological range of temperature. Furthermore, this enzyme could be activated in the cold at alkaline pH. Interestingly, enzyme concentration influenced these activities, with higher concentrations suppressing activation. Vallejo (1989) makes the case that these three variables (temperature, pH, concentration) although studied *in vitro*, have physiological significance. These relationships provide excellent opportunities for further study.

Since Slegers' (1991) review, Nambu and colleagues reported results on cyst trehalase (Nambu *et al.* 1997a) which are not easily reconciled with other findings on this enzyme; however, as they point out, there are a number of technical differences between these various studies that make comparison difficult and, in any case, are too complicated to deal with here. Nambu *et al.* (1997b) and Tanaka *et al.* (1999) obtained interesting results on trehalase genes using molecular techniques. Nithya *et al.* (1996) published preliminary findings on the acidic trehalase in crude extracts of parthenogenetic *Artemia* cysts.

Available evidence indicates that the first step in trehalose utilization is a simple hydrolysis. One of us (JSC) thinks that the seeming energetic waste of initiating this pathway by splitting trehalose into two glucose molecules is curious since there is sufficient free energy associated with breaking this glycosidic bond to allow for phosphorylation of at least one of the resulting glucose molecules. Indeed, that is the case in a number of bacteria and fungi which contain a trehalose phosphorylase, the reaction products being glucose and glucose-1-phosphate (see Eis and Nidetzky, 1999 and references therein). In cysts, the glucose produced from trehalose hydrolysis must first be phosphorylated by hexokinase, a step that requires ATP, before it can be metabolized further. *Artemia* has been fortuitous in the evolutionary acquisition of a biochemistry that is efficient and closely in tune with its natural setting, so there must be good reasons why trehalase has been selected over trehalose phosphorylase to initiate this critical pathway. If trehalose hydrolysis were taking place in embryos undergoing active aerobic metabolism one would expect to detect at least small concentrations of free glucose, but that is not the case (Vallejo, 1989; J. Clegg, unpublished results) notably in cyst populations having high viability. That result could be explained by a very tight metabolic coupling ('channelling') between trehalase and the next enzyme in the pathway (hexokinase) involving the direct transfer of glucose and avoiding its release into the aqueous phase. We have more to learn about the initiation of trehalose metabolism.

Another possibility for regulation during stress-induced metabolic transitions has been advanced by Lazou *et al.* (1994) and Lazou and Frosinis (1994). They reported that anoxia and dehydration decreased the binding of several glycolytic enzymes to 'cell particulates' and suggested that this reduction might be at least part of the regulatory mechanisms involved in metabolic control. That possibility deserves further study.

Recall that cysts are able to undergo cycles of desiccation/re-hydration (Morris, 1971) and survive prolonged anoxia (Warner *et al.* 1997) during at least the first four hours of post-diapause development (Figure 2). Consequently, this initial phase of embryonic development, *per se*, does not reduce these capabilities. Thus, if embryos are allowed to develop for 4 hours a significant amount of trehalose is utilized, but a heat shock applied at that point results in the rapid re-synthesis of trehalose to its initial level (Clegg and Jackson, 1992) ensuring that adequate concentrations of this important sugar are maintained. There are, of course, dramatic differences between emerged embryos, newly hatched nauplii and the encysted embryos that gave rise to them (Figure 2). For instance, the former two stages cannot undergo reversible desiccation, survive less than a day of anoxia, depending on the temperature (Warner *et al.* 1997) and exhibit lower thermal tolerance than cysts (Miller and McLennan, 1988a). Although there may be several reasons for the loss of these abilities, it is relevant that larvae derived from cysts contain very little trehalose, p26 and guanosine polyphosphates (see later), and even the small amounts present are soon destined to be reduced to undetectable levels. However, we consider in sections 7 and 8 that the nauplius has its own array of adaptations related to the conditions it faces in nature, including the severe threats of swimming in a hypersaline environment, sometimes very hypersaline. First we consider the unusual nucleotide pool of encysted embryos which, once again, reveals several features of adaptive significance.

6.3. NUCLEOTIDES AND THEIR ENZYMES

A lot of work has been done on this subject, much of it appearing during the 1980s (reviewed by Sillero and Sillero, 1989; McLennan, 1992). As far back as the early 1960s it was known from the seminal work of Finamore and Warner (1963) that cysts contain massive concentrations of a variety of guanine-containing nucleotides. Subsequent research from their laboratory and others showed that the most prominent member of this family is guanosine (5') tetraphospho (5') guanosine, or Gp_4G (reviewed in depth by Warner, 1992). This compound and other dinucleoside polyphosphates (including Gp_3G and Gp_2G), GDP and GTP make up about 3% of the dry weight of cysts, and almost 90% of the total nucleotide pool, which is astonishing compared with nucleotide pools in the cells of other organisms. In most other cells the adenosine nucleotides (AMP, ADP and ATP) are, by far, the most abundant, but in *Artemia* cysts they constitute only about 5% of the total nucleotide pool. In general, cellular viability is correlated closely with concentrations of ATP in the 1–5 mM range, and the balance between ATP, ADP and AMP (energy charge) has often been viewed as central to the regulation of energy metabolism. Even in the majority of organisms undergoing metabolic rate depression (MRD) due to anoxia, the concentration of ATP is usually maintained, although its turnover may be dramatically reduced (Guppy and Withers, 1999). In striking contrast, *Artemia* cysts placed under anoxia show a rapid decline in ATP concentration to almost undetectable levels (Stocco *et al.* 1972).

It has been proposed that the guanosine polyphosphates, especially Gp_4G, represent a source of free energy during anoxia, essentially taking over the role of the adenosine nucleotides (see Warner, 1992). Support for this idea came from Warner and Sonnenfeld (1990) who studied cysts from thirteen different populations of *Artemia* and found a strong correlation between cyst viability (% hatching) and the guanylate energy charge (appropriate ratios of the concentrations of several guanine-containing nucleotides). Although Stocco *et al.* (1972) observed a decline in the content of Gp_4G during anoxia, that decrease appeared to level off after 50 days with only a slight decrease during further anoxic incubation. Nevertheless, Gp_4G remained a strong candidate to be a free energy source during prolonged anoxia, and recent evidence supports that possibility (Warner and Clegg, unpublished results). If the anoxic metabolism of Gp_4G can be demonstrated and quantified it should be possible to determine if its metabolism can account for the tiny amount of heat generated by anoxic embryos (see Hand and Podrabsky, 2000). A non-trivial question arises concerning the use to which this free energy is put, but the answer is not obvious to us, based on current knowledge (see Clegg, 1997; Clegg and Jackson, 1998).

Another unusual, but closely related feature of nucleotide metabolism in *A. franciscana* is the apparent absence of the pathway required to synthesize purine rings de novo (reviewed by Finamore and Clegg, 1969; Sillero and Sillero, 1989; Warner, 1992). The evidence for that conclusion comes from nutritional studies (see Hernandorena, 1993, 1999) as well as biochemical ones. What might be the adaptive significance of the loss of a metabolic pathway that seems to be ubiquitous in other animals? This pathway is energetically expensive and involves a large number of enzymes (see Finamore and Clegg, 1969). Consequently, one can speculate that the loss of this pathway, whenever possible, would be advantageous, energetically and also at the level of protein synthesis. It is well known that *Artemia* larvae, immatures and adults feed chiefly on purine-rich algae and bacteria and, since swimming and feeding are linked (Lent, 1971) feeding is essentially constant. One can suppose that these motile stages receive adequate amounts of purines in their diets, so there is no need to synthesize them, permitting the repression or loss of the appropriate genes for this pathway, a speculation that can be tested by molecular genetics. However, exogenous purines are not available to the non-feeding encysted embryo or the early naupliar stages whose digestive tract is not functional (see Criel, 1991a, b).

What then is the source of adenine and guanine nucleotides for nucleic acid synthesis in encysted and emerged embryos, and particularly in newly hatched, non-feeding nauplii? An obvious candidate is Gp_4G, and there is substantial evidence for that possibility (reviewed by Warner, 1992). It is interesting and important to note that the nucleotides of immatures and adults are conventional, lacking the large guanosine polyphosphate pool. Thus, at least part of the adaptive significance of the highly unusual guanine-nucleotide pool seems to be its purine-storage function in non-feeding stages of the life cycle.

Research from one laboratory indicates that an active *de novo* pathway for purine ring synthesis in *Artemia* does exist (see Liras *et al.* 1992; Rotllan *et al.* 1993). We recognize this difference of opinion, but note that Warner (1992) has provided an explanation for this discrepancy in experimental results, involving bacterial contamination. Additional experiments should resolve the issue.

Coverage of purine nucleotides in *Artemia* should include mention of the long and detailed series of studies by Hernandorena on the effects of dietary purines on several aspects of brine shrimp development. Her research, carried out over the last 30 years, has implicated the metabolism of purines in several developmental processes, including pattern formation (see Hernandorena, 1993, 1999).

A number of other papers have been published on enzymes of nucleotide metabolism during the last decade and we refer the interested reader to them (Liras *et al.* 1990; Prescott *et al.* 1992; Díaz *et al.* 1993; Liu and McLennan, 1994; Sillero *et al.* 1994; Díaz and Heredia, 1996; Montero *et al.* 1996;

Cartwright and McLennan, 1999). Finally, Gp$_4$G has been shown to be a significant inhibitor of casein kinase II from *Artemia* cysts (Pype and Slegers, 1993). This could prove to be a very important observation since this major second-messenger-independent protein kinase has been implicated in the activation of stored (masked) mRNA and as a regulator of protein synthesis during development.

6.4. PROTEASES AND THEIR INHIBITORS

In his comprehensive review, Warner (1989) concluded that *Artemia* was the most studied invertebrate in terms of the involvement of proteases in embryonic development. The evidence is abundant that proteases play significant roles in such diverse processes as hatching, yolk platelet degradation, protein synthesis regulation and RNA metabolism. We previously stressed the necessity for very tight control over proteases of all types in these cysts, notably during prolonged anoxia and also when in diapause which can last for well over a year (Clegg and Jackson, 1998). Hand and colleagues found that the ubiquitin-based pathway of proteolysis was inhibited by anoxia (Anchordoguy and Hand, 1994, 1995) and recent evidence indicates that acidic pH is involved in the regulation of this system (van Breukelen and Hand, 2000) adding further importance to the pH switch as a metabolic regulator.

The first detailed study of *Artemia* proteases (Osuna *et al.* 1977) showed that these activities were hardly detectable in cyst extracts, and remained so until emergence and hatching when marked increases took place (Warner, 1989; Slegers, 1991). Subsequent research from Warner's laboratory revealed that over 90% of the measurable protease activity in cysts is due to a cysteine protease (CP) consisting of two subunits and six isoforms (Warner, 1989; Warner *et al.* 1995, 1997). Interestingly, the optimal pH for CP activity *in vitro* is acidic, and similar to pH$_i$ of anoxic cysts. However, CP is under very tight control by inhibitor proteins (CPI) whose activity is also maximal in the same pH range. Thus, these protein-protein interactions exist in harmony with the pH switch. CPI decreases at about the time of emergence, releasing CP from inhibition (Warner *et al.* 1997). The latter study also suggested that p26 might interact with CP and CPI, forming a particle of *M*r 95,000. That being the case, CP could be involved in the rapid degradation of p26 that takes place beginning at emergence and continuing through hatching (Clegg *et al.* 1994; Liang and MacRae, 1999). Detailed immunocytochemical and biochemical studies by Warner and Matheson (1998) and Warner *et al.* (1995) showed that CP is an extremely important enzyme in nauplii, being involved directly or indirectly in hatching, yolk utilization, moulting and digestion when larval feeding begins. Evidence for these functions is extensive and convincing.

Stabili *et al.* (1999) raised the possibility that the shells of cysts contain enzymes that serve as a first line of defense against invasive microorganisms. They reported trypsin-like and lysozyme-like activities in the embryos and in well-washed shell fragments. The extent to which these activities are due to contaminating bacteria and/or fungi embedded in the shells is uncertain, although the authors do not consider that to be important.

7. Emergence and Hatching

The mechanisms of emergence and hatching have been studied in detail (see Clegg and Conte, 1980; Trotman, 1991; Rosowski *et al.* 1997) and will be dealt with only briefly here. Rosowski *et al.* (1997) provide ultrastructural details about emergence and hatching, and a good discussion of the mechanisms involved. There is some evidence for the participation of metabolism in the emergence process, involving an increase in free glycerol whose accumulation might cause (or contribute to) rupture of the shell (Clegg, 1964). If that mechanism plays a significant role then it would be another adaptation at the level of physiology and biochemistry that can be related to *Artemia* ecology – in this case the variable osmolality faced by encysted embryos. No specialized region of the shell seems to be the site of emergence, although the anterior end of the pre-nauplius (E-1, Figure 2) is always the part that protrudes first.

A fascinating series of studies indicates that gravity has a powerful effect on the direction of emergence (Gouthro and Rosowski, 1994). When cysts were adhered to a stationary substrate over 75% emerged upward, and only 1% downward. The direction of light and point of cyst adhesion had no effect on these results. They and their colleagues went on to show that cysts exposed to microgravity in NASA's space shuttle **Discovery** emerged in random directions, while hypergravity studies on earth (73 g) increased the proportion of upward-emerging cysts to 91% (Rosowski *et al.* 1995). These authors provided several suggestions about the possible adaptive significance of these results, one being the need for cysts in lake sediments to emerge in the direction of the water column. Other work on this interesting effect in development, emergence and hatching during spaceflight has been published (Spooner *et al.* 1994a, b).

The identity of the gravity sensor was not established in those studies, so a recent paper by Schimek *et al.* (1999) might be of interest in this connection. They showed that sporangiophores of the fungus, *Phycomyces blakesleeanus* (which orient with respect to gravity) contain paracrystallin proteins that sediment, and lipid droplets that float upward, both at unit gravity. The authors concluded that these structures were the principal mechanisms involved in the perception of gravity. While a similar situation could conceivably operate in *Artemia* cysts, its internal structure is highly compact and dense, at least at the level of organelles (Figure 1).

The transition from the encysted embryo through emergence to the free-swimming nauplius is obviously critical to developmental success. In addition, the larva faces serious ionic and osmotic problems once it leaves the ion-impermeable shell that protected the embryo so well. What adaptations are involved in meeting those challenges to homeostasis?

8. Emerged Embryos and Instar-I Nauplius Larvae

We focus chiefly on osmotic and ionic regulation, and associated adaptations, because of their ecological relevance. Following the pioneering work of D.E. Copeland, P.C. Croghan and P.G. Smith (see Conte, 1984), a series of detailed studies throughout the 1970s by Frank Conte and colleagues conclusively established that the haemolymph of nauplii can be maintained at concentrations and compositions that are considerably lower and qualitatively different, respectively, from the external hyperosmotic medium. These results, reviewed thoroughly by Conte (1984) require that the emerged embryos and nauplii retain or take up water from the medium while eliminating excess of ions, both of these sometimes taking place against formidable electrochemical gradients. It is well established that larval osmoregulation takes place chiefly in what has been referred to as the 'neck organ', but more appropriately called the larval salt gland since its ultrastructural, physiological and biochemical features are similar to many other types of salt extruding structures in other animals (Conte, 1984).

Hypoosmotic regulation is based on the regulation of sodium and particularly chloride ions that are transported outward, an active process that requires the participation of the key enzyme Na,K-adenosine triphosphatase (Conte, 1984; Baxter-Lowe et al. 1989). That enzyme first appears in emerged embryos, then increases dramatically as the larvae hatch and pass through the first stage. The Conte laboratory demonstrated that the ion-transporting elements in the larval salt gland are chloride cells, typical of ion transporting epithelia of animals in general. Since then, studies by Sun et al. (1992) and Escalante et al. (1995) revealed the spatial and temporal expression of this enzyme's mRNA, using in situ hybridization techniques (also see Escalante et al. 1994). More recently, García-Sáez et al. (1997) carried out detailed molecular biological work on the gene coding for a subunit of Na,K-ATPase. They pointed out the importance of this enzyme to larval osmoregulation and discussed its evolutionary relationships amongst various taxa. Additional studies on *Artemia* Na,K-ATPase have been published by Fisher et al. (1986) and Salon et al. (1989) including recent molecular biological work by Saez et al. (2000). The effects of juvenile hormone and related compounds on the level of naupliar Na,K-ATPase activity have also been described (Ahl and Brown, 1991).

An interesting and ecologically relevant feature of larval osmoregulation concerns the developmental programming of Na,K-ATPase synthesis. Lee and Watts (1994) showed that increases in the amounts of this enzyme before and during the first larval stage are independent of environmental salinity, a situation much different from that in adults (Holliday et al. 1990). Larval exposure to hypo- or hypersalinity takes place suddenly at the time of emergence and hatching, and the regulatory mechanisms needed to handle this exposure are programmed into their development. This adaptive programming is similar to that involved in the accumulation of trehalose, glycerol, Gp_4G and p26 in the diapause-destined embryo.

Watts *et al.* (1996) contributed to our understanding of larval osmoregulation by examining the responses of nauplii to hypoosmotic conditions. Larvae reared in seawater (32 ppt) responded to transfer to 4 and 12 ppt by rapidly synthesizing substantial amounts of ornithine decarboxylase (ODC), the rate-limiting enzyme in polyamine synthesis. Polyamines have been implicated in a variety of basic cellular processes, including responses to altered cellular osmolality (see Watts *et al.* 1996). They proposed that increases in ODC with subsequent changes in polyamine synthesis, degradation or transport play a significant role in the short-term regulation of osmotic and ionic balance in larvae. Short-term cell volume regulation, in the face of osmotic challenge, is deserving of further study in the motile stages of *Artemia*.

Conte (Clegg and Conte, 1980) reviewed evidence indicating that increases in Na,K-ATPase are correlated with an increase in cytochrome oxidase, suggesting that changes in ion transport capability and cellular respiration are linked. Conte (1984) constructed a useful diagram that integrates pathways of glycogen breakdown, energy metabolism and yolk platelet dissolution in nauplii.

A recent contribution to our understanding of larval salt gland development is the work of Chavez *et al.* (1999). They demonstrated the expression of a homeobox gene, APH-1, during encysted embryo development, and found that its expression is restricted to the larval salt gland. It seems likely that the product of this gene, a specific kind of transcription factor, must be important to the development of the salt gland, and possibly involved with its degeneration in young immature stages. The larval salt gland works very well indeed, and its loss is accompanied by development of the more conventional osmoregulatory system of adults.

9. Adult Osmoregulation

Artemia is arguably the best osmoregulator in the animal kingdom. What are the upper limits of their abilities in this regard? At the Lake Grassmere solar salt works, New Zealand, adults were found in the final crystallizing ponds where the brine specific gravity (SG) was 1.22 at 30 °C (Marshall, 1988). Recent observations (Rawle and Trotman, unpublished) revealed that large numbers of adults had been living for at least 8 days in waters of 1.240 SG, and for 13 days (the last 5 in captivity) at 1.223. The latter had evidently been let in from a reservoir of 1.218 SG. However, brine of 1.245–1.251 proved to be lethal to all these animals. The latter range of SG corresponds to roughly 10 times the concentration of ocean seawater, well over 300 ppt, and above the point of NaCl crystallization. There is uncertainty about the longevity of these animals, and whether they can reproduce under such conditions. The analysis by Browne and Wanigasekera (2000), and the literature they cite on the study of *Artemia* from other salt lakes and ponds, suggests that the Grassmere animals may have been transported into the crystallizing ponds where they survived for a while, but were not able to live there for any length of time. These uncertainties aside, the ability of *Artemia* to live and reproduce in solutions of remarkably high ionic strength and osmotic pressure is definitely impressive (Clegg and Conte, 1980; Conte, 1984; Browne and Wanigasekera, 2000) and, to our knowledge, is unequaled by any other animal.

We refer the reader to the review by Conte (1984) and the work of Holliday *et al.* (1990) whose paper seems to be the most recent on adult osmoregulation. The basic strategy of adults involves drinking the external medium, and transferring ions and fluid from the gut into the haemolymph at about the same rate as fluid is lost across the body wall to the environment. Haemolymph ions are then transported outward. Urine is produced by the maxillary gland which, however, is unable to concentrate inorganic ions (*i.e.*, the urine is isosmotic with respect to the environment). Outward ion transport is performed chiefly by leaf-like structures (metepipodites) on the individual thoracic swimming appendages (phyllopodia). These structures house the ion-secreting epithelia and, as expected, have very high levels of Na,K-ATPase activity (Holliday *et al.* 1990). The gut and maxillary gland also contain substantial levels of this enzyme activity, reflecting their participation in ionic and osmotic regulation. Unlike larvae, levels of Na,K-ATPase in the metepipodites and other adult tissues are influenced strongly by environmental salinity. That is understandable since adults are likely to encounter wide fluctuations in their ionic and osmotic surroundings during their lifetime; larvae, on the other hand, seem to have systems designed for shorter terms.

In spite of the remarkable osmoregulatory abilities of *Artemia* adults, the physiological mechanisms they use are, on balance, quite similar to those of ordinary marine teleost fishes, including functional similarities between the chloride cells of teleost gills and those of brine shrimp metepipodites (Holliday *et al.* 1990).

10. Oxygen Transport and Consumption in Adults

Artemia haemolymph does contain cells but these are not concerned with the transport of molecular oxygen by haemoglobin which is carried in free solution. Swimming, feeding and other vital activities are seriously impaired if the dissolved oxygen of the medium falls below about 2 ml O_2/l at seawater salinity. Below 1.5 ml O_2/l, respiration falls drastically and the organism stops moving and relies on anaerobic metabolic pathways for survival.

Gilchrist (1954) examined the effects of both oxygen and salinity on haemoglobin synthesis, higher salinity and lower oxygen tension generally equating with a stimulation of haemoglobin synthesis. Respiratory rates of adults (Kuenen, 1939; Eliassen, 1952; Gilchrist, 1956, 1958) are different from those of nauplii (Kratowich, 1964; Angelovic and Engel, 1968; Conte *et al.* 1980). Furthermore, adult males and females responded differently in their oxygen consumption to the effects of salinity changes (Gilchrist, 1956, 1958), the changes being greater in females than in males. Both male and female adults respond to diminished oxygen by synthesizing more haemoglobin (Gilchrist, 1954; Spicer and El-Gamal, 1999). At about 1.0 ml O_2/l, enhanced haemoglobin production is principally in the form of HbIII, which appropriately has the highest oxygen affinity and the lowest Bohr effect (Decleir *et al.* 1980). Thus, the organism is better able to sequester scarce oxygen. Angelovic and Engel (1968), and later Decleir *et al.* (1980) and Vos *et al.* (1979), investigated the effects of temperature and found, not surprisingly, that respiration rates increased with increasing temperature.

11. Haemoglobins

11.1. PHYSIOLOGY AND CHARACTERIZATION

Physical and physiological characteristics of the *Artemia* extracellular haemoglobins (Hb) have been comprehensively reviewed by Moens (1982) and Moens *et al.* (1991). Three Hb variants exist, formed by the permutations of two different types of subunit known as α and β, permitting the dimers HbI (α,α), HbII (α,β) and HbIII (β,β). The three dimeric Hbs can be separated by ion exchange or electrophoresis and their physiological parameters have been distinguished. The key parameters of oxygen affinity (P_{50}, mm Hg) and Hill coefficient are 5.3 and 1.6 for HbI; 3.7 and 1.9 for HbII; and 1.8 and 1.6 for HbIII (1 mm Hg = 0.133 kPa). Studies on different biological strains and under different conditions have yielded different but comparable values (Moens *et al.* 1991).

Three points of interpretation stand out. First, the three *Artemia* Hbs are different from each other, which may be of physiological significance. Second, their oxygen affinity is considerably higher than, for example, mammalian Hbs in general, and is more like myoglobin (Mb); for comparison P_{50} in humans (under physiological conditions and near sea level) is 26 mm Hg for Hb and approximately 1 mm Hg for Mb. Third, and contrariwise, the *Artemia* Hbs display cooperativity which aids oxygen release. Oxygen affinity in clones and expressed isolated domains is up to an order of magnitude stronger than in the physiological polymer where allosteric cooperativity is possible, P_{50} values being in the order of 0.5 mm Hg in the isolated domains (Vandenberg and Trotman, 1999).

Physical, chemical and analytical studies (Moens, 1982) including amino acid composition, haeme and Fe^{2+} content, and estimates of *M*r by analytical centrifugation, light scattering or gel filtration, led to a consensus model comprising eight globin domains in each α or β polypeptide. Limited proteolysis with subtilisin (Moens et al. 1984) fortuitously cut the polymer at presumably exposed inter-domain bridges and gave eight fragments by chromatofocusing, strongly supporting the eight domain model. Eight globin domains of nominal Mr 16,000 (excluding heme) would dictate a total Mr of 128,000, close to values measured by Wood et al. (1981). Values of Mr 17,000–17,500 per domain derived from compositional analysis (Moens, 1982) and from domain sequences (Moens et al. 1988), both of which techniques included the inter-domain linkers, posited a total Mr of 136,000 to 140,000, but this higher value was accommodated by measurements of about 135,000–165,000 from gradient gel electrophoresis (Moens, 1982).

11.2. SEQUENCE AND STRUCTURE

It was a surprise when the cDNA for the entire molecule was sequenced (Manning et al. 1990) to find not eight but nine concatenated domains. The definitive Mr added up to 158,980 (excluding haeme) or an average of 17,664 per domain, the high value being due to inter-domain linkers of typically 14 amino acids and also unique N- and C-terminal extensions. Since initially it was not possible to assign the definitive cDNA sequence unequivocally to either α or β polypeptide its translation was called T. The complementary polypeptide was called C when sequenced later via cDNA (Matthews et al. 1998). C and T are now considered to be α and β, respectively. Recent sequence and Southern analysis (Vandenberg, 2000) suggest that four Hb polymers (T1, T2, C1 and C2) are expressed in Artemia. Protein sequencing of the native Hbs revealed that there are limitations as to which two polymers associate. HbI comprises C1C2; HbII comprises C1T2; HbIII comprises T1T2. These pairings allow the levels of these three Hbs to be regulated independently by polymer expression alone, therefore explaining the previously inconsistent developmental and hypoxia-induced expression patterns.

A working model of the molecule comprises two nine-domain rings stacked as a cylinder 9 nm tall and 14.5 nm diameter. Each ring has globin domains of nominal dimensions 4.5 nm × 3.5 nm × 2.5 nm joined by highly conserved inter-domain linkers, suggesting a symmetrical arrangement (Trotman et al. 1994). The model is consistent with an axial ratio of 1.7 measured by ultracentrifugation (De Voeght and Clauwaert, 1981) and with electron micrographs (Wood et al. 1981).

Artemia and Parartemia zietziana Hbs, on the basis of a 17% amino acid difference (too low to need correction for multiple hits) and assuming a divergence of 1% per 5 MY (Dickerson and Geiss, 1983), appear to have diverged about 85 MY ago (Coleman et al. 1998), coincidentally if not significantly about the time Australia separated from the rest of the world. The 17% difference appears, upon examination, to exemplify neutral evolution, preserving all idiosyncrasies of Artemia versus other Hbs including fluctuations of domain length and unusual substitutions characteristic of specific domains, with one crucial exception. The deletion between amino acids C1 and C4 (deemed to be C2), leaving three amino acids in place of four, is a unique mutation confined to the 9th and final domain of both T and C polymers of *Artemia*. This deletion is in the critical haeme environment and difficult to credit, but structurally plausible (Trotman *et al.* 1991). *P. zietziana* does not have this deletion in the 9th or any domain (there is only one type of polymer, P). Evidently the deletion happened in the line leading to *Artemia* but before the T and C genes diverged about 60 MY ago. None of the differences between the *Artemia* and *Parartemia* Hb sequences can be related to functional differences except for the possibility, currently under study, that the C2 deletion might render domain 9 non-functional, but in any case the *Parartemia* molecule appears to be the least corrupt and yet the least employed since it is scarcely expressed.

11.3. GENE EXPRESSION AND REGULATION

One interesting question is the physiological relevance of the three Hb variants. Adult *Artemia* may tolerate environmental extremes such as 340 g/l salinity, over 10 times the concentration of sea water, and a dissolved oxygen concentration of 1.5–2.0 ml/l, being less than half the 5 ml/l typical of surface seawater (Decleir *et al.* 1980). Do responses to environmental factors correlate with the Hb expression pattern?

Hb expression is detectable within about 2 hours of hatching under aerobic seawater-like conditions and is predominantly of the heterodimer, HbII, as expected if both polymers are equally expressed. About 8 hours after hatching, however, HbIII is detectable and may represent about 25% of total Hb (Heip *et al.* 1978a, b) until the shrimp are about 3 weeks old. At 3–4 weeks of age HbIII expression drops towards zero and HbI rises to be similar to HbII (Heip *et al.* 1978a, b), although the pattern may differ according to the strain of *Artemia* (Bowen *et al.* 1969). The steady expression of the heterodimer HbII effectively from the time of hatching means that both α and β polymers are being expressed and the changing ratios of the three variants could simply reflect an imbalance in expression of the two genes. In fact a 25:50:25 ratio of Hbs I:II:III is to be anticipated if inter-subunit affinities are equal and the observations above could be explained by expression of β exceeding α for the first three weeks, followed by a reversal.

Environmental oxygen depletion is associated with a shift of balance towards β expression as evidenced by decreased HbI and increased HbIII synthesis (Bowen *et al.* 1969), which is nicely rationalized in terms of HbIII having the highest affinity for oxygen and therefore sequestering it more efficiently. Oxygen availability decreases with increasing salinity of the habitat (Mitchell and Geddes, 1977), both directly as a solubility affect and indirectly because the increase in density with salinity leads to reduced movement and gas transfer between strata. A tertiary effect weighing in the same direction is that shallow, dense saline lagoons harbouring *Artemia* are often rich in purple pigmented microorganisms, favouring heat absorption, a higher temperature (40 °C is not uncommon) and thereby a lower oxygen tension.

Artemia hatched under hypoxic conditions (PO_2 of 10 kPa, approximately half atmospheric) respond with increased Hb expression and earlier respiratory regulation (Spicer and El-Gamal, 1999). Experimentally, organisms in a closed chamber are found to respire at a constant rate until the available oxygen concentration falls to a critical PO_2, below which this ability is lost. *Artemia* hatched hypoxically maintain respiratory regulation down to a lower oxygen tension (4 kPa) than normoxic controls (down to 4.6 kPa) but, equally interesting, the hypoxic animals achieved this ability in developmental stage 3 (about 0.9 mm length) rather than the normal stage 6, about 1.1 mm on the 19-stage developmental scale of Weisz (1946, 1947).

The effect of hypoxia on development therefore appears to be two-fold: to hasten progress through the early stages following hatching, and to upregulate Hb expression in those early stages. Hb expression appears gradually and there may be no 'point' at which it is first expressed, however by stages 3–6 haemolymph Hb concentration in hypoxically-hatched individuals, at 14–16 mg/ml, is vastly higher than the 0.6 mg/ml of normoxic controls at the same stages. Spicer and El-Gamal (1999) also noted that a shift of P_{50} by stage 6 towards a lower affinity (0.6 kPa) in normoxic controls did not occur in hypoxics, in which the affinity of about 0.4 kPa was not significantly different from stage 3. This would imply, in hypoxics, a mix of HbII and HbIII at stages 3 and 6, similar to early analyses of the Hb mix under normal conditions (Heip *et al.* 1978a,b; Bowen *et al.* 1969) whereas in normoxics the shift was towards HbI. As Spicer and El-Gamal (1999) themselves comment, the opposite, namely a higher affinity associated with hypoxia, might be more rational.

It has been assumed tacitly that *Artemia* Hb is an oxygen transporter but one of us (CNAT) wonders to what extent is this the case. As Mangum (1997) pointed out, Hb concentration can be so low in some other small invertebrates that one wonders why it is there at all. However, haemolymph Hb concentration in *Artemia* can be quite high, adult values of 8 mg/ml in the Great Salt Lake strain and as high as 23 mg/ml in a Kazakhstan strain having been reported (Decleir *et al.* 1989) but Hb levels may also be very low, and lower in males than females (Gilchrist, 1954; Bowen *et al.* 1969). The possibility that Hb is of little consequence for oxygen transport in *Artemia*, but has been diverted to some prosaic task, cannot be dismissed in view of the low profile role of the molecule in *Parartemia*, a brine shrimp similar to *Artemia* but sufficiently different to represent a separate genus. One such difference is that *Parartemia* cysts sink rather than float, which has probably contributed to the isolation of different species in different Australian salt lakes (Geddes, 1980, 1981). At least in *Parartemia zietziana*, Hb expression is a trace 0.1 mg/ml and could contribute a mere 1.7% to haemolymph dissolved oxygen capacity (Manwell, 1978). Furthermore, *Parartemia* is about twice the size of *Artemia*. Although much has been made of the ability of *Artemia* to tolerate more concentrated brines and lower oxygen tensions, *Parartemia* can survive 300 g/l salt and an oxygen tension as low as 1.8 mg/l (Williams and Geddes, 1991), not greatly different from *Artemia*.

11.4. ADDITIONAL ROLES FOR HAEMOGLOBIN ?

Possession of substantial Hb correlates with the extended environmental reach of *Artemia* but, as demonstrated by *Parartemia*, does not necessarily cause it. Could Hb have an additional role, for instance in quiescence and the emergence process?

Emergence involves escape of the embryo from its surrounding shell and involves a sudden rupture of the rigid alveolar structure of the cyst wall together with the thick cuticular layer inside, which is bounded by outer and inner cuticular membranes. Depending on the temperature, this happens about 12–18 hours after rehydration of dried, activated cysts in seawater. Piecing together clues, including the effects of respiratory inhibitors, a requirement for bicarbonate ion and the effects of bicarbonate dehydratase inhibitors, it appears that essential components of the emergence mechanism include bicarbonate-chloride exchange, a Na^+-K^+ exchange pump and oxidative phosphorylation to supply the ATP needed for ion pumping (Trotman *et al.* 1980, 1987; Trotman, 1991). The likely location of these components is the embryonal surface, presumably being part of the developmental program leading to the diapause embryo.

The proposed scenario for successful expulsion of the prenauplius centres on the hatching membrane that intimately surrounds the embryo, forming a sac. Hydrostatic pressure within this sac causes the cuticular and alveolar layers to rupture whereupon the sac, still enclosing the embryo, inflates and widens the rupture transiently to propel the embryo to the exterior. The question is how? Prior to cyst rupture the volume between the embryo and the hatching membrane is small and ion extrusion by the embryo would maintain a high osmotic pressure in that small space. The cyst wall ruptures, water enters osmotically and the hatching membrane sac starts to inflate. This is where Hb may become relevant. A momentary large increase in ion pumping activity by the embryo is essential if the osmotic pressure in the sac is to be maintained and its inflation is not to collapse. The necessary oxidative phosphorylation may be protected by a reserve of oxygen stored in Hb in case oxygen availability is marginal.

Evidence for this mechanism comes from its occasional failure. A subtle but lethal defect in emergence can lead to inflation of the hatching sac and expulsion of the prenauplius, but not quite powerfully enough to rupture the inner cuticular membrane. Such a defect occurs under conditions that would partially inhibit oxidative phosphorylation or bicarbonate-chloride exchange (Trotman *et al.* 1980, 1987; Trotman, 1991). Against this theory, it might be pointed out that Hb synthesis is not detectable until some hours after hatching is completed! However, a search for Hb in cysts and embryos by means of the immunogold technique led to the unexpected finding of dense globin immunoreactivity throughout the thickness of the cuticular layer of the cyst wall except for the inner cuticular membrane (Trotman *et al.* 1986). The cuticle is an embryonal tissue, whereas the alveolar layer (which did not reveal globin immunoreactivity) is maternal (Morris and Afzelius, 1967; Lochhead and Lochhead, 1940). Globin immunoreactivity was also present in the larval exoskeleton and gut lining (a body surface, developmentally) but not elsewhere (Trotman *et al.* 1986).

A provocative parallel exists between the probable occurrence of Hb in the *Artemia* cyst cuticular layer and its presence in the aleurone layer of plant seeds (Taylor *et al.* 1994). Ancestral Hb could have an antiquity of well over 1,000 MY (Matthews and Trotman, 1998; Trotman *et al.* 1998) if not at least 1,800 MY (Appleby, 1994) at which time oxygen may have been scarce and the role of Hb could have been more to sequester reserves for transient use. This storage role may survive in the high-affinity Hb of modern *Artemia* and *Parartemia*. We have referred previously in this review (section 3.2) to the potential value of an oxygen sink in the cyst shell, a role for which Hb could also be suited.

12. Concluding Remarks

We have chosen to focus largely on those aspects of biochemistry and physiology that can be related to the natural setting of *Artemia*, in an adaptive context whenever possible. Having said that, an attempt has also been made to cite and sometimes briefly comment upon most of the physiological and biochemical papers that have appeared since 1990, the year before the most recent book on *Artemia* was published (Browne *et al.* 1991). We apologize to authors whose papers have not been included.

Artemia truly is an extraordinary animal, its encysted embryos being an 'extremophile' of the animal kingdom, while the larvae, immatures and adults are the 'top guns' of osmoregulation. It is hard to imagine a more suitable model system amongst all the invertebrates for the study of these subjects. In the many locations where *Artemia* occurs it dominates the animal population in numbers and biomass, its most abundant animal companion in the water column being larvae of the brine fly, *Ephydra* (which, by the way, exhibit some remarkable adaptations of their own – see Herbst, 1999). The evolutionary and ecological strategy of brine shrimps has been to acquire adaptations that permit them to thrive where very few other animals can even survive, and none can complete their entire life cycle. This success seems to have been achieved very early since the Anostraca first appeared about 400 million years ago and, from the scanty fossil record (Tasch, 1963) it appears that fairy shrimps and brine shrimps were among these early crustaceans, changing very little in their morphology (and probably physiology and biochemistry as well) for hundreds of millions of years.

13. Acknowledgements

JSC acknowledges financial support from the U.S. National Science Foundation (grant MCB-9807762) and a Senior Fulbright Research Fellowship while at Ghent University, Belgium, in 1999. CNAT acknowledges financial support from the Marsden Fund administered by the Royal Society of New Zealand. The word processing skills of Diane Cosgrove and technical contributions of Susan Jackson are greatly appreciated.

14. References

Abatzopoulos, Th., Triantaphyllidis, G., Sorgeloos, P. and Clegg, J.S. (1994) Evidence for the induction of diapause by heat shock in *Artemia. Journal of Crustacean Biology* **14**, 226–230.

Ahl, J.S.B. and Brown, J.J. (1991) The effect of juvenile hormone III, methyl farnesoate and methoprene on Na / K-ATPase activity in larvae of the brine shrimp, *Artemia. Comparative Biochemistry and Physiology A* **100**, 155–158.

Anchordoguy, T.J. and Hand, S.C. (1994) Acute blockage of the ubiquitin-mediated proteolytic pathway during invertebrate quiescence. *American Journal of Physiology* **267**, R895–R900.

Anchordoguy, T.J. and Hand, S.C. (1995) Reactivation of ubiquitination in *Artemia franciscana* embryos during recovery from anoxia-induced quiescence. *Journal of Experimental Biology* **198**, 1299–1305.

Angelovic, J.W. and Engel, D.W. (1968) Interaction of gamma irradiation and salinity on respiration of brine shrimp (*Artemia salina*) nauplii. *Radiation Research* **35**, 102–108.

Appleby, C.A. (1994) The origin and functions of haemoglobin in plants. *Science Progress (Oxford)* **76**, 365–398.

Baxter-Lowe, L.A., Guo, J.Z., Bergstrom, E.E. and Hokin, L.E. (1989) Molecular cloning of the Na,K-ATPase α-subunit in developing brine shrimp and sequence comparison with higher organisms. *FEBS Letters* **257**, 181–187.

Beattie, G.M., Crowe, J.H., Lopez, A.D., Cirulli, V., Ricordi, C. and Hayek, A. (1997) Trehalose: a cryoprotectant that enhances recovery and preserves function of human pancreatic islets after long-term storage. *Diabetes* **46**, 519–523.

Beissinger, M. and Buchner, J. (1998) How chaperones fold proteins. *Biological Chemistry* **379**, 245–259.

Bond, A.D., McClelland, B.W., Einstein, J.R. and Finamore, F.J. (1972) Ascorbic acid-2-sulfate of the brine shrimp, *Artemia salina. Archives of Biochemistry and Biophysics* **153**, 207–214.

Bowen, S.T., Lebherz, H.G., Poon, M.-C., Chow, V.H.S. and Grigliatti, T.A. (1969) The hemoglobin of *Artemia salina*. I. Determination of phenotype by genotype and environment. *Comparative Biochemistry and Physiology* **31**, 733–747.

Brandsma, M., Janssen, G.M.C. and Möller, W. (1997) Termination of quiescence in Crustacea. The role of transfer RNA aminoacylation in the brine shrimp *Artemia. Journal of Biological Chemistry* **272**, 28912–28917.

Browne, R.A. and Wanigasekera, G. (2000) Combined effects of salinity and temperature on survival and reproduction of five species of *Artemia. Journal of Experimental Marine Biology and Ecology* **244**, 29–44.

Browne, R.A., Sorgeloos, P. and Trotman, C.N.A. (1991) *Artemia Biology*, CRC Press, Boca Raton, Florida.

Bryant, C. (1991) *Metazoan Life Without Oxygen*, Chapman and Hall, New York.

Busa, W.B. (1986) Mechanisms and consequences of pH-mediated cell regulation. *Annual Review of Physiology* **48**, 388–402.

Busa, W.B. and Crowe, J.H. (1983) Intracellular pH regulates transitions between dormancy and development of brine shrimp (*Artemia salina*) embryos. *Science* **221**, 366–368.

Busa, W.B. and Nuccitelli, R. (1984) Metabolic regulation via intracellular pH. *American Journal of Physiology* **246**, R409–R438.

Busa, W.B., Crowe, J.H. and Matson, G.B. (1982) Intracellular pH and the metabolic status of dormant and developing *Artemia* embryos. *Archives of Biochemistry and Biophysics* **216**, 711–718.

Carpenter, J.F. and Hand, S.C. (1986) Arrestment of carbohydrate metabolism during anaerobic dormancy and aerobic acidosis of *Artemia* embryos: determination of pH-sensitive control points. *Journal of Comparative Physiology B* **156**, 451–459.

Cartwright, J.L. and McLennan, A.G. (1999) Formation of a covalent N-epsilon-2-guanylylhistidyl reaction intermediate by the GTP:GTP guanylyltransferase from the brine shrimp *Artemia*. *Archives of Biochemistry and Biophysics* **361**, 101–105.

Chavez, M., Landry, C., Loret, S., Muller, M., Figueroa, J., Peers, B., Rentier-Delrue, F., Krauskopf, M. and Martial, J.A. (1999) APH-1, a POU homeobox gene expressed in the salt gland of the crustacean, *Artemia franciscana*. *Mechanisms of Development* **87**, 207–212.

Clegg, J.S. (1962) Free glycerol in dormant cysts of the brine shrimp *Artemia salina*, and its disappearance during development. *Biological Bulletin* **123**, 295–301.

Clegg, J.S. (1964) The control of emergence and metabolism by external osmotic pressure and the role of free glycerol in developing cysts of *Artemia salina*. *Journal of Experimental Biology* **41**, 879–892.

Clegg, J.S. (1965) The origin of trehalose and its significance during the formation of encysted dormant embryos of *Artemia salina*. *Comparative Biochemistry and Physiology A* **14**, 135–143.

Clegg, J.S. (1978) Interrelationships between water and cellular metabolism in *Artemia* cysts. VIII. Sorption isotherms and derived thermodynamic quantities. *Journal of Cellular Physiology* **94**, 123–138.

Clegg, J.S. (1986a) *Artemia* cysts as a model for the study of water in biological systems. *Methods in Enzymology* **127**, 230–239.

Clegg, J.S. (1986b) The physical properties and metabolic status of *Artemia* cysts at low water contents: the water replacement hypothesis. In A.C. Leopold (ed.), *Membranes, Metabolism and Dry Organisms*, Cornell University Press, New York, pp. 169–187.

Clegg, J.S. (1993) Respiration of *Artemia franciscana* embryos after continuous anoxia over 1-year period. *Journal of Comparative Physiology B* **163**, 48–51.

Clegg, J.S. (1994) Unusual response of *Artemia franciscana* embryos to prolonged anoxia. *Journal of Experimental Zoology* **270**, 332–334.

Clegg, J.S. (1997) Embryos of *Artemia franciscana* survive four years of continuous anoxia: the case for complete metabolic rate depression. *Journal of Experimental Biology* **200**, 467–475.

Clegg, J.S. (2001) Cryptobiosis – a peculiar state of biological organization. *Comparative Biochemistry and Physiology B* **128**, 613–624.

Clegg, J.S. and Conte, F.P. (1980) A review of the cellular and developmental biology of *Artemia*. In G. Persoone, P. Sorgeloos, O. Roels and E. Jaspers (eds.), *The Brine Shrimp Artemia*, Vol. 2, Universa Press, Wetteren, Belgium, pp. 11–54.

Clegg, J.S. and Drost-Hansen, W. (1991) On the biochemistry and cell physiology of water. In P.W. Hochachka and T.P. Mommsen (eds.), *The Biochemistry and Molecular Biology of Fishes*, Vol. 1, Elsevier Science Publishers, Amsterdam, pp. 1–23.

Clegg, J.S. and Jackson, S.A. (1992) Aerobic heat shock activates trehalose synthesis in embryos of *Artemia franciscana*. *FEBS Letters* **303**, 45–47.

Clegg, J.S. and Jackson, S.A. (1997) Significance of cyst fragments of *Artemia* sp. recovered from a 27,000 year old core taken under the Great Salt Lake, Utah, USA. *International Journal of Salt Lake Research* **6**, 207–216.

Clegg, J.S. and Jackson, S.A. (1998) The metabolic status of quiescent and diapause embryos of *Artemia franciscana* (Kellogg). *Archiv für Hydrobiologie Special Issues Advances in Limnology* **52**, 425–439.

Clegg, J.S., Zettlemoyer, A.C. and Hsing, H.H. (1978) On the residual water content of dried but viable cells. *Experientia* **34**, 734–735.

Clegg, J.S., Jackson, S.A. and Warner, A.H. (1994) Extensive intracellular translocations of a major protein accompany anoxia in embryos of *Artemia franciscana*. *Experimental Cell Research* **212**, 77–83.

Clegg, J.S., Jackson, S.A., Liang, P. and MacRae, T.H. (1995) Nuclear-cytoplasmic translocations of protein p26 during aerobic-anoxic transitions in embryos of *Artemia franciscana*. *Experimental Cell Research* **219**, 1–7.

Clegg, J.S., Drinkwater, L.E. and Sorgeloos, P. (1996) The metabolic status of diapause embryos of *Artemia franciscana* (SFB). *Physiological Zoology* **69**, 49–66.

Clegg, J.S., Willsie, J.K. and Jackson, S.A. (1999) Adaptive significance of a small heat shock/α-crystallin protein (p26) in encysted embryos of the brine shrimp, *Artemia franciscana*. *American Zoologist* **39**, 836–847.

Clegg, J.S., Jackson, S.A. and Popov, V.I. (2000a) Long-term anoxia in encysted embryos of the crustacean, *Artemia franciscana*: viability, ultrastructure, and stress proteins. *Cell and Tissue Research* **301**, 433–446.

Clegg, J.S., Jackson, S.A., Hoa, N.G. and Sorgeloos, P. (2000b) Thermal resistance, developmental rate and heat shock proteins in *Artemia franciscana*, from San Francisco Bay and southern Vietnam. *Journal of Experimental Marine Biology and Ecology* **252**, 85–96.

Clegg, J.S., Hoa, N.G. and Sorgeloos, P. (2001) Thermal tolerance and heat shock proteins in encysted embryos of *Artemia* from widely different thermal habitats. *Hydrobiologia*, in press.

Coleman, J.S., Heckathorn, S.A. and Hallberg, R.L. (1995) Heat shock proteins and thermotolerance: linking molecular and ecological perspectives. *Trends in Ecology & Evolution* **10**, 305–306.

Coleman, M., Geddes, M.C. and Trotman, C.N.A. (1998) Divergence of *Parartemia* and *Artemia* haemoglobin genes. *International Journal of Salt Lake Research* **7**, 171–180.

Conte, F.P. (1984) Structure and function of the crustacean larval salt gland. *International Review of Cytology* **91**, 45–106.

Conte, F.P., Lowry, J., Carpenter, J., Edwards, A., Smith, R. and Ewing, R.D. (1980) Aerobic and anaerobic metabolism of *Artemia* nauplii as a function of salinity. In G. Persoone, P. Sorgeloos, O. Roels and E. Jaspers (eds.), *The Brine Shrimp Artemia*, Vol. 2, Universa Press, Wetteren, Belgium, pp. 125–136.

Criel, G.R.J. (1991a) Ontogeny of *Artemia*. In R.A. Browne, P. Sorgeloos and C.N.A. Trotman (eds.), *Artemia Biology*, CRC Press, Boca Raton, Florida, pp. 155–185.

Criel, G.R.J. (1991b) Morphology of *Artemia*. In R.A. Browne, P. Sorgeloos and C.N.A. Trotman (eds.), *Artemia Biology*, CRC Press, Boca Raton, Florida, pp. 119–153.

Crowe, J.H. and Clegg, J.S. (1973) *Anhydrobiosis*, Dowden, Hutchinson and Ross, Stroudsburg, Pennsylvania.

Crowe, J.H., Crowe, L.M. and O'Dell, S.J. (1981) Ice formation during freezing of *Artemia* cysts of variable water content. *Molecular Physiology* **1**, 145–152.

Crowe, L.M., Mouradian, R., Crowe, J.H., Jackson, S.A. and Womersley, C. (1984) Effects of carbohydrates on membrane stability at low water activities. *Biochimica et Biophysica Acta* **769**, 141–150.

Crowe, J.H., Crowe, L.M., Drinkwater, L. and Busa, W.B. (1987) Intracellular pH and anhydrobiosis in *Artemia* cysts. In W. Decleir, L. Moens, H. Slegers, P. Sorgeloos and E. Jaspers (eds.), *Artemia Research and its Applications*, Vol. 2, Universa Press, Wetteren, Belgium, pp. 19–40.

Crowe, J.H., Hoekstra, F.A. and Crowe, L.E. (1992) Anhydrobiosis. *Annual Review of Physiology* **54**, 579–599.

Crowe, L.M., Reid, D.S. and Crowe, J.H. (1996) Is trehalose special for preserving dry biomaterials? *Biophysical Journal* **71**, 2087–2093.

Crowe, J.H., Carpenter, J.F. and Crowe, L.M. (1998a) The role of vitrification in anhydrobiosis. *Annual Review of Physiology* **60**, 73–103.

Crowe, J.H., Clegg, J.S. and Crowe, L.M. (1998b) Anhydrobiosis: the water replacement hypothesis. In D.S. Reid (ed.), *The Properties of Water in Foods ISOPOW 6*, Chapman and Hall, New York, pp. 440–455.

De Chaffoy, D., De Maeyer-Criel, G. and Hondo, M. (1978) On the permeability and formation of the embryonic cuticle during development *in vivo* and *in vitro* of *Artemia salina* embryos. *Differentiation* **12**, 99–109.

Decleir, W., Vos, J., Bernaerts, F. and Van den Branden, C. (1980) The respiratory physiology of *Artemia* sp. In G. Persoone, P. Sorgeloos, O. Roels and E. Jaspers (eds.), *The Brine Shrimp Artemia*, Vol. 2, Universa Press, Wetteren, Belgium, pp. 137–145.

Decleir, W., Moens, L., Slegers, H., Sorgeloos, P. and Jaspers, E. (1987) *Artemia Research and Its Applications*, Vol. 2, Universa Press, Wetteren, Belgium.

Decleir, W., Wolf, G. and De Wachter, B. (1989) Adaptation to hypoxia in *Artemia*. In A.H. Warner, T.H. MacRae and J.C. Bagshaw (eds.), *Cell and Molecular Biology of Artemia*, Plenum Press, New York, pp. 47–54.

De Graaf, J., Amons, R. and Möller, W. (1990) The primary structure of artemin from *Artemia* cysts. *European Journal of Biochemistry* **193**, 737–750.

De Herdt, E., Slegers, H. and Kondo, M. (1979) Identification and characterization of a 19-S complex containing a 27,000-Mr protein in *Artemia salina*. *European Journal of Biochemistry* **96**, 423–430.

de Jong, W.W., Caspers, G.-J. and Leunissen, J.A.M. (1998) Genealogy of the α-crystallin/small heat-shock protein super family. *International Journal of Biological Macromolecules* **22**, 151–162.

De Voeght, F. and Clauwaert, J. (1981) Structural characterization by biophysical methods of extracellular hemoglobins of *Artemia salina*. In J. Lamy (ed.), *Structure, Active Site and Function of Invertebrate Oxygen Carriers*, Marcel Dekker, New York, pp. 61–67.

Díaz, A.R. and Heredia, C.F. (1996) Purification and characterization of *Artemia* 2′–3′-cyclic nucleotide 3′-phosphodiesterase. *Biochimica et Biophysica Acta* **1290**, 135–140.

Díaz, A.R., Pérez-Grau, P. and Heredia, C.F. (1993) 2′:3′ Cyclic nucleotide 3′ phosphodiesterase in *Artemia* and other organisms. *Comparative Biochemistry and Physiology B* **104**, 275–279.

Dickerson, R.E. and Geiss, I. (1983) *Hemoglobin: Structure, Function, Evolution, and Pathology*, Benjamin Cummings, Menlo Park, California.

Drinkwater, L.E. and Crowe, J.H. (1987) Regulation of embryonic diapause in *Artemia*: environmental and physiological signals. *Journal of Experimental Zoology* **241**, 297–307.

Drinkwater, L.E. and Clegg, J.S. (1991) Experimental biology of cyst diapause. In R.A. Browne, P. Sorgeloos and C.N.A. Trotman (eds.), *Artemia Biology*, CRC Press, Boca Raton, Florida, pp. 93–117.

Dutrieu, J. (1960) Observations biochimiques et physiologiques sur le développement d'*Artemia salina*, Leach. *Archives de Zoologie Experimentale et Génerale* **99**, 1–134.

Eads, B.D. and Hand, S.C. (1999) Regulatory features of transcription in isolated mitochondria from *Artemia franciscana* embryos. *American Journal of Physiology (Regulatory, Integrative and Comparative Physiology)* **277**, R1588–R1597.

Eis, C. and Nidetzky, B. (1999) Characterization of trehalose phosphorylase from *Schizophyllum commune*. *Biochemical Journal* **341**, 385–393.

Eliassen, E. (1952) The energy metabolism of *Artemia* in relation to body size, seasonal rhythms and different salinities. *Arb. Nat. Forsch. Vehr. Riga* **11**, 1–18.

Elliott, B., Haltiwanger, R.S. and Futcher, B. (1996) Synergy between trehalose and Hsp104 for thermotolerance in *Saccharormyces cerevesiae*. *Genetics* **144**, 923–933.

Ellis, R.J. (1999) Molecular chaperones: pathways and networks. *Current Biology* **9**, R137–R139.

Ellis, R.J. and Hartl, F.U. (1999) Principles of protein folding in the cellular environment. *Current Opinion in Structural Biology* **9**, 102–110.

Escalante, R., García-Sáez, A., Ortega, M.-A. and Sastre, L. (1994) Gene expression after resumption of development of *Artemia franciscana* cryptobiotic embryos. *Biochemistry and Cell Biology* **72**, 78–83.

Escalante, R., García-Sáez, A. and Sastre, L. (1995) In situ hybridization analyses of Na, K-ATPase α-subunit expression during early larval development of *Artemia franciscana*. *Journal of Histochemistry & Cytochemistry* **43**, 391–399.

Feder, M.E. (1999) Organismal, ecological and evolutionary aspects of heat shock proteins and the stress response: established conclusions and unresolved issues. *American Zoologist* **39**, 857–864.

Feder, M.E. and Hofmann, G.E. (1999) Heat-shock proteins, molecular chaperones, and the stress response: evolutionary and ecological physiology. *Annual Review of Physiology* **61**, 243–282.

Finamore, F.J. and Warner, A.H. (1963) The occurrence of P1, P4 – diguanosine 5′-tetraphosphate in brine shrimp eggs. *Journal of Biological Chemistry* **238**, 344–348.

Finamore, F.J. and Clegg, J.S. (1969) Biochemical aspects of morphogenesis in the brine shrimp, *Artemia salina*. In G.M. Padilla, G.L. Whitson and I.L. Cameron (eds.), *The Cell Cycle. Gene-Enzyme Interactions*, Academic Press, New York, pp. 249–278.

Fisher, J.A., Baxter-Lowe, L.A. and Hokin, L.E. (1986) Regulation of Na, K-ATPase biosynthesis in developing *Artemia salina*. *Journal of Biological Chemistry* **261**, 515–519.

Frankenberg, M.M., Jackson, S.A. and Clegg, J.S. (2000) The heat shock response of adult *Artemia franciscana*. *Journal of Thermal Biology* **25**, 481–490.

García-Sáez, A., Perona, R. and Sastre, L. (1997) Polymorphism and structure of the gene coding for the α-1 subunit of the *Artemia* franciscana Na/K-ATPase. *Biochemical Journal* **321**, 509–518.

Geddes, M.C. (1980) The brine shrimps *Artemia* and *Parartemia* in Australia. In G. Persoone, P. Sorgeloos, O. Roels and E. Jaspers (eds.), *The Brine Shrimp Artemia*, Vol. 3, Universa Press, Wetteren, Belgium, pp. 57–65.

Geddes, M.C. (1981) The brine shrimp *Artemia* and *Parartemia*. *Hydrobiologia* **81**, 169–179.

Gilchrist, B.M. (1954) Haemoglobin in *Artemia*. *Proceedings of the Royal Society of London B* **152**, 118–136.

Gilchrist, B.M. (1956) The oxygen consumption of *Artemia salina* in different salinities. *Hydrobiologia* **8**, 54–65.

Gilchrist, B.M. (1958) The oxygen consumption of *Artemia salina* (L.). *Hydrobiologia* **12**, 27–37.

Gouthro, M.A. and Rosowski, J.R. (1994) Gravity-directed orientation of postgastrular development and emergence of prenauplii from cysts of *Artemia* (Anostraca). *Journal of Crustacean Biology* **14**, 722–728.

Grosfeld, H. and Littauer, U.Z. (1976) The translation *in vitro* of mRNA from developing cysts of *Artemia salina*. *European Journal of Biochemistry* **70**, 589–599.

Guppy, M. and Withers, P. (1999) Metabolic depression in animals: physiological perspectives and biochemical generalizations. *Biological Reviews of the Cambridge Philosophical Society* **74**, 1–40.

Guppy, M., Fuery, C.J. and Flanigan, J.E. (1994) Biochemical principles of metabolic depression. *Comparative Biochemistry and Physiology B* **109**, 175–189.

Hand, S.C. (1993) pHi and anabolic arrest during anoxia in *Artemia franciscana* embryos. In P.N. Hochachka, P.L. Lutz, M. Rosenthal and G. van den Thillart (eds.), *Surviving Hypoxia*, CRC Press, Boca Raton, Florida, pp. 171–185.

Hand, S.C. (1995) Heat flow is measurable from *Artemia franciscana* embryos under anoxia. *Journal of Experimental Zoology* **273**, 445–449.

Hand, S.C. (1997) Oxygen, pH_i and arrest of biosynthesis in brine shrimp embryos. *Acta Physiologica Scandinavica* **161**, 543–551.

Hand, S.C. (1998) Quiescence in *Artemia franciscana* embryos: reversible arrest of metabolism and gene expression at low oxygen levels. *Journal of Experimental Biology* **201**, 1233–1242.

Hand, S.C. and Carpenter, J.F. (1986) pH-induced metabolic transitions in *Artemia* embryos mediated by a novel hysteretic trehalase. *Science* **232**, 1535–1537.

Hand, S.C. and Hardewig, I. (1996) Downregulation of cellular metabolism during environmental stress: mechanisms and implications. *Annual Review of Physiology* **58**, 539–563.

Hand, S.C. and Podrabsky, J.E. (2000) Bioenergetics of diapause and quiescence in aquatic animals. *Thermochimica Acta* **349**, 31–42.

Hardewig, I., Anchordoguy, T.J., Crawford, D.L. and Hand, S.C. (1996) Profiles of nuclear and mitochondrial encoded mRNAs in developing and quiescent embryos of *Artemia franciscana*. *Molecular and Cellular Biochemistry* **158**, 139–147.

Hartl, F.U. (1996) Molecular chaperones in cellular protein folding. *Nature* **381**, 571–580.

Heip, J., Moens, L., Joniau, M. and Kondo, M. (1978a) Ontogenetical studies on extracellular hemoglobins of *Artemia salina*. *Developmental Biology* **64**, 73–81.

Heip, J., Moens, L. and Kondo, M. (1978b) Effect of concentrations of salt and oxygen on the synthesis of extracellular hemoglobins during development of *Artemia salina*. *Developmental Biology* **63**, 247–251.

Henle, E.S. and Linn, S. (1997) Formation, prevention and repair of DNA damage by iron/hydrogen peroxide. *Journal of Biological Chemistry* **272**, 19095–19098.

Herbst, D.A. (1999) Biogeography and physiological adaptations of the brine fly genus *Ephydra* (Diptera: Ephydridae) in saline waters of the Great Basin. *Great Basin Naturalist* **59**, 127–135.

Hernandorena, A. (1993) Guanylate requirement for patterning the postcephalic body region of the brine shrimp *Artemia*. *Roux's Archives of Developmental Biology* **203**, 74–82.

Hernandorena, A. (1999) The patterning function of purines in the brine shrimp, *Artemia*. *Comptes Rendus de l'Academie des Sciences Serie III* **322**, 289–301.

Hinton, H.E. (1968) Reversible suspension of metabolism and the origin of life. *Proceedings of the Royal Society of London B* **171**, 43–57.

Hochachka, P.W. and Guppy, M. (1987) *Metabolic Arrest and the Control of Biological Time*, Harvard University Press, Cambridge.

Hochachka, P.W., Lutz, P.L., Sick, T., Rosenthal, M. and van den Thillart, G. (1993) *Surviving Hypoxia. Mechanisms of Control and Adaptation*, CRC Press, Boca Raton, Florida.

Hofmann, G.E. and Hand, S.C. (1990a) Arrest of cytochrome-c oxidase synthesis coordinated with catabolic arrest in dormant *Artemia* embryos. *American Journal of Physiology* **258**, R1184–R1191.

Hofmann, G.E. and Hand, S.C. (1990b) Subcellular differentiation arrested in *Artemia* embryos under anoxia: evidence supporting a regulatory role for pH. *Journal of Experimental Zoology* **253**, 287–302.

Hofmann, G. and Hand, S.C. (1992) Comparison of mRNA pools in active and dormant *Artemia franciscana* embryos. *Journal of Experimental Biology* **164**, 103–116.

Hofmann, G. and Hand, S.C. (1994) Global arrest of translation during invertebrate quiescence. *Proceedings of the National Academy of the United States of America* **91**, 8492–8496.

Holliday, C.W., Roye, D.B. and Roer, R.D. (1990) Salinity-induced changes in branchial Na^+/K^+-ATPase activity and transepithelial potential difference in the brine shrimp, *Artemia salina*. *Journal of Experimental Biology* **151**, 279–296.

Hontoria, F., Crowe, J.H., Crowe, L.E. and Amat, F. (1993) Metabolic heat production by *Artemia* embryos under anoxic conditions. *Journal of Experimental Biology* **178**, 149–159.

Hottiger, T., De Virgilio, C., Hall, M.N., Boller, T. and Wiemken, A. (1994) The role of trehalose synthesis for the acquisition of thermotolerance in yeast. II. Physiological concentrations of trehalose increase the thermal stability of proteins *in vitro*. *European Journal of Biochemistry* **219**, 187–193.

Hultin, T. and Morris, J.E. (1968) The ribosomes of encysted embryos of *Artemia salina* during cryptobiosis and resumption of development. *Developmental Biology* **17**, 143–163.

Iwahashi, H., Nwaka, S., Obuchi, K. and Komatsu, Y. (1998) Evidence for the interplay between trehalose metabolism and Hsp 104 in yeast. *Applied and Environmental Microbiology* **64**, 4614–4617.

Jackson, S.A. and Clegg, J.S. (1996) The ontogeny of low molecular weight stress protein p26 during early development of the brine shrimp, *Artemia franciscana*. *Development, Growth & Differentiation* **38**, 153–160.

Jardel, J.P.L. (1986) Ultrastructural modifications associated with anhydrobiosis in the brine shrimp, *Artemia*, Ph.D. Thesis, Monash University, Australia.

Kasturi, S.R., Seitz, P.K., Chang, D.C. and Hazlewood, C.F. (1990) Intracellular water in *Artemia* cysts (brine shrimp). Investigations by deuterium and oxygen-17 nuclear magnetic resonance. *Biophysical Journal* **58**, 483–491.

Keilin, D. (1959) The problem of anabiosis or latent life: History and current concepts. *Proceedings of the Royal Society of London B* **150**, 149–191.

Kratowich, N.R. (1964) The effect of metabolic inhibitors on osmoregulation of nauplii of *Artemia salina* at various salinities. *American Zoologist* **4**, 389–390.

Kuenen, D.J. (1939) Systematical and physiological notes on the brine shrimp, *Artemia*. *Archives Néerlandaises de Zoologie* **3**, 365–445.

Kwast, K.E. and Hand, S.C. (1996a) Acute depression of mitochondrial protein synthesis during anoxia: contributions of oxygen sensing, matrix acidification and re-dox state. *Journal of Biological Chemistry* **271**, 7313–7319.

Kwast, K.E. and Hand, S.C. (1996b) Oxygen and pH regulation of protein synthesis in mitochondria from *Artemia franciscana* embryos. *Biochemical Journal* **313**, 207–213.

Kwast, K.E., Shapiro, J.I., Rees, B.B. and Hand, S.C. (1995) Oxidative phosphorylation and the re-alkalinization of intracellular pH during recovery from anoxia in *Artemia franciscana*. *Biochimica et Biophysica Acta* **1232**, 5–12.

Lavens, P. and Sorgeloos, P. (1987) The cryptobiotic state of *Artemia* cysts, its diapause deactivation and hatching: a review. In P. Sorgeloos, D.A. Bengston, W. Decleir and E. Jaspers (eds.), *Artemia Research and its Applications*, Vol. 3, Universa Press, Wetteren, Belgium, pp. 27–63.

Lazou, A. and Frosinis, A. (1994) Kinetic and regulatory properties of pyruvate kinase from *Artemia* embryos during incubation under aerobic and anoxic conditions. The effect of pH on the kinetic constants. *Comparative Biochemistry and Physiology B* **109**, 325–332.

Lazou, A., Polydoros, I. and Beis, Is. (1994) Effect of anaerobiosis and anhydrobiosis on the extent of glycolytic enzyme binding in *Artemia* embryos. *Journal of Comparative Physiology B* **164**, 306–311.

Lee, K.J. and Watts, S.A. (1994) Specific activity of Na^+ K^+ ATPase is not altered in response to changing salinities during early development of the brine shrimp, *Artemia franciscana*. *Physiological Zoology* **67**, 910–924.

Lent, C.M. (1971) Metachronal limb movements by *Artemia salina*: synchrony of male and female during coupling. *Science* **173**, 1247–1248.

Liang, P. and MacRae, T.H. (1999) The synthesis of a small heat shock/α-crystallin protein in *Artemia* and its relationship to stress tolerance during development. *Developmental Biology* **207**, 445–456.

Liang, P., Amons, R., MacRae, T.H. and Clegg, J.S. (1997a) Purification, structure and *in vitro* molecular-chaperone activity of *Artemia* p26, a small heat shock/α-crystallin protein. *European Journal of Biochemistry* **243**, 225–232.

Liang, P., Amons, R., Clegg, J.S. and MacRae, T.H. (1997b) Molecular characterization of a small heat-shock/α-crystallin protein in encysted *Artemia* embryos. *Journal of Biological Chemistry* **272**, 19051–19058.

Lion, M.B., Kirby-Smith, J.H. and Randolph, M.L. (1961) Electron-spin resonance signals from lyophilized bacterial cells exposed to oxygen. *Nature* **192**, 34–36.

Liras, A., Argomaniz, L. and Llorente, P. (1990) Presence, preliminary properties and partial purification of 6-phosphoribosylpyrophosphate amidotransferase from *Artemia* sp. *Biochimica et Biophysica Acta* **1033**, 114–117.

Liras, A., Rotllan, P. and Llorente, P. (1992) *De novo* purine biosynthesis in the crustacean *Artemia*: influence of salinity and geographical origin. *Journal of Comparative Physiology B* **162**, 263–266.

Liu, J.J. and McLennan, A.G. (1994) Purification and properties of GTP:GTP guanylyltransferase from encysted embryos of the brine shrimp *Artemia*. *Journal of Biological Chemistry* **269**, 11787–11794.

Lochhead, J.H. and Lochhead, M.S. (1940) The egg shells of the brine shrimp *Artemia*. *Anatomical Record* **78**, 75–76.

MacRae, T.H. (2000) Structure and function of small heat shock/α-crystallin proteins: established concepts and emerging ideas. *Cellular and Molecular Life Sciences* **57**, 899–913.

MacRae, T.H., Bagshaw, J.C. and Warner, A.H. (1989) *Biochemistry and Cell Biology of Artemia*, CRC Press, Boca Raton, Florida.

Malarkey, K., Coker, K.J. and Sturgill, T.W. (1998) Ribosomal S6 kinase is activated as an early event in pre-emergence development of encysted embryos of *Artemia salina*. *European Journal of Biochemistry* **251**, 269–274.

Mangum, C.P. (1997) Invertebrate blood oxygen carriers. In W.H. Dantzler (ed.), *Handbook of Physiology*, section 13, Comparative Physiology, Vol. 2, OUP, pp. 1097–1135.

Manning, A.M., Trotman, C.N.A. and Tate, W.P. (1990) Evolution of a polymeric globin in the brine shrimp *Artemia*. *Nature* **348**, 653–656.

Manwell, C. (1978) Haemoglobin in the Australian anostracan *Parartemia zietziana*: evolutionary strategies of conformity vs regulation. *Comparative Biochemistry and Physiology* A **59**, 37–44.

Marshall, C.J. (1988) Structures of the haemoglobins of the brine shrimp, *Artemia*, Ph.D. Thesis, University of Otago, New Zealand.

Matthews, C.M. and Trotman, C.N.A. (1998) Ancient and recent intron stability in the *Artemia* hemoglobin gene. *Journal of Molecular Evolution* **47**, 763–771.

Matthews, C.M., Vandenberg, C.J. and Trotman, C.N.A. (1998) Variable substitution rates of the 18 domain sequences in *Artemia* hemoglobin. *Journal of Molecular Evolution* **46**, 729–733.

McLennan, A.G. (1992) *Ap₄A and Other Dinucleoside Polyphosphates*, CRC Press, Boca Raton, Florida.

Mead, C.G. and Finamore, F.J. (1969) The occurrence of ascorbic acid sulfate in the brine shrimp, *Artemia salina*. *Biochemistry* **8**, 2652–2655.

Merchie, G., Lavens, P., Dhert, P., Dehasque, M., Nelis, H., De Leenheer, A. and Sorgeloos, P. (1995) Variation of ascorbic acid content in different live food organisms. *Aquaculture* **134**, 325–337.

Miller, D. and McLennan, A.G. (1988a) The heat shock response of the cryptobiotic brine shrimp, *Artemia*. – I. A comparison of the thermotolerance of cysts and larvae. *Journal of Thermal Biology* **13**, 119–123.

Miller, D. and McLennan, A.G. (1988b) The heat shock response of the cryptobiotic brine shrimp, *Artemia*. – II. Heat shock proteins. *Journal of Thermal Biology* **13**, 125–134.

Mitchell, B.D. and Geddes, M.C. (1977) Distribution of the brine shrimps *Parartemia zietziana* Sayce and *Artemia salina* (L.) along a salinity and oxygen gradient in a South Australian saltfield. *Freshwater Biology* **7**, 461–467.

Moens, L. (1982) The extracellular haemoglobins of *Artemia* sp.: a biochemical and ontogenetical study. *Mededelingen van de Koninklijke Academie voor Wetenschappen, Letteren en Schone Kunsten van België, Klasse der Wetenschappen* **44**, 1–21.

Moens, L., Geelen, D., Van Hauwaert, M.-L., Wolf, G., Blust, R., Witters, R. and Lontie, R. (1984) The structure of the *Artemia* sp. hemoglobin. Cleavage of the native molecule into functional units by limited subtilisin digestion. *Biochemical Journal* **223**, 801–811.

Moens, L., Van Hauwaert, M.-L., De Smet, K., Geelen, D., Verpooten, G., Van Beeumen, J., Wodak, S., Alard, P. and Trotman, C.N.A. (1988) A structural domain of the covalent polymer globin chains of *Artemia*. Interpretation of amino acid sequence data. *Journal of Biological Chemistry* **263**, 4679–4685.

Moens, L., Wolf, G., Van Hauwaert, M.-L., De Baere, I., Van Beeumen, J., Wodak, S. and Trotman, C.N.A. (1991) The extracellular hemoglobins of *Artemia*. Structure of the oxygen carrier and respiration physiology. In R.A. Browne, P. Sorgeloos and C.N.A. Trotman (eds.), *Artemia Biology*, CRC Press, Boca Raton, Florida, pp. 187–219.

Montero, C., Llorente, P., Argomaniz, L. and Menendez, M. (1996) Thermal stability of *Artemia* HGPRT: effect of substrates on inactivation kinetics. *International Journal of Biological Macromolecules* **18**, 255–262.

Morimoto, R.I., Tissičres, A. and Georgopoulos, C. (1994) *The Biology of Heat Shock Proteins and Molecular Chaperones*, Cold Spring Harbor, Laboratory Press, New York.

Morris, J.E. (1968) Dehydrated cysts of *Artemia salina* prepared for electron microscopy by totally anhydrous techniques. *Journal of Ultrastructure Research* **25**, 64–72.

Morris, J.E. (1971) Hydration, its reversibility and the beginning of development in the brine shrimp, *Artemia salina. Comparative Biochemistry and Physiology A* **39**, 843–857.

Morris, J.E. and Afzelius, B.A. (1967) The structure of the shell and outer membranes in encysted *Artemia salina* embryos during cryptobiosis and development. *Journal of Ultrastructure Research* **20**, 244–259.

Nakanishi, Y.H., Iwasaki, T., Okigaki, T. and Kato, H. (1962) Cytological studies of *Artemia salina*. I. Embryonic development without cell multiplication after the blastula stage. *Annotationes Zoologicae Japonenses* **35**, 223–228.

Nakanishi, Y.H., Okigaki, T., Kato, H. and Iwasaki, T. (1963) Cytological studies of *Artemia salina*. II. Deoxyribonucleic acid content and the chromosomes in encysted dry eggs and nauplii. *Proceedings of the Japan Academy* **39**, 306–309.

Nambu, Z., Nambu, F. and Tanaka, S. (1997a) Purification and characterization of trehalase from *Artemia* embryos and larvae. *Zoological Science (Tokyo)* **14**, 419–427.

Nambu, Z., Tanaka, S. and Nambu, F. (1997b) The expression of trehalase during post-dormant development of the brine shrimp, *Artemia*. Comparison of the two species. *Journal of the University Occupational Environmental Health (Japan)* **19**, 255–264.

Nelis, H.J., Lavens, P., Moens, L. Sorgeloos, P. and De Leenheer, A.P. (1989) Carotenoids in relationship to *Artemia* development. In T.H. MacRae, J.C. Bagshaw and A.H. Warner (eds.), *Biochemistry and Cell Biology of Artemia*, CRC Press, Boca Raton, Florida, pp. 159–190.

Nelis, H.J., Merchie, G., Lavens, P., Sorgeloos, P. and De Leenheer, P. (1994) Solid-phase extraction of ascorbic acid 2-sulfate from cysts of the brine shrimp, *Artemia franciscana. Analytical Chemistry* **66**, 1330–1333.

Nithya, M., Shanmugasundaram, G.K. and Munuswamy, N. (1996) Role of trehalase and protease during the development of the brine shrimp *Artemia parthenogenetica. Biomedical Letters* **54**, 51–58.

Nover, L. (1991) *Heat Shock Response*, CRC Press, Boca Raton, Florida.

Olson, C.S. and Clegg, J.S. (1978) Cell division during the development of *Artemia salina. Wilhelm Roux's Archives of Developmental Biology* **184**, 1–13.

Osuna, C., Ollala, A., Sillero, A., Sillero, M.A.G. and Sebastian, J. (1977) Induction of multiple proteases during the larval development of *Artemia salina. Developmental Biology* **61**, 94–103.

Packer, L. (1993) Carotenoids, Part. B, Metabolism, genetics and biosynthesis. *Methods in Enzymology* **214**, 1–468.

Perona, R., Bes, J.-C. and Vallejo, C.G. (1988) Degradation of yolk in the brine shrimp, *Artemia*. Biochemical and morphological studies on the involvement of the lysosomal system. *Biology of the Cell* **63**, 361–366.

Persoone, G., Sorgeloos, P., Roels, O. and Jaspers, E. (1980) *The Brine Shrimp Artemia*, Vol. 2, Universa Press, Wetteren, Belgium.

Prescott, M., Thorne, M.H., Milne, A.D. and McLennan, A.G. (1992) Characterization of a *bis* (5′-nucleosidyl) triphosphate pyrophosphohydrolase from encysted embryos of the brine shrimp *Artemia. International Journal of Biochemistry* **24**, 565–571.

Pype, S. and Slegers, H. (1993) Inhibition of casein kinase II by dinucleoside polyphosphates. *Enzyme & Protein* **47**, 14–21.

Richardson, A., Landry, S.J. and Georgopoulos, C. (1998) The ins and outs of a molecular chaperone machine. *Trends in Biochemical Sciences* **23**, 138–143.

Rosowski, J.R., Gouthro, M.A., Schmidt, K.K., Klement, B.J. and Spooner, B.S. (1995) Effect of microgravity and hypergravity on embryo axis alignment during postcystment embryogenesis in *Artemia franciscana* (Anostraca). *Journal of Crustacean Biology* **15**, 625–632.

Rosowski, J.R., Belk, D., Gouthro, M.A. and Lee, K. (1997) Ultrastructure of the cyst shell and underlying membranes of the brine shrimp, *Artemia franciscana* Kellogg (Anostraca), during post-encystment development, emergence and hatching. *Journal of Shellfish Research* **16**, 233–249.

Rotllan, P., Liras, A. and Llorente, P. (1993) Salvage and interconversion of purines in developing *Artemia*. *Biochimica et Biophysica Acta* **1156**, 128–134.

Rudneva, I.I. (1999) Antioxidant system of Black Sea animals in early development. *Comparative Biochemistry and Physiology C* **122**, 265–271.

Saez, A.G., Escalante, R. and Sastre, L. (2000) High DNA sequence variability at the α1 Na/K-ATPase locus of *Artemia franciscana*: Polymorphism in a gene for salt resistant organism. *Molecular Biology and Evolution* **17**, 235–250.

Salon, J., Cortas, N. and Edelman, L.S. (1989) Isoforms of Na, K-ATPase in *Artemia salina*: I. Detection by FITC binding and time course. *Journal of Membrane Biology* **108**, 177–186.

Schimek, C., Eibel, P., Grolig, F., Horie, T., Ootaki, T. and Galland, P. (1999) Gravitropism in *Phycomyces*: a role for sedimenting protein crystals and floating lipid globules. *Planta* **210**, 132–142.

Shanmugasundaram, G.K., Ramasubramanian, V. and Munuswamy, N. (1996a) α-Tocopherol in *Artemia* cysts: a report. *Aquaculture International* **4**, 377–378.

Shanmugasundaram, G.K., Rani, K. and Munuswamy, N. (1996b) Studies on the role of antioxidants in the cryptobiotic cysts of the brine shrimp, *Artemia parthenogenetica*. *Biomedical Letters* **53**, 17–22.

Sillero, A. and Sillero, M.A. (1989) Purine nucleotide metabolism in *Artemia*. In T.H. MacRae, J.C. Bagshaw and A.H. Warner (eds.), *Biochemistry and Cell Biology of Artemia*, CRC Press, Boca Raton, Florida, pp. 95–112.

Sillero, M.A.G., de Diego, A., Cerdan, S., Criel, G. and Sillero, A. (1994) Occurrence of millimolar concentrations of guanosine (5′) tetraphospho (5′) guanosine (Gp4G) in encysted embryos of *Thamnocephalus platyurus*. *Comparative Biochemistry and Physiology B* **108**, 41–45.

Singer, M.A. and Lindquist, S. (1998) Multiple effects of trehalose on protein folding *in vitro* and *in vivo*. *Molecular Cell* **1**, 639–648.

Skoultchi, A.I. and Morowitz, H.J. (1964) Information storage and survival of biological systems at temperatures near absolute zero. *Yale Journal of Biology and Medicine* **37**, 158–163.

Slegers, H. (1991) Enzyme activities through development: a synthesis of the activity and control of the various enzymes as the embryo matures. In R.A. Browne, P. Sorgeloos and C.N.A. Trotman (eds.), *Artemia Biology*, CRC Press, Boca Raton, Florida, pp. 37–73.

Somero, G.N. (1986) Protons, osmolytes and fitness of internal milieu for protein function. *American Journal of Physiology* **20**, R197–R213.

Somero, G.N. (1995) Proteins and temperature. *Annual Review of Physiology* **57**, 43–68.

Spicer, J.I. and El-Gamal, M.M. (1999) Hypoxia accelerates the development of respiratory regulation in brine shrimp – but at a cost. *Journal of Experimental Biology* **202**, 3637–3646.

Spooner, B.S., DeBell, L., Armbrust, L., Guikema, J.A., Metcalf, J. and Paulsen, A. (1994a) Embryogenesis, hatching and larval development of *Artemia* during orbital spaceflight. *Advances in Space Research* **14**, 229–238.

Spooner, B.S., Metcalf, J., DeBell, L., Paulsen, A., Noren, W. and Guikema, J.A. (1994b) Development of the brine shrimp *Artemia* is accelerated during spaceflight. *Journal of Experimental Zoology* **269**, 253–262.

Stabili, L., Miglietta, A.M. and Belmonte, G. (1999) Lysozyme-like and trypsin-like activities in the cysts of *Artemia franciscana* Kellog, 1906: Is there a passive immunity in a resting stage? *Journal of Experimental Marine Biology and Ecology* **237**, 291–303.

Stocco, D.M., Beers, P.C. and Warner, A.H. (1972) Effects of anoxia on nucleotide metabolism in encysted embryos of the brine shrimp. *Developmental Biology* **241**, 479–489.

Storey, K.B. (1998) Survival under stress: molecular mechanisms of metabolic rate depression in animals. *South African Journal of Zoology* **33**, 55–64.

Storey, K.B. and Storey, J.M. (1990) Metabolic rate depression and biochemical adaptation in anaerobiosis, hibernation and estivation. *Quarterly Review of Biology* **65**, 145–174.

Sun, W.Q. and Leopold, A.C. (1997) Cytoplasmic vitrification and survival of anhydrobiotic organisms. *Comparative Biochemistry and Physiology A* **117**, 327–333.

Sun, D.Y., Guo, J.Z., Hartmann, H.A., Uno, H. and Hokin, L.E. (1992) Differential expression of the alpha2 and beta messenger RNAs of Na, K-ATPase in developing brine shrimp as measured by in situ hybridization. *Journal of Histochemistry & Cytochemistry* **40**, 555–562.

Tanaka, S., Nambu, F. and Nambu, Z. (1999) Cloning and characterization of cDNAs encoding trehalase from post-dormant embryos of the brine shrimp, *Artemia franciscana*. *Zoological Science (Tokyo)* **16**, 269–272.

Tasch, P. (1963) XI Evolution of the Branchiopoda. *Phylogeny and Evolution of Crustacea, Museum of Comparative Zoology, Special Publication*, pp. 145–157.

Taylor, E.R., Nie, X.Z., MacGregor, A.W. and Hill, R.D. (1994) A cereal haemoglobin gene is expressed in seed and root tissues under anaerobic conditions. *Plant Molecular Biology* **24**, 853–862.

Trantham, E.C., Rorschach, H.E., Clegg, J.S., Hazlewood, C.F., Nicklow, R.M. and Wakabayashi, N. (1984) The diffusive properties of water in *Artemia* cells determined by quasi-elastic neutron scattering. *Biophysical Journal* **45**, 927–938.

Triantaphyllidis, G.V., Abatzopoulos, T.J. and Sorgeloos, P. (1998) Review of the biogeography of the genus *Artemia* (Crustacea, Anostraca). *Journal of Biogeography* **25**, 213–226.

Trotman, C.N.A. (1991) Normality and abnormality in early development. In R.A. Browne, P. Sorgeloos and C.N.A. Trotman (eds.), *Artemia Biology*, CRC Press, Boca Raton, Florida, pp. 75–92.

Trotman, C.N.A., Mansfield, B.C. and Tate, W.P. (1980) Inhibition of emergence, hatching and protein biosynthesis in *Artemia salina*. *Developmental Biology* **80**, 167–174.

Trotman, C.N.A., Wells, B., Marshall, C.J., Kean, A. and Tate, W.P. (1986) Immunoreactive haemoglobin in embryonic and naupliar *Artemia*. *Biochemistry International* **13**, 425–432.

Trotman, C.N.A., Gieseg, S.P., Pirie, R.S. and Tate, W.P. (1987) Abnormal development in *Artemia*. Defective emergence of the prenauplius with bicarbonate deficiency. *Journal of Experimental Zoology* **243**, 225–232.

Trotman, C.N.A., Manning, A.M., Moens, L. and Tate, W.P. (1991) The polymeric haemoglobin molecule of *Artemia*. Interpretation of translated cDNA sequence of nine domains. *Journal of Biological Chemistry* **266**, 13789–13795.

Trotman, C.N.A., Manning, A.M., Bray, J.A., Jellie, A.M., Moens, L. and Tate, W.P. (1994) Interdomain linkage in the polymeric hemoglobin molecule of *Artemia*. *Journal of Molecular Evolution* **38**, 628–636.

Trotman, C.N.A., Coleman, M., Jellie, A.M., Shieffelbien, D.C. and Tate, W.P. (1998) Haemoglobin: a primordial role preserved in diapause? *Archiv für Hydrobiologie Special Issues Advances in Limnology* **52**, 441–449.

Utterback, P.J. and Hand, S.C. (1987) Yolk platelet degradation in pre-emergence *Artemia* embryos: response to protons *in vivo* and *in vitro*. *American Journal of Physiology* **252**, R774–R785.

Vallejo, C.G. (1989) *Artemia* trehalase: regulation by factors that also control resumption of development. In A.H. Warner, T.H. MacLean and J.C. Bagshaw (eds.), *Cell and Molecular Biology of Artemia Development*, Plenum Press, New York, pp. 173–189.

Vallejo, C.G., Sillero, M.A.G. and Marco, R. (1979) Mitochondrial maturation during *Artemia salina* embryogenesis. General description of the process. *Cellular and Molecular Biology* **25**, 113–124.

Vallejo, C.G., Lopez, M., Ochoa, P., Manzanares, M. and Garesse, R. (1996) Mitochondrial differentiation during the early development of the brine shrimp *Artemia franciscana*. *Biochemical Journal* **314**, 505–510.

van Breukelen, F. and Hand, S.C. (2000) Characterization of ATP-dependent proteolysis in embryos of the brine shrimp, *Artemia franciscana*. *Journal of Comparative Physiology B* **170**, 125–133.

van Breukelen, F., Maier, R. and Hand, S.C. (2000) Depression of nuclear transcription and extension of mRNA half-life under anoxia in *Artemia franciscana* embryos. *Journal of Experimental Biology* **203**, 1123–1130.

Vandenberg, C.J. (2000) Expression and characterization of individual globin domains of *Artemia* haemoglobin, Ph.D. Thesis, University of Otago, New Zealand.

Vandenberg, C. and Trotman, C.N.A. (1999) Expression and purification of isolated *Artemia* globin domains for functional analysis. *Genome Diversity and Bioinformatics, 14th Symposium, Federation of Asian and Oceanian Biochemists and Molecular Biologists, Dunedin*, Abst. No. 141.

van den IJssel, P., Norman, D.G. and Quinlan, R.A. (1999) Molecular chaperones: small heat shock proteins in the limelight. *Current Biology* **9**, R103–R105.

Vos, J., Bernaerts, F., Gabriels, I. and Decleir, W. (1979) Aerobic and anaerobic respiration of adult *Artemia salina* (L.) acclimated to different oxygen concentrations. *Comparative Biochemistry and Physiology A* **62**, 545–548.

Warner, A.H. (1989) Proteases and protease inhibitors in *Artemia* and their role in the developmental process. In T.H. MacRae, J.C. Bagshaw and A.H. Warner (eds.), *Biochemistry and Cell Biology of Artemia*, CRC Press, Boca Raton, Florida, pp. 113–132.

Warner, A.H. (1992) Diguanosine and related nonadenylated polyphosphates. In A.G. McLennan (ed.), *Ap$_4$A and other Dinucleoside Polyphosphates*, CRC Press, Boca Raton, Florida, pp. 275–303.

Warner, A.H. and Sonnenfeld, M.J. (1990) Comparative study of guanosine nucleotide levels in thirteen populations of *Artemia* cysts. *Comparative Biochemistry and Physiology B* **97**, 855–860.

Warner, A.H. and Matheson, C. (1998) Release of proteases from larvae of the brine shrimp, *Artemia franciscana* and their potential role during the molting process. *Comparative Biochemistry and Physiology B* **119**, 255–263.

Warner, A.H., MacRae, T.H. and Bagshaw, J.C. (1989) *Cell and Molecular Biology of Artemia Development*, Plenum Press, New York.

Warner, A.H., Perz, M.J., Osahan, J.K. and Zielinski, B.S. (1995) Potential role in development of a major cysteine protease in larvae of the brine shrimp, *Artemia franciscana*. *Cell and Tissue Research* **282**, 21–31.

Warner, A.H., Jackson, S.J. and Clegg, J.S. (1997) Effect of anaerobiosis on cysteine protease regulation during the embryonic-larval transition in *Artemia franciscana*. *Journal of Experimental Biology* **200**, 897–908.

Watts, S.A., Yeh, E.W. and Henry, R.P. (1996) Hypoosmotic stimulation of ornithine decarboxylase activity in the brine shrimp *Artemia franciscana*. *Journal of Experimental Zoology* **274**, 15–22.

Weisz, P.B. (1946) Space time pattern of segment formation in *Artemia salina*. *Biological Bulletin* **91**, 119–140.

Weisz, P.B. (1947) The histological pattern of metameric development in *Artemia salina*. *Journal of Morphology* **81**, 45–95.

Wera, S., De Schrijver, E., Geyskens, I., Nwaka, S. and Thevelin, M. (1999) Opposite roles of trehalase activity in heat-shock recovery and heat-shock survival in *Saccharomyces cerevisiae*. *Biochemical Journal* **343**, 621–626.

Whitley, D., Goldberg, S.P. and Jordan, W.D. (1999) Heat shock proteins: a review of the molecular chaperones. *Journal of Vascular Surgery* **29**, 748–751.

Williams, W.D. and Geddes, M.C. (1991) Anostracans of Australian salt lakes, with particular reference to a comparison of *Parartemia* and *Artemia*. In R.A. Browne, P. Sorgeloos and C.N.A. Trotman (eds.), *Artemia Biology*, CRC Press, Boca Raton, Florida, pp. 351–368.

Winzerling, J.J. and Law, J.H. (1997) Comparative nutrition of iron and copper. *Annual Review of Nutrition* **17**, 501–526.

Winzor, C.L., Winzor, D.J., Paleg, L.G., Jones, P.G. and Naidu, B.P. (1992) Rationalization of the effects of compatible solutes on protein stability in terms of thermodynamic non-ideality. *Archives of Biochemistry and Biophysics* **296**, 102–107.

Wood, E.J., Barker, C., Moens, L., Jacob, W., Heip, J. and Kondo, M. (1981) Biophysical characterization of *Artemia salina* extracellular haemoglobin. *Biochemical Journal* **193**, 353–359.

Yancey, P.H., Clark, M.E., Hand, S.C., Bowlus, R.D. and Somero, G.N. (1983) Living with water stress: Evolution of osmolyte systems. *Science* **217**, 1214–1222.

CHAPTER IV

ZOOGEOGRAPHY

GILBERT VAN STAPPEN
Laboratory of Aquaculture & Artemia Reference Center
Faculty of Agricultural & Applied Biological Sciences
Ghent University, Rozier 44, B-9000 Gent
Belgium

1. Introduction

'*Artemia* is widely distributed on the five continents in many salt lakes, coastal lagoons, and solar saltworks' – this, or a variation on the same theme, is the standard opening phrase in most scientific contributions related to brine shrimp zoogeography. This chapter is no exception to this rule.

As early as 1915 Abonyi published a list of 80 *Artemia* sites, located in 21 countries. Similar lists were published by Artom (1922), with 18 sites, Stella (1933) 28 sites, and Barigozzi (1946) 29 sites. Persoone and Sorgeloos (1980) reviewed the zoogeography of *Artemia*, which included a list of about 250 sites distributed over 48 countries. Vanhaecke *et al*. (1987) published an updated list, covering 350 natural *Artemia* sites, and completed it with geographical coordinates and information about reproduction mode and species, if available. Ten years later Triantaphyllidis *et al*. (1998) compiled previous reviews, recent literature reports and personal communications into a list of about 500 *Artemia* sites, and discussed the zoogeographical data in relation to the most recent knowledge of genetics and morphometry of the genus *Artemia*. Obviously correspondence, informal communications and trip reports were an information source, as valuable as scientific articles, published in official journals.

The distribution of these sites over the continents is very uneven (see Figure 1): it is mainly a reflection of prospecting and exploration activities. As such it does not give a precise picture of the actual occurrence of *Artemia* over the globe. This is illustrated in Table 1, where the distribution of natural *Artemia* sites over the different (sub)continents is listed.

Europe (where the occurrence of *Artemia* is largely limited to the Mediterranean and Black Sea basin), the USA and Canada have seen thorough inventorization campaigns in recent decades. Information about Africa on the other hand, especially the sub-Saharan part of the continent, remains limited (the few sites, described in the last ten years, are almost all located in South-Africa). For the Indian subcontinent there are numerous scattered reports about brine shrimp occurrence, both in inland salt lakes and in coastal saltworks. Australia can be considered as a special case, since it is the home of the

Figure 1. World distribution of *Artemia* species.

Table 1. Distribution of known natural Artemia sites, grouped per (sub)continent.

Area	1980 [1]	1987 [2]	1998 [3]
Africa			
Mediterranean basin	24	31	32
sub-Sahara	9	10	15
America			
North (USA and Canada)	71	84	89
Central	18	43	57
South	22	39	48
Asia (excl. PR China and ex-USSR)	17	37	36
Australia	10	9	10
Europe (excl. ex-USSR)	62	76	116
PR China	2	4	73
ex-USSR republics	15	25	29

[1] Persoone and Sorgeloos, 1980
[2] Vanhaecke et al. 1987
[3] Triantaphyllidis et al. 1998

endemic genus *Parartemia*. Intensive exploration work has further been done in PR China (Xin et al. 1994) and for various individual countries like Chile (Gajardo et al. 1992, 1995, 1998, 1999; Gajardo, 1995), Mexico (Maeda-Martinez, 1991), Spain (Amat et al. 1995b) and – to a much more limited extent – countries of the former Soviet Union, with a focus on Ukraine, Kazakhstan and Russia, esp. South Siberia (Baitchorov and Nagorskaja, 1999). Though valuable characterization work has been done in the (ex-)USSR (Solovov and Studenikina, 1990), published in national (Russian-language) journals, this has hardly reached the international scientific community.

The exploration of new *Artemia* habitats is not exclusively stimulated by pure scientific interests in 'new' *Artemia* strains and species. The decline since 1997 of the *Artemia* cyst yields on the traditional harvesting ground, the Great Salt Lake in Utah, USA (Lavens and Sorgeloos, 2000), has intensified the search for alternative resources, especially in inland lakes that are big and productive enough to justify commercial exploitation. Competition is hard, and thus discretion is a keyword here. As a result, several sites, especially in continental Asia, are exploited occasionally or on a regular basis (with some local investment), and these cysts are being used worldwide in aquaculture. The identity or location of these sites has not yet reached the scientific literature, and attempts are seldom made to perform a systematic characterization of the respective strains. Table 2 updates the list of Triantaphyllidis et al. (1998) with the most recent information, predominantly originating from the informal circuit. Nevertheless, for some areas where competition by private companies, exploiting the *Artemia* resources, is particularly tough (like Russia and the Central Asian republics) and thus confidentiality is a top priority, the sites listed in Table 2 only very partially reflect the reality in the field.

A continued survey will necessarily lead to the discovery of many more

Artemia biotopes in different parts of the world. On the other hand, brine shrimp populations may disappear as certain biotopes (*e.g.* coastal saltworks and lagoons) are subject to human intervention and climatological fluctuations, like Lymington, UK – the population, described by Schlösser in 1755 (in Sorgeloos, 1980 and references therein); Capodistria, Slovenia (Vanhaecke *et al.* 1987); Germany (Persoone and Sorgeloos, 1980).

Lists of *Artemia* sites principally include biotopes where the population reappears each year. Additionally, temporal populations occur where *Artemia* is introduced on a seasonal basis, mostly in seasonal salt operations, *e.g.* Panama, Costa Rica, Burma, Thailand, Philippines, Vietnam, Indonesia and other places (Vanhaecke *et al.* 1987). Eventually, depending on the climatic conditions and management procedures, some of these populations may become established as feral strains; *e.g.* Cam Ranh Bay, Vietnam (Vu Do Quynh and Nguyen Ngoc Lam, 1987).

2. Ecological Aspects of *Artemia* Distribution

2.1. INTRODUCTION

Two critical factors determine the population dynamics of *Artemia* and its zoogeographical distribution: firstly, whether water body conditions allow the animals to survive throughout the year, and secondly, whether the seasonality of the environment is predictable or not (Lenz, 1987; Amat *et al.* 1995b). In the past, hypersaline environments were generally considered as habitats with simple trophic structures and low species diversity, having overall ecological characteristics largely similar all over the world (Persoone and Sorgeloos, 1980; Lenz, 1987; Lenz and Browne, 1991). This concept resulted in underestimation of the diversity in physical, chemical and biotic characteristics of *Artemia* habitats. Por (1980) thus states that 'at high salinities we find a tendency towards cosmopolitan distribution and there is basically one biotic association to be found almost everywhere'. This species association is composed of the green alga *Dunaliella*, *Artemia* sp. and *Ephydra* sp. Furthermore, *Artemia* spp. show little morphological variation, suggesting little difference among the respective populations in their interaction with the environment. Numerous laboratory studies in later years, however, indicated that *Artemia* strains vary significantly with respect to their life history traits and gene pools, presumably as a result of selective pressures in their native habitats (Browne *et al.* 1984).

In natural environments, it is difficult to detect possible synergistic or antagonistic interactions between factors affecting life history traits, such as salinity, temperature or food levels (Von Hentig, 1971; Browne, 1982; Wear *et al.* 1986; Mura, 1995b). Nonetheless a few general remarks can be made regarding the impact of some environmental factors on the *Artemia* distribution.

2.2. SALINITY

The common feature of all *Artemia* biotopes is their high salinity. Salinity is without any doubt the predominant abiotic factor determining the presence of *Artemia* and consequently limiting its geographical distribution. Salinity as crucial factor for *Artemia* presence is linked to the occurrence of predators of brine shrimp at lower and intermediate salinities (see further). Hammer *et al.* (1975) studied the salinity and presence of *Artemia* in salt lakes in the Saskatchewan region, Canada. Whereas salinity ranged from 2.4 up to 370 ppt, *Artemia* was only found in the lakes with the highest salinity, where no or very few other organisms (other zooplankton, benthic fauna or fish) were found. An analogous distribution of *Artemia* in relation to salinity has been reported by McCarraher (1970) for salt lakes in Nebraska, USA. Other variables (temperature, light intensity, primary food production) may have an influence on the quantitative aspects of the *Artemia* population, or may cause only a temporary absence of brine shrimp. On the other hand, not all highly saline biotopes are populated with *Artemia*, *e.g.* McCarraher (1972), in his preliminary bibliography of inland salt lakes, listed over 30 highly salines lakes (> 100 ppt) in the continental USA not inhabited by *Artemia*, most of them extremely saline (> 200 ppt) and/or sulphate lakes. Also in Western Australia *Artemia* is absent in several natural salt lakes with a salinity exceeding 100 ppt (Williams, 1981). In the particular case of Australia, however, competition with the endemic *Parartemia* (see further) may play a role (Geddes, 1980, 1981). In other cases – and in the absence of potential predators of brine shrimp (see further) – apparently none of the distribution vectors wind, birds, or man has allowed *Artemia* to colonise these biotopes.

Vanhaecke *et al.* (1987) have used the classification system of Thornthwaite (1948) to relate the natural distribution of *Artemia* sites to different climate types. This system is based on the calculation of the water balance of the soil taking into account precipitation and evaporation data. This study revealed that the geographical distribution of brine shrimp is limited by climatological conditions, *i.e.* no natural *Artemia* populations are found in humid climate types, and 97% of the biotopes are located in areas where yearly evaporation exceeds yearly precipitation. In humid climate types, normally human intervention is required to keep high salinity levels (*e.g.* removal of freshwater stratification layers by drainage or pumping, storage of brine in reservoirs). Additionally, Vanhaecke *et al.* (1987) outlined the potential distribution pattern of *Artemia* on a global level, stressing nevertheless the restricted relevance of this kind of extrapolation: climate types may vary significantly even within relatively short distances and specific local conditions may result in isolated microclimates suitable or unsuitable for *Artemia*. The finding of *Artemia* sites in the subhumid extreme southern part of Chile (Campos *et al.* 1996; Gajardo *et al.* 1999), however, might be an illustration of this theory. Morever, besides general annual precipitation and evaporation figures, seasonal distribution of precipitation and evaporation are also important, and more detailed climatological conditions and waterbalance data have to be used to obtain an accurate idea about the potential world distribution of *Artemia*.

Brine shrimp have been found alive in supersaturated brines at salinities as high as 340 ppt (Post and Youssef, 1977). Under these extreme conditions, however, the animals barely manage to survive and their normal physiological and metabolic functions are seriously affected. The lower salinity limit in nature is basically a function of the presence of predatory animals. Brine shrimp are rarely found in waters with salinity lower than 45 ppt, although physiologically they thrive in seawater and even in brackish waters (Persoone and Sorgeloos, 1980). Brine shrimp do not possess any anatomical defense mechanism against predation and consequently *Artemia* populations are always in danger at salinities tolerated by carnivorous species. The list of *Artemia* predators includes by definition all species populating natural seawaters and feeding on zooplankton (see further). Brine shrimp have, however, developed a very efficient ecological defense mechanism by their physiological adaptation to media with very high salinity. As such they can escape from their predators thanks to this salinity barrier (for predation by birds there is, of course, no escape in shallow waters). As a general rule, the lowest salinity at which *Artemia* is found in nature thus varies from place to place and is determined by the upper salinity tolerance level of the local predator(s). Hedgpeth (1959) mentioned that for several species of marine fish and invertebrates this level can be as high as 80 to 100 ppt; some fish species even seem to survive in salinities above 100 ppt and even up to 130 ppt (Caspers, 1957; Bayly, 1972; Kristensen and Hulscher-Emeis, 1972; Lenanton, 1977; Davis, 1978). The cyprinodont *Aphanius* in the Mediterranean and Red Sea area even tolerates salinities exceeding 150 ppt (P. Baert, pers. comm.). Mullet (*Mugil* sp.), milkfish (*Chanos* sp.), *Tilapia* and several crustacean species (*e.g. Palaemonetes*) predate heavily on brine shrimp in saltpans and salt lakes. Calanoid copepods are efficient predators of *Artemia* nauplii at lower salinities. Adults of the Anostracan Branchiopod *Branchinella spinosa* have been reported to feed on *Artemia* larvae and post-larvae (Abatzopoulos *et al.* 1999). Several categories of insects also regularly predate on brine shrimp: Odonata larvae, aquatic Coleoptera and Hemiptera (especially the families Corixidae and Notonectidae) all can withstand considerable salinities in the range 60–120 ppt (Rawson and Moore, 1944; Bayly, 1972).

Hammer and Hurlbert (1992) investigated if the absence of *Artemia* at lower salinities was determined by the presence of predators or by lower salinity itself. In laboratory tests with *A. franciscana* they found that survival was normal at salinities exceeding 38 ppt, but at lower salinities adults died and young grew very slowly. Nevertheless tank production of brine shrimp is commonly done in natural seawater (Lavens and Sorgeloos, 1996). Hammer and Hurlbert (1992) furthermore made direct laboratory observations of invertebrates capturing and feeding on brine shrimp nauplii and metanauplii, like *Cyclops* spp. (cyclopoid Copepoda), *Trichocorixa* spp., *Diaptomus* spp. and the damselfly *Enallagma*, and studied the predation rate of the respective organisms.

No optimum can clearly be defined for salinity of the *Artemia* environment. For physiological reasons this optimum must be situated towards the lower end of the salinity range, as higher ambient salinity requires higher energy costs for osmoregulation. Ambient salinity finally also plays a role in cyst metabolism, as *Artemia* cysts will only start to develop when the salinity of the medium drops below a certain threshold value, which is strain dependent (*e.g.* 85 ppt for the San Francisco Bay strain). At salinities above this threshold *Artemia* cysts will never hatch because the hydration level they reach is insufficient, which is one of the prerequisites for the onset of the hatching metabolism (Lavens and Sorgeloos, 1987). Situations of a temporary low salinity often occur in a salt lake, *e.g.* in a restricted area of inflowing freshwater, or after rainfall when for a while a freshwater layer remains on top of the heavier salt water.

2.3. IONIC COMPOSITION

Artemia can withstand environments in which the ratio of the major anions and cations may be totally different from that in seawater, and even reach extremely high or low values in comparison to natural seawater (Persoone and Sorgeloos, 1980). The Na^+/K^+ ratio, which is 28 in seawater, may range from 8 to 173 in some *Artemia* habitats; the Cl^-/CO_3^{2-} ratio, which is 137 in seawater, may decrease to 101 and reach 810 at the other extreme; the Cl^-/SO_4^{2-} ratio which is 7 in seawater has been reported to vary between 0.5 and 90 in certain *Artemia* biotopes (Cole and Brown, 1967; Bowen *et al.* 1978). Depending on the prevailing anions, *Artemia* may thus inhabit chloride, sulphate or carbonate waters and/or combinations of two or even three major anions. The most renowned example of a carbonate habitat is no doubt Mono Lake in California, USA, inhabited by a local subspecies of *A. franciscana*, assigned the name *A. monica*. This striking physiological adaptation to such extreme chemical habitats brought Cole and Brown (1967) to the conclusion that 'the ionic composition of the waters inhabited by *Artemia* varies more than that of any other aquatic metazoan'.

The ionic composition of the habitat of *Artemia* can result in ecological isolation of particular strains (Bowen *et al.* 1985, 1988) and can furthermore result in morphological differences (Hontoria and Amat, 1992b). For comparative tests, standardization of the culture medium is therefore of utmost importance (Hontoria and Amat, 1992a; Triantaphyllidis *et al.* 1997a).

2.4. TEMPERATURE

Apart from salinity temperature also affects the distribution pattern of *Artemia* (Vanhaecke *et al.* 1987). No *Artemia* is found in the cold tundra and frost climate types as the year-round prevailing extremely low temperatures preclude

Artemia development. Moreover, potential evaporation is very limited in these regions, which (with only very few exceptions) excludes the presence of highly saline biotopes. At least *A. franciscana* does not seem to survive temperatures below 5 °C for extended periods of time, unless under the form of cysts (field data from Gajardo and Beardmore, 1993, show that the temperature at the period of maximum population size in Atacama, Chile, is about 6 °C). No data are available, however, for other strains. A lot of strains are found in continental areas (continental USA and China, Central Asia, South Siberia) with extremely cold winter temperatures, but hot summer temperatures allow the hatching of the cysts and subsequent colonisation of the environment. *A. tibetiana*, however, seems to be physiologically adapted to low temperatures, as it survives in a habitat, a carbonate lake on the high plateau of Tibet, with a salinity of ± 60 ppt (Zheng, 1997), with annual temperatures fluctuating between –26 and +24 °C, and with an average annual air temperature of ± 1.6 °C, and a maximal daily water temperature in the cyst production season of ± 15 °C (Abatzopoulos *et al.* 1998; Clegg, 2001). No other *Artemia* site of higher altitude has ever been reported than this carbonate lake, situated at a height of 4490 m (Zheng, 1997). The effect of extreme height (*e.g.* more intense UV irradiation) on the local brine shrimp population still needs to be studied.

The maximum temperature that *Artemia* populations tolerate has repeatedly been reported to be close to 35 °C, a temperature often attained in the shallow tropical salterns that constitute a large part of the *Artemia* habitats. This tolerance threshold is, however, strain-dependent. Moreover, early inoculation tests in Thailand revealed that, after a certain adaptation period, brine shrimp from Macau (Brazil) survived for weeks at temperatures around 40 °C (Vos and Tunsutapanit, 1979). Physiological adaptation of San Francisco Bay (SFB) *Artemia* after a number of generations to the high temperatures (± 40 °C) in Vietnamese salt ponds has also been reported (Clegg, 2001). As for salinity, temperature optima are difficult to define and are strain-dependent; generally, however, one can state that the optimum for *Artemia* must be situated in the range 25–30 °C. The ametabolic dehydrated cysts resist a much wider temperature range, which never occurs in nature; *i.e.* the minimum being the absolute zero (–273 °C – Skoultchi and Morowitz, 1964), and the maximum close to 100 °C, be it for short exposure times (Hinton, 1954). Finally, previous work has shown that temperature also has a major impact on the hatching ability of cysts, and that this response is species specific (Vanhaecke and Sorgeloos, 1989). As explained further, this differential response is used to verify the presence of contamination of *A. franciscana* material in (commercial) cyst samples from parthenogenetic and Old World bisexual strains (Triantaphyllidis *et al.* 1994b).

2.5. BIOTIC ELEMENTS

While hypersaline environments are characterised by monocultures of *Artemia* as major zooplankton, lower and intermediate salinity habitats are populated by various groups of invertebrates. Depending on the seasonal cycle or the salinity gradient in certain biotopes (*e.g.* in man-managed saltworks) there can thus be a range where there is some degree of coexistence with brine shrimp. Numerous studies report about the faunal and floral elements of these biotopes, and the extent to which they are found in coexistence with *Artemia*. According to Bayly (1993) the absence of *Artemia* populations in certain South American lagoons is related to low salinities: *Daphniopsis* sp. (Cladocera) and *Boeckella poopoensis* (Copepoda, Centropagidae) were observed in these habitats. The latter halobiont copepod is often found in saline water up to 80 ppt, while nearly all 14 other species of *Boeckella* found in South America are confined to freshwater. In the Central Andes no lakes have both *Boeckella* and *Artemia*. In this environment salinity is thus a fairly good predictor for the presence of these crustaceans (Zuñiga *et al.* 1999). Williams *et al.* (1995) confirm that in salt lakes of the Bolivian Altiplano, at salinities exceeding 50 ppt, the only taxa surviving are *Artemia*, Ephydridae, Dolichopodidae (Insecta), and *Boeckella poopoensis*.

Artemia salina and *A. franciscana* males have been reported to clasp females of the Anostracans *Branchinella* (Mura, 1987) and *Branchinecta* (Belk and Serpa, 1992), respectively. An interesting co-occurrence of *Artemia* and *Branchinella* has been reported by Mura *et al.* (1987) in coastal lagoons and saltpans of Sardinia (Italy), where at some sites the brackishwater *Branchinella* is dominant in wintertime. When salinity levels increase, *Artemia* becomes more numerous and finally in hypersaline summer conditions *Branchinella* disappears. Another case is reported from Greece by Abatzopoulos *et al.* (1999) where *B. spinosa* was found to coexist with parthenogenetic *Artemia*. Also in Australia several Anostracan genera occur in inland saline waters below ± 50 ppt: *Parartemia* (Branchipodidae), *Branchinecta* (Branchinectidae), *Branchinella* and *Branchinectella* (Thamnocephalidae). Above 50 ppt, only the genera *Artemia* and *Parartemia* survive (Hammer, 1986; Williams and Geddes, 1991), while only *Artemia* can tolerate the highest salinities. Hammer and Forró (1992) and Hammer (1993) have investigated the zooplankton fauna of Canadese salt lakes (British Columbia, Alberta, Saskatchewan). Of the numerous crustacean species found in these habitats, none displayed the same euryhalinity as *A. franciscana*, which was found over a salinity range of 33–270 ppt. Bos *et al.* (1996) confirm that, although both *Artemia franciscana* and *Moina hutchinsoni* dominate saline waters in British Columbia, the distribution of these two taxa also varies with brine composition; *A. franciscana* is more common in the meso-hypersaline waters that are high in Mg, Ca and sulphate, and lower in pH, while *Moina* is more common in meso-hypersaline waters that are low in Mg and Ca and higher in pH. Also Rahaman *et al.* (1993) found that in solar saltworks in India in extreme conditions only a few organisms survive, the most tolerant being *Artemia*. Its absence from a number of highly saline ponds was related to food limitation, too high temperature, or saltworks management unfavourable for colonisation by brine shrimp.

Rotifers like *Brachionus* are thought to compete with *Artemia* for food, and can outcompete *Artemia* in some cases. Other food competitors (at the lower salinity range) for phytoplankton are ciliates, copepods and other Anostracan crustaceans. The brine fly *Ephydra*, often encountered in large numbers in *Artemia* biotopes, is more a benthic feeder and does not interfere with the *Artemia* food chain (Persoone and Sorgeloos, 1980).

3. Ecological Isolation of *Artemia* Strains

Artemia franciscana is distinguished from other bisexual *Artemia* species by occurring in habitats with the largest diversity in water chemistry (Bowen *et al.* 1988). The individual *A. franciscana* populations vary considerably with respect to their tolerance for waters of different ionic composition. Reproductive isolation among these populations occurs because of the intolerance for each other's natural habitat, due to differences in lake water chemistry. The extent of this reproductive isolation of *Artemia* from non- or low-chloride habitats remained uncertain for a long time, because no culture medium permitted the coexistence, and thus the cross-breeding of brine shrimp from habitats with different ionic composition (Bowen *et al.* 1985). San Francisco Bay nauplii die within a few days in the carbonate water of Mono Lake. Conversely, Mono nauplii have low viability in seawater and other high-chloride media (Bowen, 1964; Lenz, 1980). However, in spite of this reproductive isolation of some *A. franciscana* populations in nature, they can be hybridized in well-defined laboratory media (Bowen *et al.* 1985). In spite of the 'interfertility of these physiologically differentiated clusters of populations in carefully designed permissive laboratory media, they have become substantially differentiated by the criterion of adaptation to the anionic composition of their habitats', hence the concept of 'semispecies' ('incipient' species) (Bowen *et al.* 1985).

Bowen *et al.* (1988) further analyzed the tolerance of *A. franciscana* strains to different cations and anions and found that the range of concentrations encountered by each population in their habitat is narrower than the range allowing high viability in laboratory media. The 'potential niche', which is based on the genetic potential, and which was determined by laboratory tests, is thus broader than the 'realized niche', which corresponds to the ionic compositon of the lake, and the latter often doesn't occupy intermediate, but rather more marginal values of the potential range. According to Browne and Bowen (1991) the populations in inland salt lakes of North America have experienced enormous changes in environmental ion compositions during the Holocene; supposedly certain *A. franciscana* 'semispecies' would not have succeeded (yet) in following the environmental changes. Further cross-fertility tests among North American populations suggested that there is evidence of genetic barriers between *A. franciscana* populations that inhabit the same type of ionic habitats, but are geographically isolated. In this respect, the Zuni Salt Lake (New Mexico) population for instance, might be described as a subspecies of the *A. franciscana* complex that is approaching distinct speciation.

In different ionic environments, the costs of osmoregulation are expected to differ. For the Mono Lake *Artemia* the upper salinity tolerance is 160–180 ppt, which is low compared to *Artemia* tolerances for chloride waters (Dana and Lenz, 1986). The different types of lake water may also favor different phytoplankton species. For example, *Dunaliella viridis* and *D. salina* are dominant in many chloride lakes, whereas in Mono Lake *Nannochloris* dominates (Jellison and Mellack, 1993). However, both *Dunaliella* and *Nannochloris* appear to be adequate food sources for *Artemia*.

Mono Lake *Artemia* still takes a separate position within this superspecies, because of its unique egg morphology and distinctive life history traits (*e.g.* cysts sinking in water column) (Dana, 1981). These factors, combined with the ecological barrier to gene exchange with (other) *A. franciscana* justify the nomenclature of this population as *A. monica* Verrill 1869 (Browne and Bowen, 1991). Nevertheless, the degree of genetic differentiation of *A. monica* from typical *A. franciscana* populations is very small and considerably smaller than values commonly associated with between-species differences (Beardmore and Abreu-Grobois, 1983). The classic dispersal mechanism of cysts by wind or birds does not take place for Mono Lake cysts. Although Mono Lake lies on the route of migratory bird species, and there is thus opportunity for frequent input of genes from other populations (Beardmore and Abreu-Grobois, 1983), this strain is probably much more isolated than strains producing floating cysts.

Except for *A. franciscana* strains inhabiting North and South America, the tolerance of individual species to different ionic environments has seldom been studied. The heterogeneous cluster of parthenogenetic populations, sometimes though not usefully grouped under the binomen *Artemia parthenogenetica*, is also known to occur in habitats with widely diverging ionic composition. Amat *et al.* (1995b) indicate that in Spain the parthenogenetic tetraploid strains are found preferentially in habitats of athalassohaline origin, which may be related to the greater tolerance of polyploids to stressful environments (Zhang and Lefcort, 1991).

4. Coexistence of Different *Artemia* Strains

4.1. DISTRIBUTION PATTERNS OF *ARTEMIA* SPECIES

In the early eighties the knowledge about the world distribution of parthenogenetic and bisexual strains was still limited, and it was generally recognized as a striking aspect of brine shrimp distribution that reproduction was exclusively sexual in the western hemisphere, but primarily parthenogenetic in the Old World (Browne and MacDonald, 1982); the known Old World bisexual populations were limited to *A. salina* in the Mediterranean area, and *A. urmiana*, confined to Lake Urmia in Iran. Furthermore, the contrast between the widely distributed *A. franciscana* and the geographically isolated *A. persimilis*, restricted to an area near Buenos Aires, Argentina, was remarkable. Clark and Bowen (1976) had already proven reproductive isolation between *A. persimilis* and two other species, *A. franciscana* and *A. salina*.

Several authors (reviewed by Vanhaecke *et al.* 1987) searched for a correlation between the different 'types' or sibling species of *Artemia* and their geographical distribution pattern, however without success. Bowen *et al.* (1978), studied the variation in *Artemia* haemoglobins and reported that 'a search for geographical patterns and latitudinal lines of alleles also yielded negative results'. The only fixed pattern of distribution, established until now, is the contrast between the exclusively sexual reproduction in the Americas, versus the occurrence of both parthenogenetic and bisexual strains in the rest of the world (including Australia). Contrary to the hypothesis of Browne and MacDonald (1982) it appears now that there is no correlation between the mode of reproduction and the appearance of *Artemia* in either inland salt lakes or coastal salt operations. Not only *A. franciscana*, but also Old World bisexual brine shrimp are found in both types of biotopes. Furthermore neither climate nor latitude seems to have a major influence on the distribution pattern of the different species on the world level. As it is discussed also in Chapter V of this volume, substantial genetic differentiation occurs among populations from the same species (Abreu-Grobois and Beardmore, 1980, 1982).

4.2. COEXISTENCE OF BISEXUAL STRAINS

Until now, there are no known examples of natural coexistence of different bisexual species, except for the not-confirmed coexistence of *A. persimilis* and *A. salina* in Sardinia, Italy. For the Asian species *A. urmiana*, *A. sinica* and *A. tibetiana*, a number of sites have been identified, but their exact area of distribution is not known. As Triantaphyllidis *et al.* (1998) point out, it would be interesting to know if the respective bisexual species are allopatric, or if there is hybridisation in some zones.

4.3. COEXISTENCE OF DIPLOID AND POLYPLOID PARTHENO GENETIC STRAINS

As Browne and Bowen (1991) report, *Artemia* is a useful model organism for examining the adaptiveness of sexual versus asexual reproduction. In other organisms, the asexual forms in general have replaced the related sexual species from most of their geographical ranges (Browne and Halanych, 1989).

According to Amat (1983) and Amat et al. (1995b) both in plants and in other animals parthenogenetic forms have a different geographical distribution, compared to their sexual relatives, and they generally tend to be found at higher latitude and altitudes. In an analogous way, the increased frequency of polyploids in many animal species at higher latitudes is associated with their greater tolerance to cold stress and better colonising abilities (Amat et al. 1995b). Similarly the relative fitness of diploid and polyploid parthenogenetic brine shrimp populations is thought to be a function of environmental conditions (Zhang and Lefcort, 1991; Zhang and King, 1992, 1993; Zhang, 1993). In the case of parthenogenetic populations polyploidy may account for higher heterozygosity and genetic variability (Abreu-Grobois and Beardmore, 1980, 1982). Sympatric diploids and polyploids respond differently to environmental changes, with polyploids having many advantages over sympatric diploids in stressful habitats or when exposed to lethal temperatures for short periods. Polyploid Artemia would also have developed a series of life history characteristics adapting them to environments that contrast to those of their sympatric diploids (Zhang and King, 1993). Lefcort et al. (1991) explored the behavioral response to gradients of light and heat of sympatric diploid and pentaploid parthenogenetic strains from Shandong province, PR China. From these results, they predicted that warmer areas of a pond would be populated by pentaploids during the day and by diploids during the night. Zhang and Lefcort (1991) state that polyploid Artemia are found more frequently at low and high latitudes (i.e. < 25° N and > 40° N) in the Northern hemisphere of the Old World, and hence may correspond to high and low temperature habitats. For more information on evolution, genetic variation and ecology of various types of parthenogenetic and of bisexual strains, see Chapter V.

4.4. COEXISTENCE OF PARTHENOGENETIC AND BISEXUAL STRAINS

4.4.1. Competition and Niche Partitioning
In Artemia parthenogenetic strains (especially polyploid) often attain larger sizes and have faster growth rates than sexual diploid strains (Amat Domenech, 1980; Tobias et al. 1980; Vanhaecke and Sorgeloos, 1980). Evidence has been found that males subjected to a lower food regime have a lower survival rate than similarly reared females (Browne, 1980, 1982). Since fertilization is required for each new brood, high male mortality could lead to the domination of parthenogenetic populations in stressful habitats. In Sardinia both sexual and parthenogenetic populations have been found only a few kilometres apart (Stefani, 1960), the parthenogenetic population being associated with a commercial coast saltern, and the sexual population in a natural saltern located a few kilometres inland. Bowen et al. (1978) reported that parthenogenetic strains have more haemoglobin than sexual species, which might be advantageous at high salt concentration. In competition between sexual and parthenogenetic sympatric populations, the ability to respond to environmental stress may favour parthenogenesis (Browne and MacDonald,

1982). This factor, along with the theoretical two-fold increase in reproductive rate of parthenogenetic populations and the fact that only a single individual is required for colonisation (both of which may be important in many brine shrimp populations where temperature or desiccation cycles kill all but encysted individuals) may provide an explanation why parthenogenesis would be the predominant mode of reproduction in the Old World, or, at least, dominate over bisexual species in the most stressful habitats.

Temporal cycling or niche partitioning would be expected based on temperature response profiles, as described by Browne *et al.* (1988), who reported on the effects of temperature and relative fitness of bisexual and parthenogenetic populations, by studying their lifespan and reproductive characteristics at different temperatures in the range 15–30 °C. *A. salina* was least tolerant to high temperatures, while *A. parthenogenetica* had a broader tolerance. Analogously, San Francisco Bay and parthenogenetic *Artemia* from Tanggu (China) exhibited significant differences in their response to elevated salinity levels, SFB seeming a more effective coloniser at higher salinities (Triantaphyllidis *et al.* 1995).

The effect of temperature on the performance of the population helps to determine the competitive interaction between bisexual and parthenogenetic brine shrimp strains and their relative success in competition experiments. Laboratory competition experiments result in the dominance of *A. franciscana* over parthenogenetic strains on one hand, and parthenogenetic populations over *A. salina* on the other hand (Browne, 1980; Browne and Halanych, 1989). Nevertheless, different experimental conditions (in terms of temperature, salinity, food conditions) might lead to different results, while these laboratory results need also be confirmed in the field. Remarkably, in most laboratory systems with other organisms competitive exclusion requires in the order of 10 to 100 generations, while for *Artemia* exclusion occurs in only 2 or 3 generations (Lenz and Browne, 1991), which suggests a large or nearly complete niche overlap between the respective *Artemia* species.

Vanhaecke and Sorgeloos (1989) demonstrated that the bisexual species *A. persimilis* and *A. salina* resemble parthenogenetic *Artemia* in showing a limited tolerance to high temperatures. Additionally, preliminary experiments carried out with *A. sinica* from Yimeng (Inner Mongolia, PR China) and *A. urmiana* (Iran) suggest that populations of these species are less tolerant of high temperatures, compared with *A. franciscana*. A practical application of the differential response of bisexual versus parthenogenetic strains towards environmental conditions is demonstrated in the work of Triantaphyllidis *et al.* (1994b); incubation of *Artemia* cysts at high temperature (36–37 °C) suppresses hatching of cysts from a parthenogenetic population and Old World bisexuals, while at the same time allowing *A. franciscana* cysts to hatch. This approach can be used to verify the presence of contaminating foreign material from the New World in natural cyst populations of the Old World.

Several authors report that rare males, present in parthenogenetic populations, cross with bisexual *Artemia* females, like *A. urmiana* (Clark and Bowen,

1976) or *A. sinica* (Cai, 1993), thus enabling gene flow between the populations (Pilla, 1992). It would seem that in a mixed population, consisting of an asexual and a sexual species the asexual species may have a (small) competitive advantage over its sexual counterpart, by preventing some of the males from mating with females of their own species (since asexual females do not require males in order to reproduce) (Triantaphyllidis *et al.* 1994b).

4.4.2. The Iberian Peninsula
A well-documented case of (potential) coexistence of parthenogenetic and bisexual strains is the Iberian peninsula (Amat, 1979, 1980; Perez Rodriguez, 1987; Amat *et al.* 1995b), where the populations north of 39° latitude are parthenogenetic polyploids, while both sexual and parthenogenetic diploid populations are found south of that line. Some salt ponds in southern-most Spain (area of Cadiz) seem to have a mixture of both sexual and parthenogenetic strains. However, strain dominance varies seasonally, with the bisexual strains favoured under low temperature and salinity (winter–early spring) while the parthenogenetic strains dominate under warm temperature and high salinity (late spring–summer). Amat (1983) tried to link experimental results to temperature and salinity variations recorded in the natural environment, in order to infer a relationship between these two ecological factors and the composition of the natural populations. He found that bisexual individuals dominated in samples collected in the period March–June; these would have been offspring from females that developed from January to April/May, *i.e.* at a time when temperature and salinity are lowest. On the other hand, laboratory populations from cysts collected between July and February have a preponderance of parthenogenetic females, derived from females developing from June to November, *i.e.* in summer and autumn, when temperatures and salinities in the salterns attain maximum values.

The hypothesis that bisexual strains tend to occur in inland salt lakes and parthenogenetic ones in coastal salterns (Browne and MacDonald, 1982; Lenz, 1987) does not seem to be supported by the situation on the Iberian peninsula. According to field data (Amat *et al.* 1995b), there is an obvious overlap in coastal salterns, where both bisexual and parthenogenetic diploid strains are found. In Bonmatí (Santa Pola, Alicante) salterns the co-occurrence of even three different strains throughout the year was recorded. The remaining inland locations tend to show the presence of only two strains, namely bisexual or parthenogenetic tetraploid.

4.4.3. Lake Urmia, Iran
Another case of possible coexistence (not conclusively elucidated) is Lake Urmia, Iran, where – at regular time intervals – indications are found that a parthenogenetic strain may be sharing the habitat with *A. urmiana*. Günther (1890) described brine shrimp from Lake Urmia as a bisexual species, *Artemia urmiana*. This characteristic was confirmed when Clark and Bowen (1976) demonstrated the reproductive isolation of the species from other bisexual

strains. Nevertheless, Barigozzi et al. (1987) reported a population exclusively composed of parthenogenetic individuals after culturing two cyst samples in the laboratory, and proposed to cancel the species designation A. urmiana. In 1989 Azari Takami reported the coexistence of bisexual and parthenogenetic populations in the lake. He observed that the parthenogenetic strain was dominant in spring and summer but not common in autumn and winter. Barigozzi (1989) proposed to reconsider A. urmiana as a species, and Ahmadi et al. (1990) reinforced the idea that the lake sustains a mixed population. Pador (1995), who cultured animals using cysts collected from different sampling stations, found no evidence of a parthenogenetic population. Abatzopoulos et al. (1997) used cyst membrane protein composition to discriminate between different Artemia strains; they found that the electrophoresis banding pattern of A. urmiana resembled the parthenogenetic populations and they linked this to the hypothesis that parthenogenetic populations and A. urmiana have a common ancestor and the genetic distance (Nei's D) between them is smaller than for other bisexual species (Beardmore and Abreu-Grobois, 1983). This close relationship of parthenogenetic strains and A. urmiana was confirmed by AFLP fingerprinting analysis (Triantaphyllidis et al. 1997b). Van Stappen et al. (2001), studying the population dynamics of brine shrimp in Lake Urmia, concluded that 'although the occurrence of a parthenogenetic strain thus remains a possibility, the sex ratio suggests that the Artemia population in Lake Urmia is at least predominantly bisexual'. The confusion may be caused by the occurrence of parthenogenetic strains in shallow saltpans, adjacent to the lake, with a temperature and salinity regime strongly differing from the conditions prevailing in Lake Urmia itself (Agh, personal communication).

4.4.4. PR China
Also for PR China it has been suggested that the prevailing mode of reproduction in the coastal habitats in China is parthenogenesis (Xin et al. 1994). Numerous parthenogenetic populations also exist in the inland lakes and especially in the big salt lakes of Xinjiang and Qinghai provinces. A variety of bisexual populations is found in inland China, especially in the province of Inner Mongolia. In provinces with both coastal and inland Artemia sites, like Hebei province, generally parthenogenetic populations are found in the coastal areas, while inland populations are only bisexuals. It is not clear if this distribution is linked to specific ionic requirements of the respective populations (Xin et al. 1994) or determined by other factors. Nevertheless, especially in remote inland areas, still much knowledge about Artemia sites remains to be developed, so that the complete picture might look different.

5. *Artemia* and Birds: Predation and Dispersal

Artemia world distribution is discontinuous, since Artemia cannot on its own migrate from one saline biotope to another via the seas, due to the absence of any anatomical defense structure against predation by carnivorous aquatic organisms (Persoone and Sorgeloos, 1980). The principal dispersion mechanism of *Artemia* is transportation of cysts, floating at the water surface, by wind and by waterfowl, as well as deliberate human inoculation in solar salt works. Cysts stick to the body of wading or swimming birds, or are ingested together with *Artemia* biomass. There are also numerous reports of dispersal of small aquatic organisms (algae, protozoans, nematodes, rotifers) dispersed from one isolated waterbody to another, attached to the external surfaces of birds, clinging to the feathers and bill, or sticking in the mud attached to the feet. Taking into account that migrating birds can reach a speed of 50–100 km/h (Dorst, 1962), it is obvious that they can play an important role in dispersing *Artemia* over large distances (Vanhaecke *et al.* 1987).

Artemia biotopes are known as foraging grounds for many birds, in particular flamingos (*Phoenicopterus* sp.), whose feeding behaviour has been studied in several parts of the world (Rooth, 1976). Brine shrimp is considered as a highly nutritional food for avifauna (Cooper *et al.* 1984). Flamingos are associated with hypersaline biotopes throughout the world. Also for several other species of waterfowl *Artemia* constitutes an important – if not exclusive – part of their diet, as reported for different (groups of) species in various parts of the world: *e.g.* gulls (*Larus* sp.) and terns (*Chlidonias* sp.) in the French Camargue (Isenmann, 1975, 1976a, b), avocets (*Recurvirostra* sp.) and sandpipers (*Erolia* sp., *Ereunetes* sp.) (Carpelan, 1957) and different species of other wading birds (Clark and Bowen, 1976) in San Francisco Bay area; flamingos in Bonaire (Van der Meer Mohr, 1977), gulls, flamingos, avocets, stilts (*Calidris* sp.), grebes (*Podiceps* sp.) and ducks (*Anas* sp.) in Portuguese saltworks (Narciso, 1989a), phalaropes (*Phalaropus* sp.) and eared grebes (*Podiceps nigricollis*) in Mono Lake (Melack *et al.* 1983). Feeding studies of flamingo (*Phoenicopterus ruber roseus* Pallas) and the shelduck (*Tadorna tadorna* L.) in the Camargue, France, show that these birds feed not only on live brine shrimp (possibly with cysts in the uterus), but also on foam where *Artemia* cysts are concentrated (MacDonald, 1980).

The impact of predation by birds on the *Artemia* population is not always clear and figures are seldom available; often a 'strong' predation pressure on the brine shrimp population is assumed (*e.g.* predation by flamingos in Patagonian salt lakes, Campos *et al.* 1996). In Mono Lake predation by grebes causes between 8 to 80% of the mortality of *Artemia* (Cooper *et al.* 1984). In Lake Grassmere, New Zealand gulls consume 50 g live weight *Artemia*/day, which corresponds – for the existing *Larus novaehollandiae* population – to approximately 0.05 tonnes dry weight annually for this biotope (Wear and Haslett, 1987). To a certain extent there is a reciprocal benefit in the predator-prey relationship between *Artemia* and waterfowl. Birds fertilize the biotope with their guano, contributing in this way to the productivity of the ecosystem, and influencing the phytoplankton composition (Persoone and Sorgeloos, 1980).

Migrating routes of wader birds often follow coastlines, passing from one

Artemia habitat to another, both in the Americas and in Eurasia (Clark and Bowen, 1976). Royan *et al.* (1970) and Achari (1971) make the link between occurrence of *Artemia* in the Indian subcontinent and the migrating behaviour of flamingos. The link between bird migration and dispersal of resting eggs of different crustaceans was observed by Löffler (1964). He noted that '*Artemia* eggs were not damaged when passing through the intestinal tract, but even hatched and produced normal individuals', and investigated the duration of the stay of the cysts in the intestinal tract and its effect on hatching. Eggs of different groups of crustaceans (anostracans, notostracans, conchostracans, cladocerans, ostracods) were recovered from the digestive tracts of domesticated and wild ducks (*Anas* sp.) and other birds (Proctor, 1964; Proctor and Malone, 1965; Proctor *et al.* 1967) and still proved viable. Especially *Artemia* cysts survived the gut passage well, and were viable even up to 3 days after ingestion (Löffler, 1964). As the percentage of broken cysts greatly increases when passing through the gut, the percent of assimilation of organic matter in the cysts (5–36%) implies that *Artemia* cysts may serve as a food for waterfowl (MacDonald, 1980).

Several authors have documented parasitism in Anostracan species, acting as intermediate hosts for water birds predating them (Mura, 1995a). Scattered throughout the scientific literature there are reports of contamination of brine shrimp by viruses (Persoone and Sorgeloos, 1980), fungi (*Haliphthoros milfordensis*) (Overton and Bland, 1981), procaryotic intracellular symbionts (Post and Youssef, 1977), bacteria (Spirochaetes) (Tyson, 1970, 1974, 1975), and cestode flatworms.

Information is seldom given as to what extent these parasites affect *Artemia* populations, except for cestode parasitism in brine shrimp populations of Mediterranean Europe, and more specifically *Flamingolepis liguloides* and other *Flamingolepis* sp., parasites of flamingos. Other reported cestode parasitoses refer to *Eurycestus avoceti*, parasite of the avocet (*Recurvirostra avosetta*), and *Hymenolepis stellorae*, parasite of the gull *Larus ridibundus* (Gabrion and MacDonald, 1980; Robert and Gabrion, 1991). Gabrion *et al.* (1982) observed a 6 weeks-cyclic evolution of *F. liguloides* infestation in *Artemia* populations in the Camargue, southern France. Infestation seemed to occur preferentially in juveniles; the cyclic evolution was thought to be due to a premature mortality of highly infested *Artemia*. Moreover, the distribution of parasitized *Artemia* (4.25% of the population) was not uniform within the pond, and varied according to location and depth. Thiéry *et al.* (1990) found a relationship between the prevalence of the parasite in *Artemia* and the population density of flamingos across the salt marshes of the French Mediterranean coast, where parasitism in the infected populations increased with the size of *Artemia*. In infected populations, up to 22% of *Artemia* were coloured red and showed a surface swimming behaviour, making them more conspicuous to predating birds. Behavioural aspects of the host-parasite meeting strategy were further discussed by Robert and Gabrion (1991).

Amat *et al.* (1991) described cestode cysticercoids (Hymenolepididae) parasitizing the haemocoele of brine shrimp along the Mediterranean coast of Spain, and its effect on *Artemia* biology. There was little pathology in *Artemia* associated with this infestation; the bright red color and aberrant swimming behaviour of parasitized individuals was observed here as well. Mura (1995a) studied another case of cestode parasitosis in an *Artemia* population of Sardinia, Italy, and relates the prevalence of the infestation with the *Artemia* age class, and with the anatomical area of the intermediate host, infested by the cysticercoids. The decreasing trend of parasitism observed here over the years might be related to the frequency of occurrence and feeding habits of the flamingo colony in the area.

6. Australia: *Artemia versus Parartemia*

6.1. INTRODUCTION

As for many animal taxa, Australia takes a special position because of the presence of numerous animal species, endemic to the continent. While the brine shrimp *Artemia* is a major faunal element in many hypersaline biotopes throughout the world, the genus *Parartemia* (family Branchipodidae), with at least eight species with well-defined distributions, is endemic to Australia. Based on a comparative study of the haemoglobin of both genera, Coleman *et al.* (1998) localize the date of divergence of both genera at minimum 85 million years ago, a date coinciding with the geological isolation of Australia in the late Mesozoic.

It is still a subject of controversy if the genus *Artemia*, which occurs in a few sites in Australia, is recently introduced by humans or not. Related to this is the question if there is a kind of niche separation between both genera and/or if there is a tendency that the endemic *Parartemia* is outcompeted by the allochthonous *Artemia*. In contrast to *Artemia*, *Parartemia* has only been the subject of temporary commercial interest, and the study of the genus has hardly attracted any attention outside Australia (Williams and Geddes, 1991). Its aquaculture potential seems limited, compared to *Artemia*, because of the sinking behaviour of the cysts, and its limited adaptation to low oxygen levels. In view of its large size (up to 22 mm and 4 mg dry weight for males) biomass production might show some perspective, but available data are unsufficient to assess this correctly.

6.2. MORPHOLOGICAL, PHYSIOLOGICAL AND ECOLOGICAL DIFFERENCES *ARTEMIA-PARARTEMIA*

Parartemia can be distinguished from *Artemia* by its bigger size and by certain morphological characteristics (shape of male antennae and penes, female labrum and egg sac), which are close to *Branchipodopsis*, distributed in arid regions of Africa and Asia. Different *Parartemia* species show quite divergent morphological characteristics, in contrast to *Artemia* species, where the morphological differentiation between the species is rather limited. Most studies on *Parartemia* have been made in the field, and laboratory tests on feeding and reproductive behaviour are scarce (Geddes, 1981); ecological and/or physiological studies are mainly limited to the species *P. zietziana*. A possible niche separation between genera was suggested by Geddes (1981), who proposed that *Parartemia* is adapted as a sediment eater, based on observations of the feeding behaviour by Marchant and Williams (1977), while *Artemia* is generally considered as an obligate filter feeder, filtering planktonic algae. But observations (Savage and Knott, 1998) of the feeding biology of the parthenogenetic *Artemia* population from Lake Hayward in Australia showed that the local *Artemia* mainly utilized substrate-bound food resources ('surface grazing of the benthic mat'), like filamentous and unicellular bacteria and benthic diatoms, while photosynthetic phytoplankton only comprised a minor component of the diet. It is thus unclear if there exist enough differences in feeding behaviour between both genera to result in a niche separation.

Geddes (1975a, b, c) performed extensive studies on the salinity tolerance and osmotic capacities of the most commonly studied *Parartemia* species, *P. zietziana* Sayce. Laboratory tests confirmed the broad salinity-temperature survival range of the species, which is found in the field at salinities ranging between 41.5 and 300 ppt, and (after acclimation) can tolerate laboratory media with salinity as low as 3 ppt. The capacity of hypo-osmotic regulation in *Parartemia* is comparable to *Artemia*; also the ionic composition of the haemolymph is similar, as well as the physiological mechanism of hypo-osmotic regulation. All these characteristics represent a nice example of parallel evolution, since both are phylogenetically distinct Anostracan taxa (Geddes, 1975b). Like *Artemia*, *P. zietziana* survives in low O_2 concentrations, and it contains a type of haemoglobin, although, in contrast to *Artemia*, the protein is only produced in small amounts, and production doesn't increase in response to environmental stress, like oxygen stress (Coleman *et al.* 1998); this particular haemoglobin appears not to be used for oxygen transport (Manwell, 1978). *P. zietziana* has never been reported to go red at high salinity in the field or in the laboratory, which means a basic difference from *Artemia* (Geddes, 1975a).

The population dynamics of *P. zietziana* in the field show fluctuations, similar to *Artemia*: between 40 and 100 ppt the species still coexists with other organisms, but it becomes the dominant faunal element in summer in ephemeral saline lakes when the environment approaches saturation and all other animals have perished (Geddes, 1976). Subitaneous and resting eggs are produced by the females; an increase in salinity (and drop in oxygen) seems to be the controlling factor, and hatching occurs in response to a salinity drop to between 50 and 180 ppt. Geddes (1981) confirms that resting eggs of *P. zietziana* hatch at very high salinities; but in contrast to *Artemia* cysts, *Parartemia* resting eggs sink to the sediment and come in contact with inflowing water when the salt crust has been dissolved. Consequently, the dispersal capacities of *Parartemia* through cysts are very limited, compared to *Artemia*.

6.3. DISTRIBUTION OF *ARTEMIA* AND *PARARTEMIA* OVER THE AUSTRALIAN CONTINENT

Several hypotheses have been formulated with regard to the distribution and ecological requirements of *Artemia*, *Parartemia*, and their possible interaction. In an early stage of habitat exploration, these hypotheses relied on a very incomplete insight into the distribution of both genera and were thus abandoned later, when new sites were progressively discovered. Moreover, early reports about *Artemia* occurrence in Australia were thought of questionable validity, and were interpreted as possible misidentifications of *Parartemia* (Geddes, 1973). Supposedly, if the 'mainly northern hemisphere' *Artemia* were to occur in Australia, it would rather be found in the northern part of the country, by analogy with the distribution pattern of the calanoid copepods *Diaptomus* (northern hemisphere and north Australia) and *Boeckella* (south Australia). The finding of a new species, *Parartemia minuta*, in northern Australia, however, did not contribute to the corroboration of this hypothesis.

In 1979 Geddes evaluated the early literature records of Australian *Artemia* (with a variety of species names like *A. proxima*, *australis*, *westraliensis*) and confirmed the existence of a parthenogenetic population (originally described by Edward and Watson, 1959) on Rottnest Island, Western Australia. A parthenogenetic strain was also found in saltworks at Shark Bay, Western Australia, and two bisexual populations were reported from saltworks near Bowen and Rockhampton, Queensland. A 'mainly parthenogenetic' population was found in Dry Creek, South Australia. The author assumed that the '*Artemia* populations represent an indetermined number of recent introductions of that genus to Australia', and mentioned that new introductions were still made into large new solar salt fields to control algal blooms. The inoculated material consisted of Australian *Artemia* populations, already established, or foreign material, mainly from the San Francisco Bay area. Electrophoretic and genetical research (Bowen *et al.* 1978; Abreu-Grobois and Beardmore, 1980) showed the close relationship between the Queensland populations and the *A. franciscana* from San Francisco Bay or Great Salt Lake, well separated from the West-Australian parthenogenetic strains. It was not clear if the parthenogenetic populations in Western Australia had been introduced when the saltworks were established last century, and had since then spread or been inoculated along the coast. An alternative assumption was an earlier, natural dispersal from Asia (Geddes, 1980).

At this stage *Artemia* was still thought not to exist in any natural Australian salt lake, but exclusively in coastal saltworks. This contrasted well with the distribution of *Parartemia*, which seemed to have a long history on the continent, with several morphologically distinct species. Geddes (1981) reviewed the distribution of both genera again, and at that time the number of *Artemia* sites had increased to 7. The author tried to explain the zoogeographical separation between both genera as a consequence of physiological differences, and assumed that *Artemia*, in terms of its temperature preference, reproductive biology and feeding ecology, was not so well adapted to the conditions in Australian salt lakes. In Dry Creek (where an 'almost exclusively' ovoviviparous parthenogenetic *Artemia* population occurred) *Artemia* was thought to occupy ponds from which *Parartemia zietziana* is excluded, which avoids possible competition between both genera.

In later literature reports (Geddes and Williams, 1987) the number of *Artemia* and *Parartemia* sites kept growing, and a few sporadic records were made of *Artemia* in natural or semi-natural environments, including natural hypersaline lakes, both coastal and inland. As these new sites were small and isolated, it was unclear if *Artemia* had colonized these habitats by natural dispersal or by human intervention. Ephemeral salt lakes in SE and SW Australia were considered as unsuitable for *Artemia*, because they fill in winter, when water temperatures are low, and because they have low algal productivity (Geddes, 1980).

Although in the laboratory *P. zietziana* can tolerate salinities as low as 3 ppt, it is not found in a number of lakes with salinity lower than 70 ppt, in spite of the absence of predators, food competitors or food limitations. It was therefore difficult to identify any ecological factor other than salinity itself, which might explain the absence of *Parartemia* from these lakes. The absence of the genus from deeper lakes would be related to the fact that the species is confined to astatic environments with sufficient salinity fluctuations (Geddes, 1976), mostly shallow ephemeral saline water bodies, that dry out more or less annually, though occasionally longer periods may elapse between the dryings. While the genus is not found in large permanent salt lakes, it does occur in permanent artificial waters (*e.g.* in some of coastal solar ponds) where *Artemia* may also be found at higher salinities than *Parartemia*. Almost without exception, natural localities contain only one species of *Parartemia* on a continuous basis (Williams and Geddes, 1991). No simple correlations can be drawn between the major ionic composition of Australian habitats (alkaline saline lakes occur widely in southeastern and western states of Australia) and the species of *Parartemia* found in them. A very special habitat is represented by the 'acid' salt lakes of southeastern and especially southwestern Australia, inhabited by *P. contracta*, which shows substantial survival even at a pH below 3.5 (Conte and Geddes, 1988).

With certainty artificial introductions of *Artemia* have been made *e.g.* in the St. Kilda pond near Adelaide, where, according to Conte and Geddes (1988) 'thriving populations of both genera are living side by side'. The two genera are also known to co-occur in the Dry Creek Saltfields (see above). At lower salinities there is an exclusive presence of *Parartemia*, but both species co-occur in a small number of ponds of intermediate salinities. *Parartemia* generally is absent from higher salinity ponds, where *Artemia* alone occurs.

The total salinity range for *Artemia* in this habitat is 186–330 ppt, while the range for *Parartemia* is 112–258 ppt. Respiration experiments indicated that *Artemia* has a lower 'critical' oxygen concentration than the Australian species, apparently due to the presence of haemoglobin, which may result in an adaptive advantage at high salinity (and thus low dissolved oxygen concentrations). At oxygen concentrations below ±2 mg/l the metabolic rate of *Parartemia* falls and survival decreases. Beyond 300 ppt salinity the saturation level of O_2 is less than this critical value (Gilchrist, 1955; Walker et al. 1970). As the lower salinity limit of 112 ppt for *Parartemia* is well above the lower tolerance limit of this species, recorded before, the absence of *Parartemia* in the lower salinity waters may be related to fish predation, or to the insufficient astatic nature of these ponds (with salinity fluctuations limited to 5–15 ppt).

6.4. BIOCONSERVATION

From the viewpoint of biodiversity, Australia represents a special case with respect to the introduction of *Artemia*, as it is the only continent where a distinctly different halobiont Anostracan has evolved (Geddes and Williams, 1987).

The reason for the relatively limited distribution of *Artemia* in this continent may have several reasons (Geddes, 1980), *i.e.* as a result of the remoteness of the continent – which has been a barrier for other animal taxa as well – *Artemia* has arrived only recently, and probably by human action. Physiological reasons may hinder the dispersal of *Artemia*, as many of the lakes in southern Australia exist only in the winter months when temperatures may be too low to allow colonisation of the environment by *Artemia*. Finally, there may be competition with the well-established and probably well-adapted *Parartemia* species. Of course, other types of parthenogenetic and/or bisexual *Artemia* strains might react differently to the new environment, and introduction of these strains might have quite different effects on the *Parartemia* population. As long as these possible interactions are not fully clarified, several Australian scientists have argued in favour of the protection of the endemic *Parartemia* populations (Geddes and Williams, 1987).

7. 'New' *Artemia* Biotopes: Commercial Exploitation and Fundamental Research

The decline in *Artemia* cyst harvests from the Great Salt Lake since 1994–1995 has been a very strong incentive for the exploration of alternative resources. Throughout the initial period of *Artemia* exploration there have been regular warnings on international fora against the one-sided reliance of aquaculture on a single natural biotope (Sorgeloos, 1979). As a result, research efforts were made to prove the possibility of local production of *Artemia* in third world countries to cover (part of) the local market, with varying success.

Nevertheless, in spite of this diversification, world consumption of *Artemia* cysts still relied heavily on the Great Salt Lake (GSL).

In 1991 Bengtson *et al.* stressed once more that this was a risky situation, as a bad harvest year could have a major impact on provision and price of *Artemia* cysts. While GSL was still able to meet the increasing demands by intensifying and optimizing harvesting and processing procedures and raising the number of harvesting permits, a small part (10%) of the world's provision of cysts in the nineties was supplied by a few locations with a limited production or harvesting/processing capacity. These were mainly natural sites in North and Central China and South Siberia, and semi-natural or managed sites in the San Francisco Bay area, South Vietnam, Colombia and NE-Brazil (Lavens and Sorgeloos, 2000). A series of poor GSL harvests at the end of the nineties (see Chapter VI) resulted in an intensified exploration of a number of *Artemia* sites, mainly in Central Asia, *e.g.* Lake Urmia in Iran, Aibi Lake in PR China, Bolshoye Yarovoye and nearby lakes in South Siberia, several lakes in North Kazakhstan, Lake Karabogaz-Gol in Turkmenistan, and also salt lakes in Argentina (Lavens and Sorgeloos, 2000). In contrast to the GSL, however, ecological studies of the 'new' inland salt lakes are mostly nonexisting or not accessible to the international scientific community. Assessments of sustainable exploitation are thus impossible to make, which means that the risks of local shortages must be countered by diversification of the resources (Lavens and Sorgeloos, 2000).

Additionally, smaller quantities (1–20 tonnes each) of mostly good hatching quality cysts might in the future be provided on a more continuous basis from managed ponds and saltworks worldwide. Though limited in quantity, these cyst products can provide interesting opportunities for local commercial developments, where import restrictions exist. The number of recent publications about ecological studies, especially of man-managed ponds and saltworks, grows at a fast rate. While in the past these studies were few in number and restricted to certain areas (Mura, 1995b) like Australia, USA, South America, India, New Zealand, former USSR, Spain and France, a certain diversification in geographical background can be noted, though some of these areas continue to be proportionately over-represented in literature, while information about other regions has become or remains scarce. For instance, in spite of the abundance of salt lakes in the former Soviet Central Asian republics and in South Siberia (Altai area), extremely few publications have reached the international scientific audience. Solovov and Studenikina (1990) report extensively (in Russian) about the *Artemia* sites in South Siberia, including habitat description, population characteristics and estimations of standing crop, while Baitchorov and Nagorskaja (1999) describe the reproductive characteristics of 20 *Artemia* strains from Crimea, the Russian Altai area and Kazakhstan. Often recent articles refer to renewed or intensified ecological studies of populations, inventorized years ago. The majority of the most recent *Artemia* publications from different geographical origin focus on a number of well-defined geographical areas.

7.1. MEDITERRANEAN BASIN AND SOUTHERN EUROPE

Saltworks of southern Europe, the habitat of *A. salina* and several parthenogenetic strains, have been objects of field studies over several decades. Mura (1995b) provides information about the main life history traits of a bisexual strain in solar saltworks at Sant'Antioco, Sardinia, Italy, in relation to the environmental conditions. Vieira and Amat (1996) studied the fluctuations in the zooplankton community (including *Artemia* sp.) in two solar salt ponds in Aveiro, Portugal, while the changes in the trophic structure, due to human activities, in La Salada de Chiprana, Spain, the habitat of a parthenogenetic brine shrimp population, was studied by Díaz *et al.* (1998). Extensive work, focusing mainly on morphological and genetical characterization, has been performed with bisexual and parthenogenetic *Artemia* strains from the Eastern Mediterranean (Greece and Cyprus) by Abatzopoulos *et al.* (1986, 1989, 1993) and Triantaphyllidis *et al.* (1993, 1994c, 1997a, b). Sadkaoui *et al.* (2001) report on a new *Artemia* site in Morocco, although it is on the Atlantic coast. In the Black Sea basin, Pavlova *et al.* (1998) collected data on the biota and nutrients in solar ponds of saltworks near Burgas, Bulgaria.

7.2. PR CHINA

Special efforts have been devoted in recent years to the inventorization of Chinese salt lakes in general, and more specifically to *Artemia* sites. Wang (1987) reported that half of the total lake area of China consists of saline lakes (mostly in the inland provinces of Inner Mongolia, Xinjiang, Qinghai and Tibet), while Chang (1987) reported that 14 of China's 28 largest lakes (with surface area larger than 500 km^2) are saline. According to Zheng *et al.* (1993) there are more than 1,000 lakes with a salinity over 3.5 ppt in China, with a total area of more than 50,000 km^2, of which 534 have been the subject of investigation. An overview of Chinese and Mongolian salt lakes can be found in Williams (1991) and in Zheng *et al.* (1993), while Zheng (1997) focuses on the (paleo)geology and hydrochemistry of lakes from the Qinghai-Tibet Plateau. Triantaphyllidis *et al.* (1994a) present a detailed overview of literature on the occurrence and use of Chinese *Artemia*. The species endemic to continental China, *Artemia sinica*, has been described by Cai (1987, 1989) by cross-breeding a population from Yuncheng (= Xiechi) Lake, Shanxi province, with other *Artemia* species. Xin *et al.* (1994) have compiled literature data and their own findings into an updated list of Chinese *Artemia* sites, both inland lakes and coastal saltworks, within the framework of a project aiming to evaluate the potential of *Artemia* resources in the country, and to obtain, in the long run, self-sufficiency in terms of *Artemia* cysts. One of the key problems in identifying Chinese *Artemia* sites remains the high incidence of misunderstandings or mistakes when transcribing Chinese names into the Latin alphabet. In order to alleviate this problem, Xin *et al.* (1994) proposed a new, numerical, identification system for Chinese *Artemia* sites, taking into account the province and the inland or coastal location of the site.

The exploration of new *Artemia* biotopes in PR China resulted in the description of a new bisexual species, *Artemia tibetiana*, by Abatzopoulos *et al.* (1998), found in Lagkor Co, on the high plateaus of Tibet. One of the specific features of this species is the large size of its cysts (323–330 µm) and instar-I nauplii (667 µm), the biggest values ever recorded for any *Artemia* strain. This population was already described by Liu *et al.* (1998a, b), who analyzed the nutritional composition of adults, nauplii and decapsulated cysts. The advanced status of the study of Chinese *Artemia* strains and its interdisciplinary approach is illustrated by recent papers in the field of DNA fingerprinting: Sun *et al.* (1999a) applied the technique of Random Amplification of Polymorphic DNA (RAPD) to analyze the genetic relationship between different *Artemia* species, among which *A. sinica*, and one Chinese parthenogenetic strain. Further research into the genetic differentiation by means of Amplified Fragment Length Polymorphism (AFLP) of different Chinese *Artemia* species and strains showed the separate position of *A. tibetiana* from *A. sinica* (Sun *et al.* 1999b), and may suggest that the parthenogenetic populations from Chinese inland salt lakes followed a different evolutionary pathway than their coastal counterparts. Xin *et al.* (2000) used allozyme electrophoresis to assess the inter- and intra-populational genetic variation in Chinese bisexual populations (*A. sinica*), and found a significant correlation between the expected heterozygosity (to assess genetic diversity) and the surface of the habitat of the strains.

Since 1998, several *Artemia* sites have been discovered in Mongolia, though these are often small lakes or pools (unpublished data). For a limited number of sites, cyst material has become available; the characterization of these strains is still in its very initial stage.

7.3. SOUTH AMERICA

Increasing attention is equally being paid to the *Artemia* habitats of South America. Camargo (unpublished data) tries to link the genetic relationship between brine shrimp populations from northern South America (Venezuela, Colombia) and the Carribean area on one hand, and the migration routes of birds on the other. In Central America, the reproductive behaviour of two Mexican populations was studied by Correa and Tapia (1998).

However, especially the southern part of the continent (Argentina, Chile) has recently seen intensive *Artemia* research. As Gajardo *et al.* (1992, 1998) and Zuñiga *et al.* (1999) point out in their overview of the distribution of *Artemia* populations in Chile, in this area a great diversity of climates, hydrobiological and geological conditions exist on a relatively short distance, with *Artemia* sites located both at sea level and as high as 3,700–3,800 m. Vast saline plains can be found in this area with a salt composition consisting of sulphates, chlorides, nitrates, borates, carbonates and iodates, often with volcanic background. These authors describe in detail the variety of hypersaline biotopes, present in Chile with its 4,000 km coastline, like solar saltworks, salt lakes and lagoons, saline deposits and evaporitic basins ('salares'), different in geological and hydrochemical origin and characteristics, and in hydrological regime. These biotopes can be seasonal or permanent, some are exploited and others are not. Campos *et al.* (1996) performed a limnological study of a lagoon in the Torre del Paine area, Patagonia, Chile, the habitat of a local *Artemia* population. Unlike most brine shrimp sites, this area also contains semi-arid and humid *Artemia* biotopes; more details on this biotope and on the specific status of this population can be found in Gajardo *et al.* (1999). Zuñiga *et al.* (1999) discuss the reproductive mode of the Chilean *Artemia* populations in function of the stability of their environment, and identify them as *A. franciscana* by morphometric characterization and multivariate discriminant analysis, combined with cross-breeding. A cytogenetical analysis of three Argentinian populations was performed by Rodríguez Gil *et al.* (1998), who identified these strains as *A. persimilis*.

In Gajardo (1995) and Gajardo *et al.* (1995) the levels of genetic variability and the degree of interpopulational differentiation in South American *Artemia* strains are discussed. Since their introduction or dispersal rapid evolutionary changes seem to have occurred in natural South American populations of *A. franciscana*, as inferred from the accumulation of novel alleles, apparently as a response to environmental differences. Gajardo *et al.* (1998) discuss the relationship between the geographical situation of Chile and Argentina, and the genetic similarity between samples originating from this area, both *A. franciscana* and *A. persimilis*. Geographical isolation can be an effective way to constrain gene flow among natural populations of animals in general, but can be less relevant in the case of *Artemia* as cysts are naturally dispersed over long distances by waterbirds or wind, counteracting the effect of geographical barriers. Nevertheless, the Andes range, separating Chile and Argentina, can be considered as an effective barrier to gene exchange, particularly in the northern and central part of Chile where the highest altitudes are reached. Towards the south, in the area of Patagonia, the Andes tend to disappear, which would facilitate cyst exchange by bird migration (Gajardo *et al.* 1998).

8. Conservation of *Artemia* Biotopes and their Gene Pool

A first attempt in colonising natural saltwater bodies with *Artemia* was tried in the early seventies in hypersaline lagoons on Christmas Island in the Central Pacific. The trial was not successful because of low productivity and salinity of the site (Persoone and Sorgeloos, 1980). But an inoculation in 1977 in the Macau saltworks, Natal, northeastern Brazil, with nauplii hatched from 250 g of SFB cysts, was extremely successful, and brine shrimp soon spread out over 3,000 ha of saltworks (Persoone and Sorgeloos, 1980). As a result of the introduction of *A. franciscana* in solar saltworks for improved salt production and/or for harvesting cysts and biomass for use in the aquaculture industry, permanent populations of this species are nowadays found in Brazil, Australia, China, Egypt, Portugal, *etc.* Seasonal *A. franciscana* farming is practised in many tropical and subtropical countries such as the Philippines,

Thailand, Vietnam, Sri Lanka (Triantaphyllidis *et al.* 1998). However, due to the particular climatic conditions of these countries, the *Artemia* populations are not permanent and annual inoculations are required.

When *Artemia* was first spread by humans, the process of *Artemia* inoculation was considered as unambiguously positive: salt lakes were seen as having little biogeographical variability (Geddes and Williams, 1987). Now it is widely recognized that salt lakes are unique and well-balanced ecosystems, and that there are differences in their fauna between continents and regions, and even locally, according to their salinity fluctuations, water permanence, seasonality *etc.* Many salt lakes are of intermediate or small size, and this very specific environment, including the food chain with *Artemia* and waterfowl is very vulnerable for deterioration by human intervention. Threats to salt lakes are numerous, and mainly consist of desiccation by drainage or diversion of influents, or pollution. For coastal saltworks urbanization projects (industry, harbour infrastructure, tourism, expansion of residential areas), represent a supplementary threat (Sadoul *et al.* 1998), especially in industrialized or industrializing countries. If an original *Artemia* gene pool disappears, a basic element of genetic capital is destroyed as well (Persoone and Sorgeloos, 1980).

What are the possible effects of *Artemia* introduction on other organisms inhabiting natural saltwater bodies? While this practice frequently ensures social and economic benefits, particularly in developing countries, it also bears certain risks (Beardmore, 1987; Beardmore *et al.* 1997). An obvious effect is that competition with local (or near-by) strains or species of *Artemia* may occur (Geddes and Williams, 1987), which may lead to the extinction of some genotypes, or at worst, of one of the competitors. Competition experiments suggest that *A. franciscana* may outcompete others (see above; Browne, 1980; Browne and Halanych, 1989). The effect of one introduction will not remain local but may have consequences over large areas: many saltworks in northeastern Brazil are now populated with *Artemia* since the human intervention in Macau in 1977, followed by dispersion by wind and local waterbirds over an area of more than 1,000 km. Amat *et al.* (1995b) and Narciso (1989b) report about the presence of *A. franciscana* populations in the Iberian Peninsula, where this allochthonous strain, due to intentional or non-intentional inoculations (*e.g.* through hatchery effluents), in some occasions has outcompeted the local *Artemia* populations.

The resolution put forward at the 2nd International Symposium on *Artemia*, Antwerp, Belgium, in September 1985, therefore is still valid: '. . . the 2nd International Symposium resolves that all possible measures be taken to ensure that the genetic resources of natural *Artemia* populations are conserved; such measures include the establishment of gene banks (cysts), close monitoring of inoculation policies, and where possible the use of indigenous *Artemia* for inoculating *Artemia*-free waters' (Beardmore, 1987).

Another threat for *Artemia* populations worldwide are the massive harvesting practices, that have been going on for a few decades (*e.g.* in the Great Salt Lake) or that have been launched in recent years at several sites at an intensified rate in view of a menacing cyst shortage. It is largely speculative if these indiscriminate harvests actually endanger the habitat's standing crop; *Artemia* habitats differ enormously in size and in productivity. However, some degree of selection on the local gene pool (*e.g.* favouring ovoviviparous reproduction) may be imposed by systematic long-term harvesting of *Artemia* cysts.

Table 2. List of Artemia habitats.

Abbreviations used:
B: bisexual population; P: parthenogenetic population.
2, 3, 4, 5 n: indicates the ploidy level of a parthenogenetic population (2: diploid; 3: triploid, *etc.*).
A. fra: *A. franciscana*; A. mon: *A. monica*; A. per: *A. persimilis*; A. sal: *A. salina*; A. urm: *A. urmiana*; A. sin: *A. sinica*; A. tib: *A. tibetiana*
parth: parthenogenetic strain; ? : species status unknown

1. Africa

Country/State or province	Locality	Mode of reproduction	Species	References
Algeria	Chegga Oase		?	1
	Chott Djeloud		?	1
	Chott Ouargla		?	1
	Dayet Morselli		?	1
	Gharabas Lake		?	1
	Sebket Djendli		?	1
	Sebket Ez Zemouk		?	1
	Sebket Oran		?	1
	Tougourt		?	1
	Sebkha d'Arzeut		?	58
	Salin de Bethioua		?	58, 54
Cape Verde Islands	Santa Maria, Sal isl.		?	36
Egypt	Port Fouad	P	parth.	33
	Qarun Lake	P	parth.	33
	Solar Lake (Sinai)	P	parth.	1, 3
	Wadi Natron	B	A. sal	1, 3, 33
	El Max, Alexandria	P	parth.	33
	Borg el-Arab	P	parth.	33
Kenya	Elmenteita		?	1
Libya	Mandara	B	?	1
	Ramba-Az-Zallaf (Fezzan)		?	1
	Quem el Ma		?	1
	Trouna		?	1
	Gabr Aoun (Fezzan)		?	1
	Abu Kammash	B	A. sal	37
Madagascar	Salins de Diego Suarez		?	1
	Ankiembe saltworks	P (3n)	parth.	2
	Ifaty saltworks	B	A. fra	2

Table 2. Africa (Continued)

Country/ State or province	Locality	Mode of reproduction	Species	References
Morocco	Larache	P	parth.	1
	Moulaya estuary		?	1
	Oued Ammafatma		?	1
	Oued Chebeica		?	1
	Sebket Bon Areg		?	1
	Sebket Zima		?	1
	Sidi Moussa-Oualidia lagoon complex	B	?	38
Mozambique	Lagua Quissico	P	parth.	1, 13
Namibia	Vineta Swakopmund	P(2n, 4n)	parth.	1, 2
Niger	Teguidda In Tessoun		?	1
Senegal	Dakar		?	1
	Lake Kayar		?	1
	Lake Retba		?	1
South Africa	Coega Salt Flats	P	parth.	1, 3
	Swartkops	B	A. sal	1, 43
	Veldrif salterns	B	A. sal	34
	Haagestad pan, Bloemfontein	B	A. sal	58, 43
	Hopetown	B	A. sal	58, 43
Tunisia	Bekalta	B	A. sal	1
	Chott Ariana	B	A. sal	1
	Chott El Djerid		?	1
	Mégrine	B	A. sal	1
	Sebket Kowezia		?	1
	Sebket mta Moknine	B	A. sal	1
	Sebket Sidi el Hani		?	1
	Sfax	B	A. sal	1

2. Australia and New Zealand

Country/ State or province	Locality	Mode of reproduction	Species	References
New Zealand	Lake Grassmere	B	A. fra	1
	Blenheim salt lakes		?	41
Australia				
Queensland	Bowen		?	1
	Port Alma	B	A. fra	1
	Rockhampton	B	A. fra	1
South Australia	Dry Creek, Adelaide	P	parth.	1
West Australia	Dampier		?	1
	Lake Mc Leod		?	1
	Port Hedland	P	parth.	1
	Rottnest Island	P	parth.	1
	Shark Bay	P, B	parth., A. fra	1

Table 2 (Continued)

3. North America

Country/ State or province	Locality	Mode of reproduction	Species	References
Canada	Akerlund Lake	B	?	1
	Alsask Lake	B	?	1
	Aroma Lake	B	?	1
	Berry Lake	B	?	1
	Boat Lake	B	?	1
	Burn Lake	B	?	1
	Ceylon Lake	B	?	1
	Chain Lake	B	?	1
	Chaplin Lake	B	A. fra	1, 28
	Churchill	B	?	1
	Coral Lake	B	?	1
	Drybore Lake	B	?	1
	Enis Lake	B	?	1
	Frederick Lake	B	?	1
	Fusilier Lake	B	?	1
	Grandora Lake	B	?	1
	Gull Lake	B	?	1
	Hatton Lake	B	?	1
	Horizon Lake	B	?	1
	Ingebright Lake	B	?	1
	Landis Lake	B	?	1
	La Perouse	B	?	1
	Little Manitou Lake	B	A. fra	1, 28
	Lydden Lake	B	?	1
	Mawer Lake	B	?	1
	Meacham Lake	B	?	1
	Muskiki Lake	B	?	1
	Neola Lake	B	?	1
	Oban Lake	B	?	1
	Richmond Lake	B	?	1
	Shoe Lake	B	?	1
	Snakehole Lake	B	?	1
	Sybouts Lake-East	B	?	1
	Sybouts Lake-West	B	?	1
	Verlo West	B	?	1
	Vincent Lake	B	?	1
	Wheatstone Lake	B	?	1
	Whiteshore Lake	B	?	1
USA				
Arizona	Kiatuthlana Red Pond	B	A. fra	1, 28
	Kiatuthlana Green Pond	B	A. fra	1, 28
California	Carpinteria Slough	B	?	1
	Chula Vista	B	?	1
	Mono Lake	B	A. mon	1
	Moss Landing, Monterey Bay	B	A. fra	28

Table 2. North America (Continued)

Country/State or province	Locality	Mode of reproduction	Species	References
USA				
California	Owens Lake	B	?	1
	San Diego	B	?	1
	San Francisco Bay	B	A. fra	1
	San Pablo Bay	B	A. fra	1
	Vallejo West Pond	B	?	1
Hawaii	Christmas Islands	B	?	1
	Hanapepe	B	?	1
	Laysan Atoll	B	A. fra	1, 28
Nebraska	Alkali Lake	B	?	1
	Ashenburger Lake	B	?	1
	Antioch (Potash) Lake	B	A. fra	28
	Cook Lake	B	?	1
	East Valley Lake	B	?	1
	Grubny Lake	B	?	1
	Homestead Lake	B	?	1
	Jesse Lake	B	A. fra	1
	Johnson Lake	B	?	1
	Lilly Lake	B	?	1
	Reno Lake	B	?	1
	Richardson Lake	B	A. fra	1, 28
	Ryan Lake	B	?	1
	Sheridan County Lake	B	?	1
	Sturgeon Lake	B	A. fra	28
Nevada	Fallon Pond	B	A. fra	1, 28
North Dakota	Miller Lake	B	?	1
	Stink (Williams) Lake	B	?	1
New Mexico	Laguna del Perro	B	?	1
	Loving Salt Lake	B	?	1
	Quemado	B	A. fra	1
	Zuni Salt Lake	B	A. fra	1, 28
Oregon	Lake Abert	B	?	1
Texas	Cedar Lake	B	A. fra	28
	McKenzies Playa	B	?	1
	Mound Playa	B	?	1
	Playa Thahoka	B	?	1
	Raymondville	B	?	1
	Rich Playa	B	?	1
	Snow drop Playa	B	?	1
Utah	Great Salt Lake	B	A. fra	1
Washington	Cameron Lake	B	A. fra	28
	Deposit Thirteen	B	A. fra	28
	Penley Lake	B	A. fra	28
	Hot (Bitter) Lake	B	A. fra	1, 28
	Omak Plateau	B	?	1
	Soap Lake	B	?	1

Table 2 (Continued)

4. Central America and the Caribbean

Country/ State or province	Locality	Mode of reproduction	Species	References
Bahamas	Great Inagua	B	?	1
	Long Island	B	?	1
	San Salvador	B	?	1
Brit.Virgin Islands	Anegada	B	?	1
Caribbean Islands	Antigua	B	?	1
	St. Kitts	B	?	1
	St. Martin	B	?	1
	South Caicos	B	?	14
Costa Rica	Gulfo Nicova	B	?	1
	Bahia Salinas, Guanacaste	B	A. fra	4
Dominican Repuhlic	Isla Cabra	B	?	1
	Las Calderas	B	?	1
	Monte Cristi	B	?	1
	Puerto Alejandro	B	?	1
	Punta Salinas	B	?	1
Haiti	Grandes Salines	B	A. fra	1
Mexico				

*Baja California
    Norte*
	Ejido Nueva Odisea,
	  San Quintin
		B	A. fra	6

	San José	B	?	71

*Baja California
    Sur*
	Pichilingue
	  (San Juan Nepomuceno)
		B	A. fra	6

	Guerrero Negro	B	A. fra	6
	Isla del Carmen	B	A. fra	6
Campeche	Real de las Salinas		?	71
Coahuila	Cuatro Ciénegas de Carranza	B	?	6
Chiapas	Laguna del Mar Muerto	B	?	6
	La Joya	B	?	6
	Buenavista	B	?	6
	Los Palos	B	?	6
	Solo Dios	B	?	6
	Carretas	B	?	6
	Pereyra	B	?	6
	Chanchuto	B	?	6
	Panzacola	B	?	6
Estado de Mexico	Ecatepec de Morelos, Texcoco	B	A. fra	6
Oaxaca	Ponds W. Salina Cruz	B	?	6
	Las Coloradas		?	71
San Luis Potosí	Salinas de Hidalgo	B	?	6
Sinaloa	Bahia de Ceuta	B	A. fra	6
	Ahome, Bahia de Ohuira		?	71
Sonora	Salina, Tres Hermanos, Laguna de Yavaros	B	A. fra	6

Table 2. Central America and the Caribbean (Continued)

Country/ State or province	Locality	Mode of reproduction	Species	References
Tamaulipas	El Barranco, Altamira		?	71
Yucatán	San Crisanto	B	A. fra	6
	Celestun	B	?	7
	Chuburna	B	?	7
	Xtampu	B	?	7
	Las Coloradas	B	?	7
Netherlands Antilles	Aruba	B	?	1
	Bonaire Duinmeer	B	A. fra	1
	Gotomeer	B	?	1
	Pekelmeer	B	?	1
	Martinus	B	?	1
	Slagbaai	B	?	1
	Curaçao Fuik	B	?	1
	Rifwater	B	?	1
Nicaragua	Salinas Grandes, Leon	B	A. fra	5
Puerto Rico	Bahia Salinas	B	A. fra	1
	Bogueron	B	?	1
	Cabo Rojo	B	?	1
	La Parguera	B	?	1
	Ponce	B	?	1
	Tallaboa salterns	B	A. fra	1, 28

5. South America

Country/ State or province	Locality	Mode of reproduction	Species	References
Argentina	Salinas Chicas, Bahía Blanca, Buenos Aires	B	A. per	76
	Salinas Grandes de Hidalgo, La Pampa	B	A. per	1, 76
	(Laguna de) Mar Chiquita, Cordoba	B	?	1
	Salina Colorada Chica (Salitral El Chancho), La Pampa	B	A. per	44, 76
	Salinas Grandes de Anzoatequi, La Pampa		?	44, 50
	Piena de Salinas		?	50
	Laguna Callaqueo, La Pampa	B	A. per	58, 76
	Salitral Negro		?	58
	Valcheta, Rio Negro		?	50

Table 2. South America (Continued)

Country/ State or province	Locality	Mode of reproduction	Species	References
Bolivia	Laguna Canapa	B	?	1
	Lake Chulluncani	B	?	1
	Lake Hedonia	B	?	1
	Lake Poopo	B	?	1
	Papel Pampa, Oruro		?	53
Brazil	Acaraí, Ceará	B	A. fra	80
	Aracati, Ceará	B	A. fra	1, 80
	Areia Branca, Rio Grande do Norte	B	A. fra	80
	Cabo Frio, Rio de Janeiro	B	A. fra	1, 80
	Camocim, Ceará	B	A. fra	80
	Galinhos, Rio Grande do Norte	B	A. fra	80
	Guamaré, Rio Grande do Norte	B	A. fra	80
	Grossos, Rio Grande do Norte	B	A. fra	80
	Icapuí, Cear·	B	A. fra	1, 80
	Macau, Rio Grande do Norte	B	A. fra	1, 80
	Mundaú, Ceará	B	A. fra	1, 80
	São Bento do Norte, Rio Grande do Norte	B	A. fra	80
Chile	Salar de Surire (Región I)	B	A. fra	8, 82
	Playa Yape (Iquique, Región I)	B	A. fra	8, 82
	Salar de Pintados (Región I)	B	A. fra	8, 82
	Salar de Llamara (Región II)	B	A. fra	8, 82

	Salar de Atacama (laguna
	    Cejas y Tebenquiche,
	    Región II)
		B	A. fra	8, 9

	Puerto Viejo (Copiapo, Región III)	B	A. fra	8, 82
	La Pampilla (Coquimbo, Región IV)	B	A. fra	8, 82
	Palo Colorado (Los Vilos, Región IV)	B	A. fra	8, 82
	Yali (Región V)	B	A. fra	82, 83
	Salinas de Cahuil (Pichilemu, Región VI)	B	A. fra	8, 82
	Salinas de Constitución (Región VII)	B	A. fra	8, 82
	Laguna Amarga, Torres del Paine (Región XII)	B	A. per.	42, 67
	Laguna El Cisne (Región XII)	B	?	82

Table 2. South America (Continued)

Country/ State or province	Locality	Mode of reproduction	Species	References
Colombia	Bahia Chengue, Tayrona	B	A. fra	57
	Bahia Hondita	B	A. fra	57
	Galerazamba	B	A. fra	1, 57
	Kangaru	B	A. fra	57
	Manaure	B	A. fra	1, 57
	Pozos Colorados	B	A. fra	57
	Pusheo	B	A. fra	57
	Salina Cero	B	A. fra	57
	Warrego	B	A. fra	57
Ecuador	Galapagos (S. Salvador)	B	A. fra	1, 28
	Pacoa	B	A. sp	1
	Salinas	B	A. sp	1
Peru	Ancash	B	A. fra	76
	Caucato	B	?	1
	Chicama	B	?	1
	Chilca	B	?	1
	Chimus	B	A. fra	76
	Estuario de Virrila	B	?	1
	Guadalupe	B	?	1
	Pampa de Salinas	B	?	1
	Pampa Playa Chica	B	?	1
	Puerto Huarmey	B	?	1
	Tumbes	B	?	1
	Hierba Blanca		?	58
Venezuela	Boca Chica	B	?	1
	Coya Sal	B	?	1
	Coche	B	?	1
	Coro Coastline	B	?	1
	Cumaná		?	58
	La Orchila	B	?	1
	Las Aves	B	?	1
	Los Roques	B	?	1
	Maracaibo Lake		?	52
	Port Araya	B	?	1
	Tucacas	B	?	1

6. Asia

Country/ State or province	Locality	Mode of reproduction	Species	Reference
Abu Dhabi	Al Wathba Lake		?	60
PR China				
Liaoning	Jinzhou	P	parth.	10
	Yingkou	P (2, 4, 5n)	parth.	10, 15

Table 2. Asia (Continued)

Country/ State or province	Locality	Mode of reproduction	Species	References
PR China				
Liaoning	Dongjiagou	P (2n)	parth.	10, 11
	Pulandian	P (2n)	parth.	10, 11
	Lushun	P (2, 4, 5n)	parth.	10, 15
	Fuzhouwan	P	parth.	10
Hebei	Nanpu	P (2n)	parth.	10, 11
	Luannan	P	parth.	10
	Daqinghe	P	parth.	10
	Huanghua	P (2n)	parth.	10, 27
	Shangyi	B	A. sin	10, 15
	Zhangbei	B	A. sin	10, 15
	Kangbao	B	A. sin	10, 15
Tianjin	Hangu	P (2n)	parth.	10, 11
	Tanggu	P (2, 4, 5n)	parth.	10, 25
Shandong	Chengkou	P (2n)	parth.	10, 26
	Yangkou	P (2n)	parth.	10, 11
	Dongfeng	P (2, 5n)	parth.	10, 15
	Gaodao	P	parth.	10
	Xiaotan	P	parth.	10
	Nanwan	P	parth.	10
	Jimo	P	parth.	10
Jiangsu	Xuyu	P	parth.	10
	Xu Wei		?	77
	Lianyungang	P	parth.	10
Zhejiang	Zhanmao	P	parth.	10
	Shunmu	P	parth.	10
	Zhujiajian	P	parth.	10
Fujian	Shanyao	P	parth.	10
	Xigang	P	parth.	10
	Huian	P	parth.	10
	Tong An		?	58
Guangdong		P	parth.	10
Hainan	Dongfang	P	parth.	10
	Yinggehai	P (2, 4, 5n)	parth.	10, 15
Xinjiang	Aibi	P (2, 4n)	parth.	10, 11, 15, 26
	Dabancheng	P (2, 3, 4, 5n)	parth.	10, 26
	Balikun	P (2, 4n)	parth.	10, 11
	Aletai	B	?	10
	Alaerke	B	?	79
	Lu Yanchi	P	parth.	79
	Dingshan	P	parth.	79
	Bai	P	parth.	79
	Toulekule		?	79
	Qijiaojing	P	parth.	79
	Aya Kekumu		?	79
	Aqi Kekule	P	parth.	79
	Jing Yu	B	?	79

Table 2. Asia (Continued)

Country/ State or province	Locality	Mode of reproduction	Species	References
PR China				
Tibet	Lagkor Co	B	A. tib	78
	Bong Co		?	75
	Bozi Co		?	75
	Yanjing	B	?	10
	Shenzha	B	?	10
	Bange		?	10
	Gaize		?	10
	Geji		?	10
	Zhangchaka		?	10
	Wumacuo		?	10
	Jibuchaka		?	10
	Dongcuo		?	10
Qinghai	Gahai	P (2n)	parth.	10, 11
	Xiaocaidan	P	parth.	10
	Dacaidan	P	parth.	10
	Suban	P	parth.	10
	Keke	P (4n)	parth.	10, 15
	Chaka	P	parth.	10
	Tuosu	P	parth.	10
	Wu Lan		?	58
Gansu	Gaotai	B	?	10
Inner Mongolia	Haolebaoji = Haolebaoqing	B	A. sin	10, 11
	Haotongyin Chagan(nor)	B	A. sin	10, 11
	Taigemiao Chagan(nor)	B	A. sin	10, 11
	E(r)ji(nor)	B	A. sin	10, 11
	Beidachi	B	A. sin	10, 11
	Jilantai	B	A. sin	10, 11
	Wuqiangi	B	A. sin	10, 11
	S(h)anggendalai(nor)	B	A. sin	10, 11
	Dage(nor)	B	A. sin	10, 11
	Bayan(nor)	B	A. sin	10, 11
	Zhunsaihan(nor)	B	A. sin	10, 11
	Erendabus(h)en	B	A. sin	10, 11
	Chagan(nor)	B	A. sin	10, 11
	Huhetaolergai	B	A. sin	10, 11
	Hangjinqi	B	A. sin	10, 11
	Yanhai		?	75
	Dahahu		?	75
	Subei(nor)	B	?	58
Ningxia			?	10
Shaanxi	Dingbian		?	10
Shanxi	Yuncheng = Xiechi	B	?	10, 15, 74
India				
Rajasthan	Didwana		?	12
	Sambhar Lake		?	12
Gujarat	Gulf of Kutch	P	parth.	12

Table 2. Asia (Continued)

Country/ State or province	Locality	Mode of reproduction	Species	References
India				
Gujarat	Balamba salterns	P	parth.	12
	Mithapur	P	parth.	12
	Jamnagar		?	12
Maharashtra	Vadala (Bombay)		?	12
	Bhayander	P	parth.	12
	Bahinder		?	12
Madras	Kelambakkam		?	12
	Vedaranyam		?	12
Tamil Nadu	Veppalodai (Tuticorin)		?	12
	Pattanamaruthur		?	12
	Spic Nagar		?	12
	Thirespuram		?	12
	Karsewar Island		?	12
	Saltwater springs	P	parth.	12
Kanyakumari	Thamaraikulam	P	parth.	12
Iraq	Abu-Graib, Baghdad	P	parth.	1
	Basra		?	1
	Dayala		?	1
	Mahmoodia		?	1
Iran	Urmia Lake	B	A. urm	1
	Schor-Gol	P	parth.	1, 84
	Shurabil		?	1
	Athlit		?	1
	Incheh Lake	P	parth.	45, 84
	Maharlu Lake	P	parth.	46, 84
	Tashk Lake	P	parth.	46, 84
	Qom Lake	P	parth.	84
	Bakhtegan Lake	P	parth.	84
	Varmal lagoon	P	parth.	84
	Nough Lake	P	parth.	84
	lagoons around Urmia Lake	P	parth.	84
Israel	Eilat North	P	parth.	1
	Eilat South		?	1
Japan	Chang Dao		?	1
	Tamano		?	1
	Yamaguchi	P	parth.	1
Kazakhstan	Burlyu	P	parth.	32
	Cherbakty		?	72
	Karashuk	P	parth.	48
	Kolibek		?	73
	Mangyshlak peninsula		?	32
	Maraldi	P	parth.	32
	Seyten	P	parth.	32
	Teke	B	?	47
	Zhalauly		?	73

Table 2. Asia (Continued)

Country/ State or province	Locality	Mode of reproduction	Species	References
Kuwait			?	1
Mongolia	Shar Burd		?	68
	Davsan Suizh	B	?	68
	Tuhum Lake	B	?	68
Pakistan	Korangi Creek saltworks, Karachi	P	parth.	16
Saudi Arabia	Sabkhat al Fasl		?	61
	Dhahran		?	61
South Korea	Pusan		?	1
Sri Lanka	Bundala		?	1
	Hambantota		?	1
	Palavi		?	1
	Putallam	P	parth.	1
Syria	Salina Djeroud		?	61
	Palmyra		?	61
	Adana		?	61
Taiwan	Peinan Salina		?	1
	Beimen	B	?	10
Turkey	Aci Göl/Lake, Burdur		?	81
	Ayvalik, Balikesir		?	1, 17, 81
	Camalti, Izmir	P	parth.	1, 17, 81
	Bulak (= Bolluk) Gölü/Lake, Konya		?	58, 81
	Tuz Gölü/Lake, Central Anatolia	P	parth.	1, 59, 81
	Meke Salt Lake (Meke Tuz Konya, Karapinar Gölü)		?	17, 81
	Gökçeada, Imroz		?	18, 81
Turkmenistan	Karabogaz Gol	P	parth.	47
Uzbekistan	Karshi, Kashkadarya	P	parth.	37, 40
	Cape Aktumsyk, Aral Sea	B	?	37, 40
	Navruz, Syrdarya	P	parth.	37, 40

7. Europe

Country/ State or province	Locality	Mode of reproduction	Species	References
Bulgaria	Burgas	P	parth.	1
	Pomorye		?	1
Croatia	Secovlje, Portoroz	P (4n)	parth.	24
	Strunjan	P	parth.	24

Table 2. Europe (Continued)

Country/State or province	Locality	Mode of reproduction	Species	References
Cyprus	Akrotiri Lake		?	1
	Larnaka Lake	B	A. sal	1
France	Aigues Mortes	P	?	1
	Carnac-Trinité sur Mer		?	1
	Guérande-le Croisic	P	parth.	1
	La Palme		?	1
	Lavalduc	P	parth.	1
	Mesquer-Assérac		?	1
	Porte La Nouvelle		?	1
	Salin de Berre	P	parth.	1
	Salin de Fos		?	1
	Salin de Giraud	P	parth.	1
	Salins d'Hyères		?	1
	Salin des Pesquiers		?	1
	Sète	P	parth.	1
	Sète-Villeroy (Languedoc)	B	?	35
	Villeneuve (Languedoc)	B	?	35
Greece	Citros, Pieria	P (4n)	parth.	19, 20, 21
	Megalon Embolon, Thessaloniki	P (4n)	parth.	19, 20, 21
	Kalloni, Lesbos	P (4n)	parth.	22
	Polychnitos, Lesbos	P (4n)	parth.	22
	Messolonghi	P	parth.	1
	Milos	P	parth.	1
Italy	Torre Colimena, Taranto	P	parth.	65
	Cervia, Ravenna	P (4n)	parth.	1, 31
	Commachio, Ferrara	P (4n)	parth.	1, 31
	Margherita di Savoia, Foggia	P (2, 4n)	parth.	1, 31
	Tarquinia, Viterbo	B	A. sal	1, 31
	Quartu (= sal. di Quartu, sal. di Poetto, sal. di Spiaggia, San Bartolomeo) Cagliari, Sardinia	B	A. sal	1, 3
	Carloforte, Sardinia	B	A. sal	1, 3
	Sant' Antioco, Sardinia	B	A. sal	1, 3
	Santa Gilla, Sardinia	P (2n)	parth.	1, 31
	Simbirizzi, Sardinia		?	56
	Su Pallosu, Oristano, Sardinia	B	A. sal	55
	Mari Ermi, Oristano, Sardinia	B	A. sal	62
	Sale Porcus, Oristano, Sardinia	B	A. sal	62
	Notteri, Cagliari, Sardinia	B	A. sal	64
	Siracusa, Sicily		?	1
	Torre Nubia, Trapani, Sicily	B	A. sal	1, 31, 62, 63

Table 2. Europe (Continued)

Country/ State or province	Locality	Mode of reproduction	Species	References
Italy	Isola Longa, Marsala, Trapani, Sicily	B	A. sal	62, 63
	Salina Granda, Trapani, Sicily	B	A. sal	62, 63
Portugal	Castro Marim		?	58
	Alcochete	P	parth.	1
	Tejo estuary		?	1
	Sado estuary		?	1
	Ria de Aveiro		?	1
	Ria de Faro		?	1
Romania	Lake Techirghiol	P	parth.	23
	Lacul Sârat, Brâila	P	parth.	23
	Movila Miresii		?	23
	Baia Baciului	P	parth.	23
	Baia Neagrâ, SP	P	parth.	23
	Baia Verde I, SP	P	parth.	23
	Baia Verde II, SP	P	parth.	23
	Baia Verde III, SP	P	parth.	23
	Baia Rosie, SP	P	parth.	23
	Telega Bâi	P	parth.	23
	Telega II	P	parth.	23
	Telega III	P	parth.	23
	Ocna Sibiului	P?	?	23
	Sovata	P?	?	23
Russia	Belenkoye		?	69
	Bolshoye Shklo	P	parth.	32
	Bolshoye Yarovoye	P	parth.	32
	Buazonsor	P	parth.	32
	Bura Lake		?	69
	Karachi Lake	B	?	32
	Kuchukskoye	P	parth.	32
	Kulundinskoye	P	parth.	32
	Kurichye	P	parth.	32
	Maloye Yarovoye	P	parth.	32
	Medvezhoye (Bear Lake)		?	58
	Mirabilit	P	parth.	32
	Mormishanskoye 1	P	parth.	32
	Mormishanskoye 2	P	parth.	32
	Petuchovo	P	parth.	32
	Schekulduk	P	parth.	32
	Solyenoye	B	?	32
	Tanatar	B	?	32
	Tinaki Lake		?	58
	Volga delta, Astrakhan		?	70
Spain				
Alava	Añana	P (4n)	parth.	30

Table 2. Europe (Continued)

Country/ State or province	Locality	Mode of reproduction	Species	References
Spain				
Albacete	Petrola	P(4n)	parth.	30
	Pinilla	P (4n)	parth.	30
Alicante	Bonmati, S. Pola	B, P (2, 4n)	A. sal/parth.	29, 30
	Bras de Port, S. Pola	B	A. sal	29
	Calpe	P (2n)	parth.	29, 30
	La Mata	P (2n)	parth.	30
	Molina del Segura	B	A. sal	29
	Salinera Espanola, S. Pola	B	A. sal	29
	Villena	B	A. sal	29
Balearic Islands	Salinera Espanola, Formentera	B	A. sal	29
	Salinera Espanola, Ibiza	B	A. sal	29, 30
	Campos del Puerto, Mallorca	B	A. sal	29, 30
Burgos	Poza de la Sal	B	A. sp	29
Cadiz	Sanlucar de Barrameda	P	parth.	29
	Dos hermanas	B, P (2n)	A. sal/parth.	29, 30
	San Eugenio	B, P (2n)	A. sal/parth.	29, 30
	San Felix	B	A. sal	29, 30
	San Fernando	B	A. sal	29
	San Juan	B, P	A. sal/parth.	29
	San Pablo	B, P	A. sal/parth.	29
	Santa Leocadia	B, P	A. sal/parth.	29
	Barbanera	B	A. sal	29, 30
Canary islands	Janubio, Lanzarote	P (2n)	parth.	29
Cordoba	Encarnacion	P (4n)	parth.	29
	Puente Montilla	P (4n)	parth.	29
Guadalajara	Armalla	P (4n)	parth.	29
	Imon	P (4n)	parth.	29, 30
	Olmeda	P (4n)	parth.	29, 30
	Rienda	P (4n)	parth.	29
Huelva	Ayamonte	P (2n)	parth.	29
	Lepe	P (2n)	parth.	29
	Isla Cristina	P (2n)	parth.	29
	San Juan del Puerto	B	A. sal	29
Huesca	Rolda	P	parth.	29, 30
	Peralta de la Sal	P	parth.	29
Jaen	San Carlos	B	A. sal	29
	Don Benito	B	A. sal	29
Malaga	Fuente de Piedra	B, P (2n, 4n)	A. sal/parth.	29, 30
Murcia	San Pedro del Pinatar	B	A. sal	29, 30
	Jumilla	B	A. sal	29, 30
	sal. Punta Galera	B	A. sal	29
	sal. Catalana	B	A. sal	29
Soria	Medinaceli	P (4n)	parth.	29, 30
Tarragona	Delta del Ebro	P (4n)	parth.	29, 30
Teruel	Arcos de las Salinas	P (4n)	parth.	29

Table 2. Europe (Continued)

Country/ State or province	Locality	Mode of reproduction	Species	References
Zaragoza	Salada de Chiprana	P(4n)	parth.	29, 30, 39
	Bujaraloz	P (4n)	parth.	29, 30
Ukraine	Bolshoye Otar Moynakskyoe		?	32
	Dscharylgach		?	32
	Ghenicheskoye		?	32
	Kujalnicsky liman	P	parth.	32
	Popovskoye (= Ojburgskoye)	P	parth.	32
	Sakshoye		?	32
	Sasyk Lake		?	32
	Shtormovoye	B	?	32
	Tchokrakskoye	B	?	32
	Tobechikskoye	P	parth.	32
	Sivashskoye	B	?	49
Yugoslavia	Ulcinj, Montenegro	P	parth.	24

References:
1: Vanhaecke et al. (1987); 2: Triantaphyllidis et al. (1996); 3: Own observations; 4: Odio (1991a); 5: Odio (1991b); 6: Maeda-Martinez (1991); 7: Torrentera and Dodson (1995); 8: Gajardo (1995); 9: Gajardo and Beardmore (1993); 10: Xin et al. (1994); 11: Thomas (1995); 12: Peter Maryan, personal communication; 13: Sousa (1994); 14: Mclean (1994); 15: Triantaphyllidis et al. (1994a); 16: Shah and Qadri (1992); 17: Koray (1988); 18:Cihan Coru, personal communication; 19: Abatzopoulos et al. (1986); 20: Abatzopoulos et al. (1989); 21: Abatzopoulos et al. (1993); 22: Triantaphyllidis et al. (1993); 23: Liliana Pana, personal communication; 24: Petrovic (1991); 25: Triantaphyllidis et al. (1995); 26: Pilla (1992); 27: Triantaphyllidis et al. (1994b); 28: Bowen et al. (1988); 29: Amat Domenech (1980); 30: Amat et al. (1995a); 31: Baratelli et al. (1990); 32: Baitchorov and Nagorskaja (1999); 33: Peter Baert, personal communication; 34: Amat et al. (1995b); 35: Thiéry and Robert (1992); 36: Lanna Cheng, personal communication; 37: Mohammed El-Magsudi, personal communication; 38: Sadkaoui et al. (2001); 39: Díaz et al. (1998); 40: E. Kreuzberg-Mukhina, personal communication; 41: Clive Trotman, personal communication; 42: Gajardo et al. (1998); 43: Tom Hecht, personal communication; 44: Carlos Noel, personal communication; 45: Makhdoomi, personal communication; 46: Gholamreza Fayazi, personal communication; 47: Howard Newman, personal communication; 48: Interryba, personal communication; 49: Nina Khanaichenko, personal communication; 50: Julio Luis Siri, personal communication; 51: Italo Salgado, personal communication; 52: Wim Tackaert, personal communication; 53: Ricardo Sahonero Irahola, personal communication; 54: Abdelkrim Aziz, Farid Ziri, personal communication; 55: Mura (1993); 56: Mura (1986); 57: William Camargo, personal communication; 58: available in cyst bank ARC; 59: Basbug and Demirkalp (1997); 60: Aspinall and Hellyer (1999); 61: Peter Hogarth, personal communication; 62: Graziella Mura, personal communication; 63: Francesco Catania, personal communication; 64: Mura (1999); 65: Mura et al. (1999); 67: Campos et al. (1996); 68: Manchin Erdenabat, personal communication; 69: Galina Tsareva, personal communication; 70: Nick Aladin, personal communication; 71: Th. Castro Barrera, personal communication; 72: Pavel Grinchenko, personal communication; 73: Leonid Velichko, personal communication; 74: Sun Yi et al. (1999a, b); 75: Xin Naihong, personal communication; 76: Cohen et al. (1999); 77: George Triantaphyllidis, personal communication; 78: Abatzopoulos et al. (1998, 2002); 79: Ren et al. (1996); 80: Marcos Camara, personal communication;81: Yasemin Basbug Saygi, personal communication; 82: Gonzalo Gajardo, personal communication; 83: De Los Rios and Zuñiga (2000); 84: Agh et al. (2001)

9. Acknowledgements

Over the past three decades, our studies with the brine shrimp *Artemia* have been supported by research contracts from the Belgian National Science Foundation, the Belgian Administration for Development Cooperation, the Belgian Ministry of Science Policy, *Artemia* Systems NV and INVE Aquaculture NV, Belgium.

10. References

Abatzopoulos, T.J., Kastritsis, C.D. and Triantaphyllidis, C.D. (1986) A study of karyotypes and heterochromatic associations in *Artemia*, with special reference to two N. Greek populations. *Genetica* **71**, 3–10.

Abatzopoulos, T., Karamanlidis, G., Léger, P. and Sorgeloos, P. (1989) Further characterization of two *Artemia* populations from Northern Greece: biometry, hatching characteristics, caloric content and fatty acid profiles. *Hydrobiologia* **179**, 211–222.

Abatzopoulos, T., Triantaphyllidis, C. and Kastritsis, C. (1993) Genetic polymorphism in two parthenogenetic *Artemia* populations from Northern Greece. *Hydrobiologia* **250**, 73–80.

Abatzopoulos, T.J., Triantaphyllidis, G.V., Beardmore, J.A. and Sorgeloos, P. (1997) Cyst membrane protein composition as a discriminant character in the genus *Artemia*. (International Study on *Artemia* LV). *Journal of the Marine Biological Association of the United Kingdom* **77**, 265–268.

Abatzopoulos, T.J., Zhang, B. and Sorgeloos, P. (1998) *Artemia tibetiana*: preliminary characterization of a new *Artemia* species found in Tibet (People's Republic of China). International Study on *Artemia*. LIX. *International Journal of Salt Lake Research* **7**, 41–44.

Abatzopoulos, T.J., Brendonck, L. and Sorgeloos, P. (1999) First record of *Branchinella spinosa* (Milne-Edwards) (Crustacea: Branchiopoda: Anostraca) from Greece. *International Journal of Salt Lake Research* **8**, 351–360.

Abatzopoulos, T.J., Kappas, I., Bossier, P., Sorgeloos, P. and Beardmore, J.A. (2002) Genetic characterization of *Artemia tibetiana* (Crustacea: Anostraca). *Biological Journal of the Linnean Society* **75**, 333–344.

Abonyi, A. (1915) Experimentelle Daten zum Erkennen der *Artemia*-Gattung. *Zeitschrift für Wissenschaftliche Zoologie* **114**, 95–168.

Abreu-Grobois, F.A. and Beardmore, J.A. (1980) International Study on *Artemia*. II. Genetic characterization of *Artemia* populations – an electrophoretic approach. In G. Persoone, P. Sorgeloos, O. Roels and E. Jaspers (eds.), *The Brine Shrimp Artemia*, Vol. 1, Universa Press, Wetteren, Belgium, pp. 133–146.

Abreu-Grobois, F.A. and Beardmore, J.A. (1982) Genetic differentiation and speciation in the brine shrimp *Artemia*. In C. Barigozzi (ed.), *Mechanisms of Speciation*, Alan R. Liss Inc., New York, USA, pp. 345–376.

Achari, G.P.K. (1971) Occurrence of the brine shrimp *Artemia salina*, in Karsewar Island of Tuticorin, Gulf of Marmar. *Indian Journal of Fisheries* **18**, 196.

Agh, N., Sorgeloos, P., Abatzopoulos, T., Razavi Rouhani, S.M. and Lotfi, G.V. (2001) *Artemia* resources in Iran. In *Book of Abstracts of International Workshop on Artemia* (Abst. No. 11), Urmia University, Urmia, Iran.

Ahmadi, M.R.M., Leibovitz, H. and Simpson, K. (1990) Characterization of Uromiah Lake *Artemia* (*Artemia uromiana*) by electrofocusing of isozyme patterns. *Comparative Biochemistry and Physiology* **95**, 115–118.

Amat, F. (1979) Diferenciación y distribución de las poblaciones de *Artemia* (Crustaceo, Branquiopodo) de España, Ph.D. Thesis, Universidad de Barcelona, Spain.

Amat Domenech, F. (1980) Differentiation in *Artemia* strains from Spain. In G. Persoone, P.

Sorgeloos, O. Roels and E. Jaspers (eds.), *The Brine Shrimp Artemia*, Vol. 1, Universa Press, Wetteren, Belgium, pp. 19–39.

Amat, F. (1983) Zygogenetic and parthenogenetic *Artemia* in Cadiz sea-side salterns. *Marine Ecology Progress Series* **13**, 291–293.

Amat, F., Gozalbo, A., Navarro, J.C., Hontoria, F. and Varó, I. (1991) Some aspects of *Artemia* biology affected by cestode parasitism. *Hydrobiologia* **212**, 39–44.

Amat, F., Barata, C. and Hontoria, F. (1995a) A Mediterranean origin for the Veldrif (South Africa) *Artemia* Leach population. *Journal of Biogeography* **22**, 49–59.

Amat, F., Barata, C., Hontoria, F., Navarro, J.C. and Varó, I. (1995b) Biogeography of the genus *Artemia* (Crustacea, Branchiopoda, Anostraca) in Spain. *International Journal of Salt Lake Research* **3**, 175–190.

Artom, C. (1922) Nuovi datti sulla distribuzione geografica e sulla biologia delle due specie (microperenica e macroperenica) del genere *Artemia*. *Atti della Accademia Nazionale dei Lincei Rendiconti* **31**, 529–532.

Aspinall, S. and Hellyer, P. (1999) The history and development of Al Wathba Lake, Abu Dhabi. *Tribulus* **9**, 22–25.

Azari Takami. G. (1989) Two strains of *Artemia* in Urmia Lake (Iran). *Artemia Newsletter* **13**, 5.

Baitchorov, V.M. and Nagorskaja, L.L. (1999) The reproductive characteristics of *Artemia* in habitats of different salinity. *International Journal of Salt Lake Research* **4**, 287–291.

Baratelli, L., Varotto, V., Badaracco, G., Mura, G., Battaglia, B. and Barigozzi, C. (1990) Biological data on the brine shrimp *Artemia* living in the Italian saltworks. *Atti della Accademia Nazionale dei Lincei Rendiconti, Scienze Fisiche e Naturali, Serie 9*, **1**, 45–53.

Barigozzi, C. (1946) Über die geografische Verbreitung der Mutanten von *Artemia salina* Leach. *Arch. Julius Klaus-Stift.* **21**, 479–482.

Barigozzi, C. (1989) The problem of *Artemia urmiana*. *Artemia Newsletter* **14**, 14.

Barigozzi, C., Varotto, V., Baratelli, L. and Giarrizzo, R. (1987) The *Artemia* of Urmia Lake (Iran): mode of reproduction and chromosome numbers. *Atti della Accademia Nazionale dei Lincei Rendiconti-Classe di Scienze Fisiche, Matematiche e Naturali* **81**, 87–90.

Basbug, Y. and Demirkalp, F.Y. (1997) A note on the brine shrimp *Artemia* in Tuz Lake (Turkey). *Hydrobiologia* **353**, 45–51.

Bayly, I.A.E. (1972) Salinity tolerance and osmotic behavior of animals in athalassic saline and marine hypersaline waters. In R.F. Johnston (ed.), *Annual Review of Ecology and Systematics*, Vol. 3, Annual Reviews Inc., Palo Alto, CA, USA, pp. 233–268.

Bayly, I.A.E. (1993) The fauna of athalassic saline waters in Australia and the Altiplano of South America: comparisons and historical perspectives. *Hydrobiologia* **267**, 225–231.

Beardmore, J.A. (1987) Concluding remarks for Symposium Session I: Morphology, Ecotoxicology, Radiobiology, Genetics. In P. Sorgeloos, D.A. Bengtson, W. Decleir and E. Jaspers (eds.), *Artemia Research and its Applications*, Vol. 1, Universa Press, Wetteren, Belgium, pp. 345–346.

Beardmore, J.A. and Abreu-Grobois, F.A. (1983) Taxonomy and evolution in the brine shrimp *Artemia*. In G.S. Oxford and D. Rollinson (eds.), *Protein Polymorphism: Adaptive and Taxonomic Significance*, The Systematics Association Special Volume No. 24, Academic Press, London, New York, pp. 153–164.

Beardmore, J.A., Mair, G.C. and Lewis, R.I. (1997) Biodiversity in aquatic systems in relation to aquaculture. *Aquaculture Research* **28**, 829–839.

Belk, D. and Serpa, L. (1992) First record of *Branchinecta campestris* (Anostraca) from California and casual observations of males of *Artemia* clasping females of *Branchinecta*. *Journal of Crustacean Biology* **12**, 511–513.

Bengtson, D.A., Léger, P. and Sorgeloos, P. (1991) Use of *Artemia* as a food source for aquaculture. In R.A. Browne, P. Sorgeloos and C.N.A. Trotman (eds.), *Artemia Biology*, CRC Press, Boca Raton, Florida, pp. 255–285.

Bos, D.G., Cumming, B.F., Watters, C.E. and Smol, J.P. (1996) The relationship between zooplankton, conductivity and lake-water ionic composition in 111 lakes from the Interior Plateau of British Columbia, Canada. *International Journal of Salt Lake Research* **5**, 1–15.

Bowen, S.T. (1964) The genetics of *Artemia salina*. IV. Hybridization of wild populations with mutant stocks. *Biological Bulletin* **126**, 333–344.

Bowen, S.T., Durkin, J.P., Sterling, G. and Clark, L.S. (1978) *Artemia* hemoglobins: genetic variation in parthenogenetic and zygogenetic populations. *Biological Bulletin* **155**, 273–287.

Bowen, S.T., Fogarino, E.A., Hitchner, K.N., Dana, G.L., Chow, V.H.S., Buoncristiani, M.R. and Carl, J.R. (1985) Ecological isolation in *Artemia*: population differences in tolerance of anion concentrations. *Journal of Crustacean Biology* **5**, 106–129.

Bowen, S.T., Buoncristiani, M.R. and Carl, J.R. (1988) *Artemia* habitats: ion concentrations tolerated by one superspecies. *Hydrobiologia* **158**, 201–214.

Browne, R.A. (1980) Competition experiments between parthenogenetic and sexual strains of the brine shrimp, *Artemia salina*. *Ecology* **61**, 471–474.

Browne, R.A. (1982) The costs of reproduction in brine shrimp. *Ecology* **63**, 43–47.

Browne, R.A. and MacDonald, G.H. (1982) Biogeography of the brine shrimp, *Artemia*: distribution of parthenogenetic and sexual populations. *Journal of Biogeography* **9**, 331–338.

Browne, R.A. and Halanych, K.M. (1989) Competition between sexual and parthenogenetic *Artemia*: a re-evaluation (Branchiopoda, Anostraca). *Crustaceana* **57**, 57–71.

Browne, R.A. and Bowen, S.T. (1991) Taxonomy and population genetics of *Artemia*. In R.A. Browne, P. Sorgeloos and C.N.A. Trotman (eds.), *Artemia Biology*, CRC Press, Boca Raton, Florida, pp. 221–235.

Browne, R.A., Sallee, S.E., Grosch, D.S., Segreti, W.O. and Purser, S.M. (1984) Partitioning genetic and environmental components of reproduction and life-span in *Artemia*. *Ecology* **65**, 949–960.

Browne, R.A., Davis, L.E. and Sallee, S.E. (1988) Effects of temperature and relative fitness of sexual and asexual brine shrimp *Artemia*. *Journal of Experimental Marine Biology and Ecology* **124**, 1–20.

Cai, Y. (1987) Observations on parthenogenetic and bisexual brine shrimp from the People's Republic of China. In P. Sorgeloos, D.A. Bengtson, W. Decleir and E. Jaspers (eds.), *Artemia Reseach and its Applications*, Vol. 1, Universa Press, Wetteren, Belgium, pp. 227–232.

Cai, Y. (1989) A redescription of the brine shrimp (*Artemia sinica*). *The Wasmann Journal of Biology* **47**, 105–110.

Cai, Y. (1993) 'OP1' *Artemia* and its crossing experiments. *Journal of Ocean University of Qingdao* **23**, 52–58.

Campos, H., Soto, D., Parra, O., Steffen, W. and Aguero, G. (1996) Limnological studies of Amarga Lagoon, Chile: a saline lake in Patagonian South America. *International Journal of Salt Lake Research* **4**, 301–314.

Carpelan, L.H. (1957) Hydrobiology of Alviso salt ponds. *Ecology* **38**, 375–390.

Caspers, H. (1957) Black Sea and Sea of Azov. In J.W. Hedgpeth (ed.), *Marine Ecology and Paleoecology*, Vol. 1, Geological Society of America Memoirs 67, pp. 801–890.

Chang, W.Y.B. (1987) Large lakes of China. *Journal of Great Lakes Research* **13**, 235–249.

Clark, L.S. and Bowen, S.T. (1976) The genetics of *Artemia salina*. VII. Reproductive isolation. *The Journal of Heredity* **67**, 385–388.

Clegg, J.S., Hoa, N.V. and Sorgeloos, P. (2001) Thermal tolerance and heat shock proteins in encysted embryos of *Artemia* from widely different thermal habitats. *Hydrobiologia* **466**, 221–229.

Cohen, R.G., Amat, F., Hontoria, F. and Navarro, J.C. (1999) Preliminary characterization of some Argentinian *Artemia* populations from La Pampa and Buenos Aires provinces. *International Journal of Salt Lake Research* **8**, 329–340.

Cole, G.A. and Brown, R.J. (1967) Chemistry of *Artemia* habitats. *Ecology* **48**, 858–861.

Coleman, M., Geddes, M.C. and Trotman, C.N.A. (1998) Divergence of *Parartemia* and *Artemia* haemoglobin genes. *International Journal of Salt Lake Research* **7**, 171–180.

Conte, F.P. and Geddes, M.C. (1988) Acid brine shrimp: metabolic strategies in osmotic and ionic adaptation. *Hydrobiologia* **159**, 190–200.

Cooper, S.D., Winkler, D.W. and Lenz, P.H. (1984) The effect of grebe predation on a brine shrimp population. *Journal of Animal Ecology* **53**, 51–64.

Correa, F. and Tapia, O. (1998) Reproductive behavior of *Artemia franciscana* (Kellogg, 1906) from San Quintin, Baja California, Mexico. *Ciencias Marinas* **24**, 295–301.

Dana, G.L. (1981) Comparative population ecology of the brine shrimp *Artemia*, M.Sc. Thesis, San Francisco State University, CA, USA.

Dana, G.L. and Lenz, P.H. (1986) Effects of increasing salinity on an *Artemia* population from Mono Lake, California. *Oecologia* **68**, 428–436.

Davis, J.S. (1978) Biological communities of a nutrient enriched salina. *Aquatic Botany* **4**, 23–42.

De Los Rios, P. and Zuñiga, O. (2000) Biometric comparison of the frontal knob in American populations of *Artemia* (Anostraca, Artemiidae). *Revista Chilena de Historia Natural* **73**, 31–38.

Díaz, P., Guerrero, M.C., Alcorlo, P., Baltanas, A., Florin, M. and Montes, C. (1998) Anthropogenic perturbations to the trophic structure in a permanent hypersaline shallow lake: La Salada de Chiprana (north-eastern Spain). *International Journal of Salt Lake Research* **7**, 187–210.

Dorst, J. (1962) *The migrations of birds*, Windmill Press Ltd, Kingwood, UK.

Edward, D.H. and Watson, J.A.L. (1959) Freshwater and brackish water swamps of Rottnest Island. *J. Proc. R. Soc. Western Australia* **42**, 85–86.

Gabrion, C. and MacDonald, G. (1980) *Artemia* sp. (Crustacé, Anostracé), hôte intermédiaire d'*Eurycestus avoceti* Clark, 1954 (Cestode Cyclophyllide), parasite de l'avocette en Camargue. *Annales de Parasitologie* **55**, 327–331.

Gabrion, C., MacDonald-Crivelli, G. and Boy, V. (1982) Dynamique des populations larvaires du cestode *Flamingolepis liguloides* dans une population d'*Artemia* en Camargue. *Acta Oecologica/Oecologia Generalis* **3**, 273–293.

Gajardo, G. (1995) Genetical and phenotypical characterization of *Artemia* strains from South America: relevance for aquaculture. In J. Calderon and P. Sorgeloos (eds.), *Memorias II Congreso Ecuatoriano de Acuicultura*.

Gajardo, G. and Beardmore, J.A. (1993) Electrophoretic evidence suggests that the *Artemia* found in the Salar de Atacama, Chile, is *A. franciscana* Kellogg. *Hydrobiologia* **257**, 65–71.

Gajardo, G., Wilson, R. and Zuñiga, O. (1992) Report on the occurrence of *Artemia* in a saline deposit of the Chilean Andes (Branchiopoda, Anostraca). *Crustaceana* **63**, 169–174.

Gajardo, G., da Conceição, M., Weber, L. and Beardmore, J.A. (1995) Genetic variability and interpopulational differentiation of *Artemia* strains from South America. *Hydrobiologia* **302**, 21–29.

Gajardo, G., Colihueque, N., Parraguez, M. and Sorgeloos, P. (1998) International Study on *Artemia*. LVIII. Morphologic differentiation and reproductive isolation of *Artemia* populations from South America. *International Journal of Salt Lake Research* **7**, 133–151.

Gajardo, G., Mercado, C., Beardmore, J.A. and Sorgeloos, P. (1999) International Study on *Artemia*. LX. Allozyme data suggest that a new *Artemia* population in southern Chile (50°29'S; 73°45'W) is *A. persimilis*. *Hydrobiologia* **405**, 117–123.

Geddes, M.C. (1973) A new species of *Parartemia* (Anostraca) from Australia. *Crustaceana* **25**, 5–12.

Geddes, M.C. (1975a) Studies on an Australian brine shrimp, *Parartemia zietziana* Sayce (Crustacea: Anostraca) – I. Salinity tolerance. *Comparative Biochemistry and Physiology* **51**, 553–559.

Geddes, M.C. (1975b) Studies on an Australian brine shrimp, *Parartemia zietziana* Sayce (Crustacea: Anostraca) – II. Osmotic and ionic regulation. *Comparative Biochemistry and Physiology* **51**, 561–571.

Geddes, M.C. (1975c) Studies on an Australian brine shrimp, *Parartemia zietziana* Sayce (Crustacea: Anostraca) – III. The mechanisms of osmotic and ionic regulation. *Comparative Biochemistry and Physiology* **51**, 573–578.

Geddes, M.C. (1976) Seasonal fauna of some ephemeral saline waters in Western Victoria with particular reference to *Parartemia zietziana* Sayce (Crustacea: Anostraca). *Australian Journal of Marine and Freshwater Research* **27**, 1–22.

Geddes, M.C. (1979) Occurrence of the brine shrimp *Artemia* (Crustacea: Anostraca) in Australia. *Crustaceana* **36**, 225–228.

Geddes, M.C. (1980) The brine shrimp *Artemia* and *Parartemia* in Australia. In G. Persoone, P. Sorgeloos, O. Roels and E. Jaspers (eds.), *The Brine Shrimp Artemia*, Vol. 3, Universa Press, Wetteren, Belgium, pp. 57–65.

Geddes, M.C. (1981) The brine shrimp *Artemia* and *Parartemia*. Comparative physiology and distribution in Australia. *Hydrobiologia* **81**, 169–179.

Geddes, M.C. and Williams, W.D. (1987) Comments on *Artemia* introductions and the need for conservation. In P. Sorgeloos, D.A. Bengtson, W. Decleir and E. Jaspers (eds.), *Artemia Research and its Applications*, Vol. 3, Universa Press, Wetteren, Belgium, pp. 19–26.

Gilchrist, B.M. (1955) Haemoglobin in *Artemia*. *Proceedings of the Royal Society of London B* **143**, 136–146.

Günther, R.T. (1890) Crustacea. In R.T. Günther (ed.), Contributions to the natural history of Lake Urmi, NW Persia and its neighbourhood. *Journal of the Linnean Society (Zoology)* **27**, 394–398.

Hammer, U.T. (1986) *Saline Lake Ecosystems of the World*, Dr. W. Junk Publishers, Dordrecht, The Netherlands.

Hammer, U.T. (1993) Zooplankton distribution and abundance in saline lakes of Alberta and Saskatchewan, Canada. *International Journal of Salt Lake Research* **2**, 111–132.

Hammer, U.T. and Forró, L. (1992) Zooplankton distribution and abundance in saline lakes of British Columbia, Canada. *International Journal of Salt Lake Research* **1**, 65–80.

Hammer, U.T. and Hurlbert, S.H. (1992) Is the absence of *Artemia* determined by the presence of predators or by lower salinity in some saline waters? In R.D. Robarts and M.L. Bothwell (eds.), *Aquatic Ecosystems in Semi-arid Regions: Implications for Resource Management*, N.H.R.I. Symposium Series 7 (Environment), Saskatoon, Canada.

Hammer, U.T., Haynes, R.C., Heseltine, J.M. and Swanson, S.M. (1975) The saline lakes of Saskatchewan. *Internationale Vereinigung für Theoretische und Angewandte Limnologie Verhandlungen* **19**, 589–598.

Hedgpeth, J.W. (1959) Some preliminary considerations of the biology of inland mineral waters. *Archivio di Oceanografia e Limnologia* **11**, 111–141.

Hinton, H.E. (1954) XXVIII – Resistance of the dry eggs of *Artemia salina* (L.) to high temperatures. *The Annals and Magazine of Natural History* **7**, 158–160.

Hontoria, F. and Amat, F. (1992a) Morphological characterization of adult *Artemia* (Crustacea, Branchiopoda) from different geographical origins. Mediterranean populations. *Journal of Plankton Research* **14**, 949–959.

Hontoria, F. and Amat, F. (1992b) Morphological characterization of adult *Artemia* (Crustacea, Branchiopoda) from different geographical origins. American populations. *Journal of Plankton Research* **14**, 1461–1471.

Isenmann, P. (1975) Observations sur la mouette pygmée (*Larus minutus*) en Camargue de 1971 à 1974. *Terre Vie* **29**, 77–78.

Isenmann, P. (1976a) Contribution à l'étude de la biologie de la réproduction et de l'étho-écologie du goéland railleur, *Larus genei*. *Ardea* **64**, 48–61.

Isenmann, P. (1976b) Observations sur la guifette noire *Chlidonias niger* en Camargue. *Alauda* **44**, 319–327.

Jellison, R. and Melack, J.M. (1993) Algal photosynthetic activitiy and its response to meromixis in hypersaline Mono Lake, California. *Limnology and Oceanography* **38**, 818–837.

Koray, T. (1988) *Artemia* news from Turkey. *Artemia Newsletter* **9**, 8–9.

Kristensen, I. and Hulscher-Emeis, T.M. (1972) Factors influencing *Artemia* populations in Antillean Salinas. *Studies Fauna Curaçao* **39**, 87–111.

Lavens, P. and Sorgeloos, P. (1987) The cryptobiotic state of *Artemia* cysts, its diapause deactivation and hatching: a review. In P. Sorgeloos, D.A. Bengtson, W. Decleir and E. Jaspers (eds.), *Artemia Research and its Applications*, Vol. 3, Universa Press, Wetteren, Belgium, pp. 27–63.

Lavens, P. and Sorgeloos, P. (1996) Manual on the production and use of live food for aquaculture. *FAO Fisheries Technical Paper No. 361*.

Lavens, P. and Sorgeloos, P. (2000) The history, present status and prospects of the availability of *Artemia* cysts for aquaculture. *Aquaculture* **181**, 397–403.

Lefcort, H., Zhang, L. and King, C.E. (1991) Distributions of diploid and pentaploid brine shrimp *Artemia parthenogenetica* in an illuminated thermal gradient. *Canadian Journal of Zoology* **69**, 2461–2465.

Lenanton, R.C.J. (1977) Fishes from the hypersaline water of the stromatolite zone of Shark Bay, Western Australia. *Copeia* **2**, 387–390.

Lenz, P.H. (1980) Ecology of an alkali-adapted variety of *Artemia* from Mono Lake, California, USA. In G. Persoone, P. Sorgeloos, O. Roels and E. Jaspers (eds.), *The Brine Shrimp Artemia*, Vol. 3, Universa Press, Wetteren, Belgium, pp. 79–96.

Lenz, P.H. (1987) Ecological studies on *Artemia*: a review. In P. Sorgeloos, D.A Bengtson, W. Decleir and E. Jaspers (eds.), *Artemia Research and its Applications*, Vol. 3, Universa Press, Wetteren, Belgium, pp. 5–18.

Lenz, P.H. and Browne, R.A. (1991) Ecology of *Artemia*. In R.A. Browne, P. Sorgeloos and C.N.A. Trotman (eds.), *Artemia Biology*, CRC Press, Boca Raton, Florida, pp. 237–253.

Liu Junying, Zheng Mianping and Luo Jian (1998a) A study of *Artemia* in Lagkor Co, Tibet. I. Biological feature. *Journal of Lake Sciences* (Abst.) **10**(2).

Liu Junying, Zheng Mianping and Luo Jian (1998b) A study of *Artemia* in Lagkor Co, Tibet. II. Nutrient. *Journal of Lake Sciences* (Abst.) **10**(3).

Löffler, H. (1964) Vogelzug und Crustaceenverbreitung. *Zoologischer Anzeiger (suppl.)* **27**, 311–316.

MacDonald, G. (1980) The use of *Artemia* cysts as food by the flamingo (*Phoenicopterus ruber roseus*) and the shellduck (*Tadorna tadorna*). In G. Persoone, P. Sorgeloos, O. Roels and E. Jaspers (eds.), *The Brine Shrimp Artemia*, Vol. 3, Universa Press, Wetteren, Belgium, pp. 97–104.

Maeda-Martinez, A.M. (1991) Distribution of species of Anostraca, Notostraca, Spinicaudata, and Laevicaudata in Mexico. *Hydrobiologia* **212**, 209–219.

Manwell, C. (1978) Haemoglobin in the Australian anostracan *Parartemia zietziana*: evolutionary strategies of conformity vs regulation. *Comparative Biochemistry and Physiology A* **59**, 37–44.

Marchant, R. and Williams, W.D. (1977) Field measurements of ingestion and assimilation for the Australian brine shrimp *Parartemia zietziana* Sayce (Crustacea: Anostraca). *Australian Journal of Ecology* **2**, 379–390.

McCarraher, D.B. (1970) Some ecological relations of fairy shrimp in alkaline habitats of Nebraska. *American Midland Naturalist* **84**, 59–68.

McCarraher, D.B. (1972) A preliminary bibliography and lake index of the inland mineral waters of the world. *FAO Fish. Circ. No 146*.

Mclean, W.K. (1994) *Artemia* in South Caicos, Caribbean. *Larviculture and Artemia Newsletter* **33**, 28.

Melack, J.M., Lenz, P.H. and Cooper, S.D. (1983) The ecology of Mono Lake. *National Geographic Society Research Reports*.

Mura, G. (1986) SEM morphological survey on the egg shell in the Italian Anostracans (Crustacea, Branchiopoda). *Hydrobiologia* **134**, 273–286.

Mura, G. (1987) Occurrence of *Artemia* in solar saltworks and coastal brine ponds in Sardinia, Italy. *Journal of Crustacean Biology* **7**, 697–703.

Mura, G. (1993) Seasonal distribution of *Artemia salina* and *Brachinella spinosa* in a saline astatic pond in South West Sardinia, Italy (Anostraca). *Crustaceana* **64**, 172–191.

Mura, G. (1995a) Cestode parasitism (*Flamingolepis liguloides* Gervais 1847 Spassky & Spasskaja 1954) in an *Artemia* population from southwestern Sardinia. *International Journal of Salt Lake Research* **3**, 191–200.

Mura, G. (1995b) An ecological study of a bisexual *Artemia* population from Sant'Antioco solar saltworks (south-western Sardinia, Italy). *International Journal of Salt Lake Research* **3**, 201–219.

Mura, G. (1999) Current status of the Anostraca in Italy. *Hydrobiologia* **405**, 57–65.
Mura, G., Filauri, A. and Palmegiano, G.B. (1987) A survey of *Artemia* and *Branchinella* populations in coastal lagoons and salt pans of Sardinia (Italy). In P. Sorgeloos, D.A. Bengtson, W. Decleir and E. Jaspers (eds.), *Artemia Research and its Applications*, Vol. 3, Universa Press, Wetteren, Belgium, p. 151.
Mura, G., Ferreri, D. and Belmonte, G. (1999) Prima segnalazione di *Branchinella spinosa* Milne Edwards 1840 (Crustacea, Branchiopoda, Anostraca) per l'Italia peninsulare. *Thalassia Salentina* **23**, 9–6.
Narciso, L. (1989a) The importance of salt ponds for the preservation of marine birds in Portugal. *Cyanopica* **4**, 368–372.
Narciso, L. (1989b) The brine shrimp *Artemia* sp.: an example of the danger of introduced species in aquaculture. *European Aquaculture Society Special Publication* **10**, 183–184.
Odio, R. (1991a) *Artemia* in Costa Rica. *Larviculture and Artemia Newsletter* **19**, 39.
Odio, R. (1991b) *Artemia* in Nicaragua. *Larviculture and Artemia Newsletter* **19**, 39.
Overton, S.V. and Bland, C.E. (1981) Infection of *Artemia salina* by *Haliphthoros milfordensis*: a scanning and transmission electron microscope study. *Journal of Invertebrate Pathology* **37**, 249–257.
Pador, E. (1995) Characterisation of the *Artemia urmiana* Gunther 1900 from Lake Urmia, Iran, M.Sc. Thesis, Vrije Universiteit Brussel and Laboratory of Aquaculture-*Artemia* Reference Center, University of Ghent, Belgium.
Pavlova, P., Markova, K., Tanev, S. and Davis, J.S. (1998) Observations on a solar saltworks near Burgas, Bulgaria. *International Journal of Salt Lake Research* **7**, 357–368.
Perez Rodriguez, J.M. (1987) Cyst production of *Artemia* in salt ponds in southeastern Spain. In P. Sorgeloos, D.A. Bengtson, W. Decleir and E. Jaspers (eds.), *Artemia Research and its Applications*, Vol. 3, Universa Press, Wetteren, Belgium, p. 215.
Persoone, G. and Sorgeloos, P. (1980) General aspects of the ecology and biogeography of *Artemia*. In G. Persoone, P. Sorgeloos, O. Roels and E. Jaspers (eds.), *The Brine Shrimp Artemia*, Vol. 3, Universa Press, Wetteren, Belgium, pp. 3–24.
Petrovic, A. (1991) The karyotype of the parthenogenetic *Artemia* (Crustacea) from Secovlje, Yugoslavia. *Genetica* **83**, 289–291.
Pilla, E.J.S. (1992) Genetic differentiation and speciation in Old World *Artemia*, Ph.D. Thesis, University College of Swansea, Wales, UK.
Por, F.D. (1980) A classification of hypersaline waters based on trophic criteria. *Marine Ecology Progress Series* **1**, 121–131.
Post, F.J. and Youssef, N.N. (1977) A prokaryotic intracellular symbiont of the Great Salt Lake brine shrimp *Artemia salina* (L.). *Canadian Journal of Microbiology* **23**, 1232–1236.
Proctor, V.W. (1964) Viability of crustacean eggs recovered from ducks. *Ecology* **45**, 656–658.
Proctor, V.W. and Malone, C.R. (1965) Further evidence of the passive dispersal of small aquatic organisms via the intestinal tract of birds. *Ecology* **46**, 728–729.
Proctor, V.W., Malone, C.R. and Devlaming, V.L. (1967) Dispersal of aquatic organisms: viability of disseminules recovered from the intestinal tract of captive killdeer. *Ecology* **46**, 672–676.
Rahaman, A.A., Somamma, E. and Ambikadevi, M. (1993) Hydrobiology of some solar salt works in India. *International Journal of Salt Lake Research* **2**, 1–15.
Rawson, D.S. and Moore, J.E. (1944) The saline lakes of Saskatchewan. *Canadian Journal of Research* **22**, 141–201.
Ren, M., Guo,Y., Wang, J., Su, R., Li, H. and Ren, B. (1996) *Survey of Artemia Ecology and Resources in Inland Salt Lakes in Northwest China*, Heilongjiang Science and Technology Press, Harbin, PR China.
Robert, F. and Gabrion, C. (1991) Cestodoses de l'avifaune camarguaise. Rôle d'*Artemia* (Crustacea, Anostraca) et stratégies de rencontre hôte-parasite. *Annales De Parasitologie Humaine et Comparée* **66**, 226–235.
Rodríguez Gil, S., Papeschi, A.G. and Cohen, R.G. (1998) Mitotic and meiotic chromosomes of *Artemia* (Branchiopoda) from populations of La Pampa Province, Argentina. *Journal of Crustacean Biology* **18**, 36–41.

Rooth, J. (1976) Ecological aspects on the flamingos in Bonaire. In E.A.V. Seasan, W. Booi, I. Kristensen and H.A.M. Dekruijf (eds.), *Ecology Conference on Flamingos, Oil Pollution and Reefs*, Sept. 75. Netherlands Antilles Nat. Parks Found. 'STINAPA' No. 11, Curaçao.

Royan, J.P., Navaneethakrishnan, P. and Selvaraj, A.M. (1970) Occurrence of *Artemia salina* in southern India. *Current Science* **39**, 14.

Sadkaoui, F., Moncef, M., Sif, J. and Van Stappen, G. (2001) *Artemia salina* in El Jadida saltworks, Morocco. *International Journal of Salt Lake Research* (in press).

Sadoul, N., Walmsley, J. and Charpentier, B. (1998) Les salins, entre terre et mer. In J. Skinner and A.J. Crivelli (eds.), *Conservation des Zones Humides Méditerranéennes*, no. 9, Tour du Valat, Arles, France.

Savage, A. and Knott, B. (1998) *Artemia parthenogenetica* in Lake Hayward, Western Australia. II. Feeding biology in a shallow, seasonally stratified, hypersaline lake. *International Journal of Salt Lake Research* **7**, 13–24.

Shah, A. and Qadri, R.B. (1992) Natural *Artemia* in Pakistan. *Larviculture and Artemia Newsletter* **23**, 26–27.

Skoultchi, A.I. and Morowitz, H.J. (1964) Information storage and survival of biological systems at temperatures near absolute zero. *Yale Journal of Biology and Medicine* **37**, 158–163.

Solovov, V.P. and Studenikina T.L. (1990) Rachok Artemiya v ozerah zapadnoy Sibiri: morfologiya, ekologiya, perspektivy hozyaystvennovo ispol'zovaniya (The brine shrimp *Artemia* in lakes of southern Siberia: morphology, ecology, perspectives for economic exploitation – in Russian) Novosibirsk 'Nauka', Siberian Department.

Sorgeloos, P. (1979) The brine shrimp, *Artemia salina*: a bottleneck in mariculture. In T.V.R. Pillay and W.A. Dill (eds.), *FAO Technical Conference on Aquaculture*, Kyoto 1976, Fishing News Books Ltd, Farnham, England, UK, pp. 321–324.

Sorgeloos, P. (1980) Life history of the brine shrimp *Artemia*. In G. Persoone, P. Sorgeloos, O. Roels and E. Jaspers (eds.), *The Brine Shrimp Artemia*, Vol. 1–3, Universa Press, Wetteren, Belgium, pp. XIX–XXIII.

Sousa, M.I. (1994) *Artemia* in Mozambique. *Larviculture and Artemia Newsletter* **33**, 27.

Stefani, R. (1960) L'*Artemia salina parthenogenetica* a Cagliari. *Rivista di Biologia* **53**, 463–490.

Stella, E. (1933) Phenotypical characteristics and geographical distribution of several biotypes of *Artemia salina* L. *Zeitschrift für Induktive Abstammungs und Vererbungslehre* **65**, 412–446.

Sun Yi, Yi-Cheng Zhong, Wen-Qin Song, Run-Sheng Zhang and Rui-Yang Chen (1999a) Detection of genetic relationships among four *Artemia* species using randomly amplified polymorphic DNA (RAPD). *International Journal of Salt Lake Research* **8**, 139–147.

Sun Yi, Wen-Qin Song, Yi-Cheng Zhong, Run-Sheng Zhang, Abatzopoulos T.J. and Rui-Yang Chen (1999b) Diversity and genetic differentiation in *Artemia* species and populations detected by AFLP markers. *International Journal of Salt Lake Research* **8**, 341–350.

Thiéry, A., Robert, F. and Gabrion, C. (1990) Distribution des populations d'*Artemia* et de leur parasite *Flamingolepis liguloides* (Cestode, Cyclophyllidea), dans les salins du littoral méditerranéen français. *Canadian Journal of Zoology* **68**, 2199–2204.

Thiéry, A. and Robert, F. (1992) Bisexual populations of the brine shrimp *Artemia* in Sète-Villeroy and Villeneuve saltworks (Languedoc, France). *International Journal of Salt Lake Research* **1**, 47–63.

Thomas, K.M. (1995) Genetic variation and differentiation in Asian populations of *Artemia*, Ph.D. Thesis, University College of Swansea, Wales, UK.

Thornthwaite, C.W. (1948) An approach toward a rational classification of climate. *Geographical Review* **38**, 55–94.

Tobias, W.J., Sorgeloos, P., Roels, O.A. and Sharfstein, B.A. (1980) International Study on *Artemia*. XIII. A comparison of production data of 17 geographical strains of *Artemia* in the St. Croix Artificial Upwelling-Mariculture System. In G. Persoone, P. Sorgeloos, O. Roels and E. Jaspers (eds.), *The Brine Shrimp Artemia*, Vol. 3, Universa Press, Wetteren, Belgium, pp. 383–392.

Torrentera, L. and Dodson, S.I. (1995) Morphological diversity of populations of *Artemia* (Branchiopoda) in Yucatan. *Journal of Crustacean Biology* **15**, 86–102.

Triantaphyllidis, G.V., Abatzopoulos, T.J., Sandaltzopoulos, R.M., Stamou, G. and Kastritsis, C.D. (1993) Characterization of two new *Artemia* populations from two solar saltworks of Lesbos island (Greece): biometry, hatching characteristcs and fatty acid profile. *International Journal of Salt Lake Research* **2**, 59–68.

Triantaphyllidis, G.V., Bo Zhang, Lixiang Zhu and Sorgeloos, P. (1994a) International Study on *Artemia*. L. Review of the literature on *Artemia* from salt lakes in the People's Republic of China. *International Journal of Salt Lake Research* **3**, 93–104.

Triantaphyllidis, G.V., Pilla, E.J.S., Thomas, K.M., Abatzopoulos, T.J., Beardmore, J.A. and Sorgeloos, P. (1994b) International Study on *Artemia*. LII. Incubation of *Artemia* cyst samples at high temperatures reveals mixed nature with *Artemia franciscana* cysts. *Journal of Experimental Marine Biology and Ecology* **183**, 273–282.

Triantaphyllidis, G.V., Katinakis, P.K. and Abatzopoulos, T.J. (1994c) Changes in abundant proteins: intrapopulation and interpopulation study of four parthenogenetic *Artemia* populations from Northern Greece. *Cytobios* **77**, 137–146.

Triantaphyllidis, G.V., Poulopoulou, K., Abatzopoulos, T.J., Pérez, C.A.P. and Sorgeloos, P. (1995) International Study on *Artemia* XLIX. Salinity effects on survival, maturity, growth, biometrics, reproductive and lifespan characteristics of a bisexual and a parthenogenetic population of *Artemia*. *Hydrobiologia* **302**, 215–227.

Triantaphyllidis, G.V., Abatzopoulos, T.J., Miasa, E. and Sorgeloos, P. (1996) International Study on *Artemia*. LVI. Characterization of two *Artemia* populations from Namibia and Madagascar: cytogenetics, biometry, hatching characteristics and fatty acid profiles. *Hydrobiologia* **335**, 97–106.

Triantaphyllidis, G.V., Criel, G.R.J., Abatzopoulos, T.J. and Sorgeloos, P. (1997a) International Study on *Artemia*. LIV. Morphological study of *Artemia* with emphasis to Old World strains. II. Parthenogenetic populations. *Hydrobiologia* **357**, 155–163.

Triantaphyllidis, G.V., Criel, G.R.J., Abatzopoulos, T.J., Thomas, K.M., Peleman, J., Beardmore, J.A. and Sorgeloos, P. (1997b) International Study on *Artemia*. LVII. Morphological and molecular characters suggest conspecificity of all bisexual European and North African *Artemia* populations. *Marine Biology* **129**, 477–487.

Triantaphyllidis, G.V., Abatzopoulos, T.J. and Sorgeloos, P. (1998) Review of the biogeography of the genus *Artemia* (Crustacea, Anostraca). *Journal of Biogeography* **25**, 213–226.

Tyson, G.E. (1970) The occurrence of a spirochete-like organism in tissues of the brine shrimp *Artemia salina*. *Journal of Invertebrate Pathology* **15**, 145–147.

Tyson, G.E. (1974) Ultrastructure of a spirochete found in tissues of the brine shrimp, *Artemia salina*. *Archives of Microbiology* **99**, 281–294.

Tyson, G.E. (1975) Phagocytosis and digestion of spirochetes by amebocytes of infected brine shrimp. *Journal of Invertebrate Pathology* **26**, 105–111.

Van der Meer Mohr, C.G. (1977) Field trip to the salinas of Bonaire. Guide to geological excursions in Curaçao, Bonaire and Aruba. Stimapa, Doc. Ser. 2 Neth. Antilles Nat. Parks Found., pp. 76–87.

Vanhaecke, P. and Sorgeloos, P. (1980) International Study on *Artemia*. IV. The biometrics of *Artemia* strains from different geographical origin. In G. Persoone, P. Sorgeloos, O. Roels and E. Jaspers (eds.), *The Brine Shrimp Artemia*, Vol. 3, Universa Press, Wetteren, Belgium, pp. 393–405.

Vanhaecke, P. and Sorgeloos, P. (1989) International Study on *Artemia*. XLVII. The effect of temperature on cyst hatching, larval survival and biomass production for different geographical strains of brine shrimp *Artemia* spp. *Annales de la Société Royale Zoologique de Belgique* **119**, 7–23.

Vanhaecke, P., Tackaert, W. and Sorgeloos, P. (1987) The biogeography of *Artemia*: an updated review. In P. Sorgeloos, D.A. Bengtson, W. Decleir and E. Jaspers (eds.), *Artemia Research and its Applications*, Vol. 1, Universa Press, Wetteren, Belgium, pp. 129–155.

Van Stappen, G., Fayazi, G. and Sorgeloos, P. (2001) International Study on *Artemia*. LXIII. Field study of the *Artemia urmiana* (Günther, 1890) population in Lake Urmiah, Iran. *Hydrobiologia* **466**, 133–143.

Vieira, N. and Amat, F. (1996) Fluctuation in the zooplankton community in two solar salt ponds, Aveiro, Portugal. *International Journal of Salt Lake Research* **4**, 327–333.

Von Hentig, R. (1971) Einfluss von Salzgehalt und Temperatur auf Entwicklung, Wachstum, Fortpflanzung und Energiebilanz von *Artemia salina*. *Marine Biology* **9**, 145–182.

Vos, J. and Tunsutapanit, A. (1979) Detailed report on *Artemia* cysts inoculation in Bangpakong, Chachoengsao Province, FAO/UNDP Field Document THA/75/008.

Vu Do Quynh and Nguyen Ngoc Lam (1987) Inoculation of *Artemia* in experimental ponds in central Vietnam: an ecological approach and a comparison of three geographical strains. In P. Sorgeloos, D.A. Bengtson, W. Decleir and E. Jaspers (eds.), *Artemia Research and its Applications*, Vol. 3, Universa Press, Wetteren, Belgium, pp. 253–269.

Walker, K.F., Williams, W.D. and Hammer, U.T. (1970) The Miller method for oxygen determination applied to saline lakes. *Limnology and Oceanography* **15**, 814–815.

Wang, H. (1987) The water resources of lakes in China. *Chinese Journal of Oceanology and Limnology* **5**, 263–280.

Wear, R.G. and Haslett, S.J. (1987) Studies on the biology and ecology of *Artemia* from Lake Grassmere, New Zealand. In P. Sorgeloos, D.A. Bengtson, W. Decleir and E. Jaspers (eds.), *Artemia Research and its Applications*, Vol. 3, Universa Press, Wetteren, Belgium, pp. 101–126.

Wear, R.G., Haslett, S.J. and Alexander, N.L. (1986) Effects of temperature and salinity on the biology of *Artemia franciscana* Kellogg from Lake Grassmere, New Zealand. 2. Maturation, fecundity, and generation times. *Journal of Experimental Marine Biology and Ecology* **98**, 167–183.

Williams, W.D. (1981) The limnology of saline lakes in Western Victoria. *Hydrobiologia* **81/82**, 233–259.

Williams, W.D. (1991) Chinese and Mongolian saline lakes: a limnological overview. *Hydrobiologia* **210**, 39–66.

Williams, W.D. and Geddes, M.C. (1991) Anostracans of Australian salt lakes, with particular reference to a comparison of *Parartemia* and *Artemia*. In R.A. Browne, P. Sorgeloos and C.N.A. Trotman (eds.), *Artemia Biology*, Boca Raton, Florida, pp. 351–368.

Williams, W.D., Carrick, T.R., Bayly, I.A.E., Green, J. and Herbst, D.B. (1995) Invertebrates in salt lakes of the Bolivian Altiplano. *International Journal of Salt Lake Research* **4**, 65–77.

Xin Naihong, Sun Juju, Bo Zhang, Triantaphyllidis, G.V., Van Stappen, G. and Sorgeloos, P. (1994) International Study on *Artemia*. LI. New survey of *Artemia* resources in the People's Republic of China. *International Journal of Salt Lake Research* **3**, 105–112.

Xin Naihong, Audenaert, E., Vanoverbeke, J., Brendonck, L., Sorgeloos, P. and De Meester, L. (2000) Low among-population genetic differentiation in Chinese bisexual *Artemia* populations. *Heredity* **84**, 238–243.

Zhang, L. (1993) Adaptive significance of polyploidy in brine shrimp (*Artemia parthenogenetica*), Ph.D. Thesis, Oregon State University, USA.

Zhang, L. and Lefcort, H. (1991) The effects of ploidy level on the thermal distributions of brine shrimp *Artemia parthenogenetica* and its ecological implications. *Heredity* **66**, 445–452.

Zhang, L. and King, C.E. (1992) Genetic variation in sympatric populations of diploid and polyploid brine shrimp (*Artemia parthenogenetica*). *Genetica* **85**, 211–221.

Zhang, L. and King, C.E. (1993) Life history divergence of sympatric diploid and polyploid populations of brine shrimp *Artemia parthenogenetica*. *Oecologia* **93**, 177–183.

Zheng Mianping (1997) *An Introduction to Saline Lakes on the Qinghai-Tibet Plateau*. In H.J. Dumont and M.J.A. Werger (eds.), Kluwer Academic Publishers, Dordrecht, The Netherlands.

Zheng Mianping, Tang Jiayou, Liu Junying and Zhang Fasheng (1993) Chinese saline lakes. *Hydrobiologia* **267**, 23–36.

Zuñiga, O., Wilson, R., Amat, F. and Hontoria, F. (1999) Distribution and characterization of Chilean populations of the brine shrimp *Artemia* (Crustacea, Branchiopoda, Anostraca). *International Journal of Salt Lake Research* **8**, 23–40.

CHAPTER V

EVOLUTION AND SPECIATION

GONZALO GAJARDO
Laboratory of Genetics & Aquaculture
Department of Basic Sciences
Universidad de Los Lagos
P.O. Box 933, Osorno
Chile

THEODORE J. ABATZOPOULOS
Department of Genetics, Development and Molecular Biology
Faculty of Sciences, Aristotle University of Thessaloniki
541 24 Thessaloniki
Greece

ILIAS KAPPAS and JOHN A. BEARDMORE
School of Biological Sciences
University of Wales Swansea
Singleton Park, Swansea SA2 8PP
Wales, United Kingdom

1. Introduction

The brine shrimp *Artemia* comprises a group of bisexual and parthenogenetic, morphologically similar, species very likely to have diverged from an ancestral form living in the Mediterranean area some 5.5 million years ago (Abreu-Grobois and Beardmore, 1982; Abreu-Grobois, 1987; Badaracco *et al.* 1987), though this estimate, based on allozymes, could be somewhat conservative as compared to that based on mitochondrial DNA (Perez *et al.* 1994). At that time the area was, according to geological indications, the only place in the world exhibiting for significant periods of time the very high salinity required for *Artemia* to thrive (Abreu-Grobois, 1987; Badaracco *et al.* 1987). The hypothesis of the Mediterranean as the centre of radiation for *Artemia* is also supported by the diversity of *Artemia* types currently found in the area, *i.e.* bisexuality and parthenogenesis on the one hand, together with diploidy and polyploidy on the other (see below).

The brine shrimp is a highly favourable organism for evolutionary studies as a range of key factors driving speciation in most organisms are observed in *Artemia* (see Table 1). Speciation in the genus should be regarded as a complex, multidimensional process involving a variety of environmental and genomic factors. Although *Artemia* could be an example of a life form restricted to the limited range of ecological conditions often described for

Table 1. Potential for evolutionary studies of the brine shrimp *Artemia*

- The genus comprises bisexual (Old and New World) and parthenogenetic forms (Old World).
- There are morphological similar species and superspecies, *i.e.* clusters of incipient species, representing a range in degrees of reproductive isolation.
- Two reproductive strategies: ovoviviparity (nauplii), oviparity (cysts). The first predominates in permanent lakes, whilst cyst production is associated with unstable, stressful habitats.
- Great variation in water characteristics/ionic composition which can cause ecological isolation between some populations (*e.g. A. monica*).
- Diapause cysts resistant to adverse conditions are suited for passive dispersal (wind, water birds), and allow comparative studies of populations from different origin.
- The disjunct nature (island-like) of hypersaline lakes promotes interpopulation genetic differences.
- Drastic changes in effective population sizes, caused by extinction/ recolonisation cycles, favour genetic drift.
- Clines in the content of heterochromatin (repetitive DNA). Evolution of certain species (*A. franciscana*) would involve amplification of repetitive sequences.
- Great life history plasticity and easy rearing under laboratory conditions (single or in large populations).
- The nuclear DNA content of *A. franciscana* is known (1.6×10^9 bp per haploid genome), a significant part (59%) is single copy, moderately repeated (20%) and highly repetitive (15%). The complete mitochondrial DNA sequence is also known.

hypersaline environments, it is extremely successful within this range, achieving high population sizes and tolerating large environmental variations (Abreu-Grobois and Beardmore, 1982; see, also, chapters III and IV of this volume).

Artemia is one of the best known aquatic organisms, and is considered to be a paradigmatic crustacean that can help to fill gaps in knowledge in evolutionary and comparative biology of arthropods and closely related groups (Marco *et al.* 1991). As a matter of fact, branchiopods have existed since at least the Devonian period (about 410 million years) and share a constellation of primitive features, whilst the Order Anostraca goes back to the lower Cretaceous (Fryer, 1987). Some authors have emphasised its importance as a model organism for evolutionary studies referring to *Artemia* as a sort of 'aquatic *Drosophila*' (Abreu-Grobois and Beardmore, 1982; Marco *et al.* 1991; Gajardo and Beardmore, 2001).

Since the last comprehensive review of *Artemia* taxonomy and population genetics (Browne and Bowen, 1991), new field and laboratory data have been produced using morphological, karyological and DNA markers, and new species have been reported. This chapter focuses on novel ideas on species and speciation that could be deduced or integrated from data accumulated so far. Speciation is still a highly debated topic of evolutionary genetics, and this is evident from the many species concepts (Templeton, 1989; Avise, 1994; Gosling, 1994; Hey, 2001) and speciation modes (Turelli *et al.* 2001) currently existing. We note for example, the limitations presented by the use of the Biological Species Concept (BSC – Mayr, 1969) in *Artemia*, a genus with sexual and asexual forms.

2. *Artemia* Species and Genome Characterization

Any discussion on the genetics of speciation should start with the observation that species are real entities in nature and not human artifacts (Coyne, 1992; Coyne and Orr, 2000). This requires us to understand how evolution, which is a continuous process, can produce the morphological and genetically discontinuous groups known as species (Coyne, 1992). Although the species concept is essential for a proper description of natural diversity, biologists have failed so far in generating consensus on how species should be identified. Indeed, at least twenty-four species concepts are currently recognized (Hey, 2001), a variety often associated with the researcher experience, the general knowledge of the group under study, the amount and quality of the available material and data, and the organism biology (Wiley, 1981). According to the BSC, the most widely used concept today, morphological gaps and underlying genetic change would be consequence of the reproductive isolation separating members of different species, in concert with selection and genetic drift. Geographical or physical barriers between conspecific populations must exist (first stage of the allopatric mode of speciation) for reproductive isolation to evolve as a byproduct of the genetic differences accumulated over time in a local environment (see Gajardo *et al.* 1998). The pre-mating isolation mechanisms developed in allopatry should be reinforced while in sympatry (second stage).

The identification of bisexual *Artemia* species has been established by combination of cross-breeding tests, morphological differentiation, cytogenetics, allozyme studies, nuclear and mitochondrial (mtDNA) DNA sequencing. With the exception of cross-mating, all these techniques have also contributed to identifying the parthenogenetic types collectively and controversially described as *A. parthenogenetica* by Barigozzi (1974) as well as to gain insights on the population structure, origin and amount of clonal diversity.

Notwithstanding the above, *Artemia* taxonomists have long been puzzled by the problem of identifying populations and/or species since the brine shrimp was first described. Even more, the phylogenetic relationship of population and/or species that comprise the genus are still a mater of discussion (for extensive reviews on names see Belk and Brtek, 1995; Triantaphyllidis *et al.* 1997b). In 1755 Schlösser described the brine shrimp based on material collected from the solar saltworks near Lymington, England (no longer extant) referred by Kuenen and Baas-Becking (1938) (in Sorgeloos, 1980); Linnaeus in 1758 classified it as *Cancer salinus*, and Leach in 1818 renamed the brine shrimp as *Artemia salina* (in Artom, 1931). Since then, the binomen *A. salina* has caused tremendous confusion worldwide (Sorgeloos and Beardmore, 1995). Very often authors have named any brine shrimp as *A. salina*, and even for some time the name *A. tunisiana* was used instead (see Trianatphyllidis *et al.* 1997b). *Artemia salina* is now recognised as a valid name for the only bisexual species found in the Mediterranean area (Mura, 1990; Triantaphyllidis *et al.* 1997b).

The differentiation of seven bisexual species, defined primarily by the criterion of laboratory reproductive isolation, and many parthenogenetic forms are currently acknowledged. Present in the Old World are the parthenogenetic forms (with different levels of ploidy, and also found in Asia and Australia), the bisexuals *A. salina* Leach 1819 (Mediterranean area) (in Triantaphyllidis *et al.* 1997b), *A. urmiana* (Günther, 1890) (Iran), *A. sinica* (Cai, 1989) (PR China), *Artemia* sp. (Pilla and Beardmore, 1994) (Kazakhstan), and *A. tibetiana* (Abatzopoulos *et al.* 1998, 2002) (Tibet). Endemic to the New World are *A. persimilis* (Piccinelli and Prosdocimi, 1968) (Argentina) and *A. franciscana* (Kellogg, 1906) (North, Central and South America), with *A. monica* being a special case of a population otherwise closely resembling typical *A. franciscana* but described for an ecologically unique habitat (Mono Lake, carbonate rich) which is highly selective for many populations (Bowen *et al.* 1985; Browne and Bowen, 1991; Lenz and Browne, 1991). According to the current distribution paradigm, *A. persimilis* would be geographically restricted to a few sites in Argentina (Triantaphyllidis *et al.* 1998). However, the recent finding of the species in southern Chile at Laguna Amarga, Torres del Paine (50° latitude South) (Gajardo *et al.* 1999) challenges the existing distribution map. This population exhibits peculiar genomic characteristics, some of them typical of *A. persimilis* such as diploid number (2n = 44) and Nei's (Nei, 1972) genetic distance between populations (as compared to an *A. franciscana* sample). However, the chromocentre frequency of this Chilean population corresponds to that of *A. franciscana*. Gajardo and collaborators (unpublished data) believe this could be a hybrid population, and it is likely that this eventual condition have led some authors to describe it, on morphological traits, as *A. franciscana* (Zuñiga *et al.* 1999) and later as *A. persimilis* (De Los Rios and Zuñiga, 2000). Recently, Papeschi *et al.* (2000) have provided evidence, based on cytogenetic data, that *A. franciscana* is found in Argentina and seems to hybridise in nature with *A. persimilis*. Following Sorgeloos and Beardmore (1995), Triantaphyllidis *et al.* (1997b) and Gajardo *et al.* (1998), later in this chapter we emphasise the need, whenever possible, of a multi-trait approach to identify species.

2.1. KARYOLOGY

The euploid chromosome number in most bisexual *Artemia* species is 2n = 42 (*i.e. A. franciscana, A. salina, A. sinica, A. tibetiana*) while it has been found to be 44 in *A. persimilis* (Barigozzi, 1974). The ploidy level in the parthenogenetic *Artemia* populations varies from diploidy (2n = 42), triploidy (3n = 63), tetraploidy (4n = 84) to pentaploidy (5n = 105) while heteroploidy and aneuploidy are very common phenomena at least in nauplii (Goldschmidt, 1952; Iwasaki, 1969; Barigozzi, 1974; Barigozzi and Baratelli-Zambruni, 1983; Abatzopoulos *et al.* 1986, 1987; Abreu-Grobois, 1987). In polyploid parthenogenetic *Artemia* meiosis is thought to be totally suppressed (apomixis), and so the process resembles mitosis. In diploid asexual populations, diploidy is achieved by the fusion of the first and second meiotic division products (automixis) (for extensive reviews see Barigozzi, 1974, 1989).

Chromosome analysis in *Artemia* species is a rather difficult task as chromosome spreads are not easily obtainable especially in polyploid cells, and this becomes more difficult when chromosomes are interconnected by filamentous heterochromatic bridges (provided that these formations are not artefacts). Moreover, nearly all chromosomes lack centric constrictions (diffused centromeres) and the homologous chromatids are not usually conspicuous due to supercoiling or spiralization (Stefani, 1963; Abatzopoulos *et al.* 1986; Barigozzi, 1989). Because of these problems, *Artemia* karyotypes or karyograms have been described only very occasionally (Mitrofanov *et al.* 1976, 1982; Abatzopoulos *et al.* 1986, 1987). Even more, the lack of a centromeric constriction makes the statistical evaluation (essential for comparisons of chromosome length between different populations and/or species) particularly difficult. Giemsa C-banding karyotypes have been constructed only for *A. franciscana*, and distinct heterochromatic regions can be seen in populations that exhibit also distinct chromocentres (Abatzopoulos *et al.* 1987).

Chromocentres are heavily stained heterochromatic areas with highly repetitive DNA (repeated in the order of 6×10^5 copies per haploid genome of *A. franciscana*) family of the type Alu I (110 bp), also named satellite I, in the interface nuclei I (Barigozzi *et al.* 1987). This genome trait varies between and within species, and is correlated to some extent with genetic differentiation based on Nei's distances (Abreu-Grobois and Beardmore, 1982; Colihueque and Gajardo, 1996) as well as with the amount of repetitive DNA. For example, *A. franciscana* shows the highest concentration of repetitive DNA and also shows the highest chromocentre frequency. No such structures have been reported in Old World species such *A. salina* and *A. urmiana*, nor in parthenogenetic populations. New World species, in particular *A. franciscana*, would then represent a derived state likely to have arisen by sequence amplification from an original common ancestor by a mechanism yet not fully understood (unequal crossing over according to Badaracco *et al.* 1987). Differences in chromocentre numbers and staining pattern in *A. franciscana* would be an indication of chromosome reorganisations, either through heterochromatin modification or differential amounts of repetitive DNA, or both of these. This change does not seem to imply reduced hybrid viability or sterility (see Rieseberg, 2001) as the reported existence of hybrids (with *A. persimilis*) might suggest. Moreover, the observed correlation between chromocentre frequency and latitude (Gajardo *et al.* 2001b) suggests a causative involvement in speciation, and so gives support to those who previously considered heterochromatin as a dynamic element during speciation (Barigozzi *et al.* 1987) and directly adaptive in certain environments (White, 1978). The colonising capacity and actual wider distribution of *A. franciscana* in the New World, as compared to *A. persimilis*, would reflect such capabilities.

Several workers have presented data on the chromosomes and chromocentres of *Artemia* species as a way to understand the role of chromosomal evolution in the speciation pattern observed in the brine shrimp (Goldschmidt, 1952; Barigozzi and Tosi, 1959; Stefani, 1963; Beardmore and Abreu-Grobois, 1983; Abatzopoulos et al. 1986; Abreu-Grobois, 1987; Barigozzi, 1989; Sun et al. 1995; Colihueque and Gajardo, 1996; Rodríguez Gil et al. 1998; Papeschi et al. 2000; Gajardo et al. 2001b).

The best material for cytogenetic analyses is squashes of instar-I nauplii (i.e. immediately after cyst hatching – Barigozzi, 1989). As development progresses, larger nuclei appear as a product of cell polyploidization which may originate from endoreduplication (Abatzopoulos et al. 1986; Freeman and Chronister, 1988). Late prophase and/or early metaphase are the most suitable stages for the study of *Artemia* chromosomes. As metaphase progresses the chromosomes become more and more condensed and non-identifiable. Just before the beginning of anaphase *Artemia* chromosomes form a configuration of non-random associations. Later they move together to the poles of the spindle. All these facts suggest a possible connection between non-random arrangement of chromosomes with heterochromatin. Although the significance of this observation remains obscure for parthenogenetic organisms, it may be related to the function of the genetic material during interphase and/or its condensation in preparation for division (Avivi and Feldman, 1980; Abatzopoulos et al. 1986).

The *Artemia* genome has been mainly studied in *A. franciscana* in material from San Francisco Bay (Marco et al. 1991). The species has a nuclear DNA content of about 1.6×10^9 bp/haploid genome, representing approximately 70 Mbp per average chromosome. 59% of the nuclear genome is single copy, 20% moderately repeated and 15% is highly repetitive DNA, with a relative low G+C content (32.4%). With regard to extra-nuclear DNA, the complete mitochondrial DNA of *A. franciscana* has been sequenced. It has a total length of 15,822 bp, with a G+C content of 35%, and encodes 37 genes in a highly compact organisation, i.e. with few intergenic spacings. The *Artemia* mtDNA is very similar in size to those of sea urchin and various insects, particularly *Drosophila* (i.e., 16S rRNA and the protein genes are located in same positions and orientations) and relatively conserved, in contrast to some other invertebrate species (Batuecas et al. 1987; Perez et al. 1994; Valverde et al. 1994; Garesse et al. 1997).

2.2. ALLOZYME DIVERGENCE

The investigation of micro- and macro-evolutionary processes in *Artemia* has been mostly tackled, at the molecular level, by means of protein electrophoresis. This cost-effective method has provided a substantial amount of data on the extend and relevance of polymorphism in both natural and inoculated populations around the world. Allozyme surveys have allowed us to monitor spatial and temporal changes in allele frequencies, to know the intraspecific population structure and inter-species differences, as well as to evaluate the importance and relative magnitude of the various evolutionary forces behind. Indeed, a comprehensive picture of the evolutionary history of *Artemia*, based on phylogenetic inferences derived from population genetic studies is now available. The main features of this picture, which considers estimates of Nei's genetic distance (D), have been outlined by Abreu-Grobois (1987). Briefly, the two major geographical separations (Old *versus* New World species) show the greatest differentiation (D = 1.497 – 1.952). A quite similar amount of divergence (D = 1.073) is observed between *A. franciscana* and *A. persimilis*. Within the Old World, genetic distance values for inter-specific comparisons range from 0.254 to 0.808 (Abreu-Grobois, 1983; Pilla, 1992; Thomas, 1995). Finally, conspecific populations of *A. franciscana* show mean genetic distances of D = 0.126, though some populations in South America, for example, exhibit genetic distances well above this value revealing high inter-population divergence and, perhaps, incipient speciation stages (Gajardo *et al.* 1995). It should be mentioned, however, that the standard errors associated with these values are often relatively high (1/2 or 1/3 of the genetic distance estimates). The body of allozyme data gathered since 1980 convincingly supports the idea that micro-evolutionary changes in *Artemia* (Old and New World species) are almost invariably associated with adaptation to local habitats, gene flow mediated by migration and dispersal, and predictability and/or seasonality of the environment. On a macro-scale, the effect of certain genetic phenomena (*i.e.* genome amplification) and the raising of any form of barriers to gene exchange complete the evolutionary scenario reconstructed so far for *Artemia* (see Table 1). The realisation of the advantages of *Artemia* for the study of evolutionary mechanisms and its importance in aquaculture, has led to new approaches for the analysis of variability.

2.3. DNA MARKERS

Over the last decade, modern techniques have brought into play new molecular markers, *i.e.* RFLPs, RAPDs, AFLPs (Perez *et al.* 1994; Badaracco *et al.* 1995; Triantaphyllidis *et al.* 1997b; Sun *et al.* 1999a,b) (Table 2). Despite some trade-offs in the number of loci amenable to analysis, these markers convey greater sequence of information than allozymes and their utility has boosted our knowledge of the evolutionary mechanisms operating at the different levels of hierarchy.

It is probably too early to judge whether the conclusions drawn from those recent approaches are congruent with our knowledge accumulated through traditional techniques. At first sight, they seem to be in line with the results from allozymes, though further studies are needed in order to make meaningful comparisons. In this interim period, however, new findings could refine our views on the ways different molecules evolve, the role certain genome sectors could play in the speciation process, and at the same time would eventually lead to the identification and mapping of genes potentially useful to aquaculture.

With regard to the bisexual species identified by allozymes, their specific

Table 2. Genetic markers used for species identification and/or characterisation in *Artemia*

1. **Allozymes**	A protein solution is electrophoresed through a gel matrix. An enzyme-specific reaction reveals allozymes coding-loci and alleles (migrate according to charge) at a given locus.
2. **RFLPs** (Restriction Fragment Length Polymorphisms)	DNA is cut with restriction enzymes, electrophoresed (agarose gels), blotted to membranes, and probed with cloned radiolabelled DNA that binds to a single locus. Alleles differing in the presence or absence of nearby restriction sites will produce different fragment sizes.
3. **AFLPs** (Amplified Fragment Length Polymorphisms)	The methods are similar to RFLPs but more time-consuming and expensive. Detecting a higher number of loci and polymorphism is significantly higher.
4. **RAPDs** (Randomly Amplified Polymorphic DNA)	An arbitrary oligonucleotide of about 10 bases used in a PCR reaction will usually anneal well enough to serve as both forward and reverse primer at 3–10 sites. The products are electrophoresed through agarose and stained. Bands present in one individual may not be present in another for a variety of reasons, chiefly variation in the primer annealing sites. RAPD is a rapid, precise and sensitive method of detection of nucleotide variation. Good for taxonomic investigations (populations, species, genera).

status is confirmed by DNA markers. *Artemia persimilis* is highly divergent from all other species (Perez et al. 1994). Genetic distance values (RAPDs method – Badaracco et al. 1995) of *A. franciscana* from *A. salina*, *A. sinica* and *A. persimilis* are 1.061, 1.105 and 1.165 respectively. *Artemia persimilis* is more distant from *A. salina* (1.287) and *A. sinica* (Cai, 1989) from China (1.257), whilst *A. salina* and *A. sinica* are separated by a distance of 1.173. A similar data set of genetic differences, though biased with regard to the bisexual species considered (*A. urmiana*, *A. sinica*, *A. franciscana*), has been reported by Sun et al. (1999a). In general, these differences are closely parallel with those obtained by use of protein electrophoresis and morphological traits.

The fact that *A. persimilis* is the most divergent species led Perez et al. (1994) and Badaracco et al. (1995) to consider it a close relative (a primitive species) to the original ancestor. From this primitive species (condition also backed by the very low chromocentre numbers) originated, at different times, *A. salina*, *A. sinica* and *A. franciscana*. The origin of *A. franciscana* is, however, somewhat confusing since it is the only species showing satellite DNA (Alu I), a trait not seen in the Eurasian populations.

The antiquity of parthenogenesis in *Artemia* contrasts to what is observed in other invertebrates and vertebrates as well, where asexual reproduction is rare and more recent. The maintenance of this reproductive mode in European *Artemia* (sometimes seasonally coexisting with bisexuals) whilst not present in the New World, is a challenging problem from an evolutionary viewpoint.

3. Genetic Variation, Ecology and Evolution

Genetic variability is one of the main constraints on evolutionary change, hence critical early stages in speciation and/or adaptation are the availability and evolution of genetic differences between population (Coyne, 1992). Documenting genetic variation and interpreting its evolutionary relevance has been, without doubt, one of the most intellectually demanding aspects of evolutionary biology since Harris (1966), and Lewontin and Hubby (1966) provided a quick, reasonably inexpensive and reproducible method for detecting genetically determined variation (see Ward and Grewe, 1995 for a review). Protein or allozyme electrophoresis (Table 2), the predominant tool used so far, has provided the most extensive survey of intra-specific diversity, population structure and a way to investigate whether phenotypic population differentiation is underlain by genetic divergence (see Mitton, 1997; and Moller and Swaddle, 1997 for recent reviews). However, this technique allows us to compare only a handful of randomly selected loci and hence a small portion of the genome (protein-coding loci), whilst a significant part (most in non-coding regions) is not amenable to analysis. Furthermore, allozyme loci are expected to be, at best, only loosely linked to key genes (Quantitative Trait Loci, QTL) or gene groups directly related with speciation such as, for example, genes known to cause pre- and post-mating reproductive isolation (Coyne, 1992; Coyne and Orr, 2000). Not surprisingly molecular methods have extended the regional analysis of the genome and have become essential tools for evolutionary geneticists (Queller *et al.* 1993; Bossart and Prowell, 1998) as well as for the management and conservation of genetic resources (Sweijd *et al.* 2000).

Natural populations are large reservoirs of genetic variation. This is now an unquestionable fact for most plant and animal species, including commercially important finfish and shellfish (Ward and Grewe, 1995). *Artemia* species are not an exception. Populations of bisexual species around the world have been extensively screened electrophoretically by Beardmore and co-workers (Abreu-Grobois and Beardmore, 1982; Abreu-Grobois, 1987; Gajardo and Beardmore, 1993; Pilla and Beardmore, 1994; Gajardo *et al.* 1995, 1999) showing that on average between 31 and 54% of the enzyme loci screened are polymorphic, and a randomly chosen individual is heterozygous at 9 to 21% of the loci surveyed. Hence levels of genetic variability for *Artemia* are among the highest within crustaceans and are comparable to the mean obtained from a variety of invertebrate species (see reviews by Abreu-Grobois, 1987, and Browne and Bowen, 1991).

3.1. HETEROZYGOSITY AND LIFE HISTORY TRAITS

Amid the controversy generated after the proposal by Kimura in 1968 (Kimura, 1983) of the neutral theory of molecular evolution, a plethora of studies were directed towards the investigation of the long suspected adaptedness of genetic variation. Since then, many authors have provided convincing evidence of a positive correlation between heterozygosity and fitness-related traits. Variables such as growth rate, viability, fecundity, feed conversion, metabolic efficiency, from a wide range of animal and plant species, were intensively analysed (see reviews by Mitton, 1997; and Moller and Swaddle, 1997). Even certain DNA sequences of unknown function were shown to correlate with adaptive features (Yang and Bielawski, 2000). Negative associations were also reported (Booth *et al.* 1990), however due to an unhealthy tendency of withholding negative results the exact proportion of such findings in the published literature is possibly an underestimate of reality.

The recognition of the advantages of brine shrimp as an experimental organism and the importance of this kind of research in aquaculture, led to the investigation of the same problem in *Artemia*. Heterozygosity in *Artemia franciscana* has been associated with increased fecundity, and the quality or type of offspring (Gajardo and Beardmore, 1989). The brine shrimp produces young as larvae or as encysted embryos resistant to unfavourable conditions, and studies with *A. franciscana* from Great Salt Lake in Utah (Gajardo and Beardmore, 1989) have shown a positive correlation between the percentage of zygotes produced as cysts and the level of heterozygosity (determined electrophoretically) in the mother. More heterozygous females also produce more zygotes, have more broods and start to reproduce at a younger age than less heterozygous females. The encysted embryos are energetically more expensive to produce than are nauplii, and so the greater reproductive success achieved by more heterozygous individuals may be postulated to be due to their better utilisation of energy. Thus, under similar experimental conditions, the more heterozygous females are able to allocate more energy to the production of offspring as cysts than are less heterozygous females. Laboratory studies have also shown increased male mating success with heterozygosity in a population of *A. franciscana* from Colombia (Zapata *et al.* 1990). Males heterozygous for three and four loci exhibited higher fitness values (2.99 and 2.55, respectively) in comparison to homozygous genotypes with an arbitrarily assigned fitness value of 1 (see also for a review Mitton *et al.* 1997). Heterozygosity is also correlated with type of habitat (*i.e.* water ionic composition) in *A. franciscana*, in such a way that higher heterozygosities are seen in non-chloride waters (Abreu-Grobois and Beardmore, 1982). In *A. sinica* heterozygosity is positively correlated ($P = 0.04$) with larger surface area of the water body (Naihong *et al.* 2000). Although a causal explanation for this finding was not provided, it may be a consequence of the greater environmental heterogeneity of large lakes. Alternatively, population sizes may differ greatly and affect levels of genetic variation.

Besides natural populations, inoculation schemes provide excellent opportunities for the study of fitness-related traits and their association to genetic variability. They comprise a microcosm of the various and complex biotic and abiotic interactions operating in the field and can be viewed as a sort of 'natural laboratory'. In the genus, the species almost exclusively used for introductions around the world has been *A. franciscana*. Its great phenotypic plasticity, genetic richness and extended gene pool have made it the perfect candidate for that. An extensively studied case over the last few years has been the inoculation of *A. franciscana* (San Francisco Bay, SFB) into Vietnam. The question whether cyst production and other life-history traits are correlated to individual heterozygosity has been prompted on grounds of theoretical considerations, evidence in other organisms and utility to aquaculture. Five fitness-related attributes were investigated across five different strains (Kappas, 2001). The strains included a batch of the source population (SFB), the Vietnamese strain (VC, present in the area since 1982) and year classes (Y1 to Y4) originated after re-inoculation of SFB and sampling in four subsequent years. The number of cysts produced per female was significantly different among strains (one-way ANOVA, P = 0.0057). Of the remaining four fitness traits, significant differences were found for the number of broods per female (Kruskal-Wallis, P = 0.0185) and the number of encysted broods (Kruskal-Wallis, P = 0.0003), but not for the number of nauplii or total zygotes produced. The level of multi-locus heterozygosity was positively correlated with the number of cysts in the Y3 strain (Kendall'τ = 0.537, P = 0.009) and negatively so with the number of broods in the Y4 strain (Kendall'τ = –0.416, P = 0.039). In the same study, correlations for number of nauplii, zygotes and broods per female in the VC strain also approached the borderline of significance (P values ranged from 0.053 to 0.069). In a follow-up study (Kappas, 2001), three strains (SFB, VC, Y1) were cultured in 80 ppt salinity and two temperatures (26 and 30 °C). Two-way ANOVA of the number of cysts (ranked data) revealed a highly significant (P = 0.00053) reduction of cyst production at 30 °C. Pronounced interactions (P < 10^{-5}) were observed for the number of nauplii, zygotes and the rank of broods. Briefly, the pattern that emerged was of decreasing encystment and offspring production and increasing ovoviviparity and reproduction at higher temperature. Finally, when correlation coefficients to individual heterozygosity were computed significance was not established. Temperature is thought to be one of the major factors that shape the gene pool in an area like Vietnam. Of course, other factors such as food availability, population density, etc., could greatly influence the genetic makeup. However, temperature fluctuations can trigger a chain reaction of the events and, consequently, are more or less a significant component of the biotic and abiotic interplay responsible for genotype representation in a particular strain. One striking feature of *Artemia* is its high potential for fast genetic changes after colonisation or introduction. For example, some alleles (detected electrophoretically) present in the Macau (Brazil) population are absent in a sample of San Francisco Bay, the probable founder population (Gajardo *et al.* 1995). Macau is a fast changing habitat due to commercial exploitation. Drastic changes are likely to be imposed on the gene pool and consequently have an effect on a certain phenotypic response, for example cyst production or temperature tolerance (Tackaert and Sorgeloos, 1991). Concerning the latter, the Brazilian population (*A. franciscana*) shows increased survivorship at temperatures above 30 °C as compared to San Francisco Bay.

The opposite is seen in an *A. franciscana* population (Gajardo and Beardmore, 1993) from the Atacama Desert in northern Chile, reputedly one of the driest places in the world. This population is reported to tolerate temperatures as low as 6 °C (Gajardo *et al.* 1992). A similar case would be likely to occur with regard to low temperature tolerance in Chilean Patagonia (Gajardo *et al.* 1999), where *Artemia* has been found in a very unusual environment subject to specific climatic conditions. Rapid evolutionary changes are not uncommon in populations colonising novel environments, but they should end up in reproductive isolation to be important to speciation as Orr and Smith (1998) have pointed out.

Some conclusions can be drawn from these and other similar studies in the genus. Firstly, the investigation of fitness-related traits that deal with reproduction in general, should be expanded to include other important variables (*i.e.* brood size, pre- and post-reproductive period, life span etc.). The benefits of a more integrated picture and a better understanding of the genetic variation underlying key traits are worth the labour involved. Secondly, there may be discrepancies in the results from introduced and natural populations, making it difficult to find a pattern. Therefore, management and sampling practices in inoculation schemes should be carefully considered in future genetic studies to which they can be a critical part of the interpretation. Thirdly, every strain is a snapshot of the overall genetic variability, and hence integrated larger bodies of data are required prior to suggesting that a particular observation may be applicable to other genetic systems. In *Artemia*, accumulation of data over the last decade on heterozygosity-fitness associations points to the fact that the organism joins the list of other animals and plants where such correlations have been established. The applicability of such phenomena to *Artemia* aquaculture is intuitively valuable. Finally, new molecular tools will certainly expand the study on the significance of genetic variation to other portions of the *Artemia* genome (nuclear and extra-nuclear). In this way not only biologically relevant questions will likely be answered, but also aquaculture management and conservation will benefit from sound genetic recommendations. The avoidance of inbreeding and mixing of genetically divergent populations is, for example, a first step. From an economic point of view, suitable lakes could be carefully managed to become competitive suppliers of cyst material and thus render the market, to some extent, independent of cyclic abundance of cysts from a sole producer (*i.e.* poor harvests in the Great Salt Lake, USA).

3.2. BISEXUAL *VS* PARTHENOGENETIC TYPES

The species gene pool in *Artemia* consists of two components depending on the type of species considered: the sexual (bisexual species) and the asexual gene pool (parthenogenetic types). The branching of parthenogenetic types from bisexual *Artemia* (*A. salina* line) would be older than five million years (Abreu-Grobois and Beardmore, 1982; Perez *et al.* 1994). According to Singh

(2000), the sexual gene pool would involve genes affecting mating behaviour, gametogenesis and fertilisation system, and would have a significant role in speciation by influencing the evolution of reproductive isolation and, comparatively a minor effect on survival. The asexual gene pool, which is made up of genes affecting development, differentiation and metabolic functions, would have a major role in the survival of organisms and a minor role in the evolution of reproductive isolation. These two gene pools would have great impact on the genetic changes occurring in early stages of speciation.

Parthenogenetic *Artemia* populations exhibit great clonal diversity, as it is evident from morphology (Hontoria and Amat, 1992; Triantaphyllidis *et al.* 1997a), and cytological and allozyme studies (Abreu-Grobois and Beardmore, 1982; Abreu-Grobois, 1987; Abatzopoulos *et al.* 1993). Parthenogenetic types exhibit different ploidy levels (diploid, triploid, tetraploid, pentaploid), and this is likely to affect the gene pool and development of organisms as a consequence of cytological, biochemical and physiological changes produced. Such changes would suit them to conditions that are beyond the limits of diploid parents. In fact, asexual *Artemia* types are usually geographically segregated from sexual forms, and are predominantly found in low and high latitudes ($< 25°$ N and $> 40°$ N), whilst bisexuals occur in temperate regions (latitudes $35-40°$ N). This sort of niche partitioning (Zhang and Lefcort, 1991), or geographical parthenogenesis, is explained by assuming that both types of genotypes (sexual, asexual) are qualitatively different in phenotypic response. At least three hypotheses would explain alternative phenotype responses (see Parker Jr. and Niklasson, 2000), *i.e.* (1) 'general purpose genotype', (2) 'elevated ploidy' and (3) 'frozen niche variation'. In the first case, selection among clones allows those with a wide tolerance range to survive over many generations, whereas selection in sexual populations cannot produce single genotypes adapted to all possible environments. In the second, polyploid genotypes are better able to withstand environmental stress, whereas in the 'frozen niche' hypothesis multiclonal populations utilise the available niche better than broadly adapted sexual populations, making them more resistant to competitive exclusion by sexual ancestors. Co-occurrence or sympatry of parthenogenetic types and bisexuals is observed in Spain, Italy and in Central and North China and probably in Iran. Parthenogenetic types tend to predominate in more disturbed, stressful conditions of salinity, temperature and food availability (see Browne and Bowen, 1991; Lenz and Browne, 1991).

4. Population Structure and Pattern of Speciation in *Artemia*

The main avenue of research on the genetics of species formation has been the determination of the amount of genetic change that accompanies speciation. A direct strategy involved the assay of populations appearing to be in various stages of the speciation process. On the other hand, surveying populations belonging to different species has been an indirect way of assessing the sum of genetic differences accumulated subsequent to, as well as during the speciation process itself (Avise, 1976). The work of Ayala (Ayala and Tracey, 1974) on the *Drosophila willistoni* complex of populations made significant inroads on this problem and reveal rather distinct patterns of speciation even within a single genus.

In *Artemia*, speciation has been tackled by comparing genetic relationships among the different species. According to the geographic mode of speciation, a species is dissected in two parts by a physical barrier preventing gene flow between them. If enough time has elapsed since separation, populations will diverge as a result of adaptation to the prevailing environmental conditions or genetic drift. Pre- and post-zygotic reproductive isolation will develop between the populations as a consequence of the accumulated genetic differences. Many authors have viewed the speciation process (though not necessarily its products) as a rather unexceptional continuation of the microevolutionary processes generating geographic population structure, with the added factor of the evolutionary acquisition of intrinsic reproductive isolation (see Avise, 1994 and references therein). It is therefore crucial to determine whether the raw materials for speciation could, occasionally, be found in the dynamics of intra-specific population differentiation.

Artemia population structure resembles to a certain extent the meta-population structure model (Hanski, 1991), hence a relatively rapid shaping and reshaping of the genetic structure of *Artemia* populations would be expected at different geographical scales. The island-like nature of *Artemia* hypersaline environments and the occurrence of successive cycles of extinction/recolonisation promote interpopulation genetic differentiation, and hence the differentiation of locally adapted populations as a consequence of restricted gene flow, selection to the local environmental and genetic drift. Not surprisingly, *Artemia* populations are highly heterogeneous genetically and display great variability in life-history traits throughout their distribution. In fact, high levels of genetic sub-structuring, inferred from substantial F_{ST} values, have been reported for conspecific *A. franciscana* populations ($F_{ST} = 0.24$). Genetic sub-structuring is also highest among South American populations, derived either from genetic distances (Gajardo *et al.* 1995) or F_{ST} values (= 0.38) (Gajardo *et al.* 2001b).

Although *Artemia* populations exhibit many of the characteristics which render them prone to speciation, interpopulation genetic differentiation is not always observed as expected. In populations of *A. sinica* from China, genetic differentiation is not related to geographic distance and this is thought to be related to habitat size. The argument used to explain this, is that effective gene flow is often much lower than dispersal and discrepancy between the two increases with time, as the establishment success of immigrants reduces in habitats in which competition with resident conspecifics is severe (De Meester, 1996). On the other hand, low levels of genetic differentiation between populations observed in this work are attributed to the great dispersal capacity of *Artemia*, a factor often counterbalancing the disjunct nature of habitats. The various evolutionary consequences of dispersal need to be addressed in *Artemia*. Since these are difficult to test in the field, the brine shrimp offers advantages as a suitable laboratory organism.

Notwithstanding the above, the variety of ways by which new species can emerge and the differential impact that forces like natural selection and genetic drift can have on the genome, illustrate the need to approach the problem in a multidisciplinary way. A more thorough understanding of the evolutionary history of the genus could be gained by collectively analysing the diverse data sets through, for example, the employment of sophisticated statistical approaches. Although still in their infancy, meta-analytical methods are beginning to address questions ranging from physiology to phylogeny (Sokal and Rohlf, 1995; Hillis *et al.* 1996). Studies that incorporate molecular, morphological and other markers will provide better descriptions and interpretations of the biological diversity. A step towards this direction has been the determination by Abatzopoulos and co-workers (Abatzopoulos *et al.* 1998, 2002) of the specific status of *Artemia tibetiana*. Biometrics of cysts and nauplii, cytogenetics, cross-breeding laboratory tests, allozyme electrophoresis and RAPD analysis have been used to provide conclusive evidence for the taxonomic status of *Artemia* samples from Tibet. In another study, Triantaphyllidis *et al.* (1997b) utilised morphological and AFLP markers to resolve the controversy over the binomen *Artemia salina*. A similar multi-trait approach has been used to understand the speciation process in the New World species *A. franciscana* and *A. persimilis* (Gajardo *et al.* 1998, 1999, 2001; Gajardo and Beardmore, 2001).

Multidisciplinary surveys have the advantage of treating species as units in concordance with a population genetics view (a species shares a gene pool), and since speciation itself is a multidimensional process, an integrated approach is likely to reveal the multiple complexities observed in the highly hierarchical biological systems. The lessons we have learned over the last forty years from studies on speciation suggests that, it is surely a mistake to expect genetic divergence to be manifested uniformly in all kinds of characteristics or involve a particular class of the genome (Dobzhansky, 1976). It is probably true to say that, research on aspects of speciation has raised more questions than it has provided answers. However, reliable diagnosis of species requires flexibility as well as pragmatism. Parthenogenetic forms have a specific status by virtue of mode of reproduction although all clones are isolated from each other, and different ploidy levels are grouped under the same binomen. A better fit could be attained by a definition focusing on the monophyletic origin of the various parthenogenetic types within the species. Similar judgements could be applied in the case of *Artemia franciscana* and its close relative *Artemia monica*.

What do the results just discussed tell us about the genetics of species differences in *Artemia*? The available markers have provided so far good framework for assessing the specific status and evolutionary divergence among

Artemia populations and species. But studies dealing with the nature of interspecific differences have received relatively low priority. Under BSC (see earlier in this chapter), it is essential to understand the number of genes involved in phenotypic differences between species or, to put it the other way around, to understand the number of genes responsible for reproductive isolation. Although many *Artemia* traits could be thought to affect reproductive isolation (*e.g.* frontal knob), the causal relationships and the underlying genetic changes have not been investigated so far, as in *Drosophila* and some other species (Orr, 2001). Lack of examples of allopatric *Artemia* populations coming together in sympatry, where reinforcement of reproductive isolation should take place, is a probable explanation for the difficulties in tackling this topic.

4.1. A CASE STUDY OF A SUPERSPECIES: *A. FRANCISCANA*

Artemia franciscana is the best studied *Artemia* species, and is an interesting study case of evolution in progress. The species, originally found in North America and the Caribbean, has extended its range to South America, where it was introduced either for economic reasons (Brazil, Peru) or naturally dispersed (wind or water-birds such as flamingos). It is regarded as a 'superspecies', *i.e.* clusters of incipient species in *status nascendi*, as some ecologically isolated populations are reproductively isolated in nature (Browne and Bowen, 1991). Much of the taxonomy and population genetics of *Artemia*, including *A. franciscana*, has been reviewed elsewhere (Abreu-Grobois and Beardmore, 1980, 1982; Abreu-Grobois, 1987; Browne and Bowen, 1991). The combination of genetic markers (Table 2) and morphological, karyological and reproductive data currently available for the species, allows to evaluate the speciation mode most adequately describing the differentiation pattern observed, either allopatric, reinforcement, divergence-with-gene-flow or bottleneck (see Coyne, 1992; Rice and Hostert, 1993; Turelli *et al.* 2001). Lewontin (2000) has stressed the importance of considering both the genetics and phenotypic dimension to understand how evolutionary forces build population differences.

The evaluation of different traits in *A. franciscana* population from Chile and South America has permitted us to study the relationship between population structure, potential for divergence and degree of morphological and/or genetic change, all important aspects to understanding how genetic and phenotypic variation affect the speciation process in *Artemia*. Morphological (Mahalanobis distances), electrophoretic (Nei's genetic distances) and karyological (diploid and chromocentre numbers) data have been gathered for populations representing *A. franciscana* and *A. persimilis*, the two closely related sibling species found in the New World (Gajardo and Beardmore, 1993; Gajardo *et al.* 1995; Colihueque and Gajardo, 1996; Gajardo *et al.* 1998, 2001a). Generally, a good correspondence between the genetic and morphological dimension exists, though this is not always the case with several highly divergent *A. franciscana* populations. Evidence on the intra-specific allozymic divergence also comes from studies of inoculation of *Artemia* strains in different areas of the world. Geographic differentiation can be readily accessed in a known time-scale and this could shed more light to the factors involved in the process of micro-evolution in the genus. The inoculation of *Artemia* in Vietnam serves as a good example of the great evolutionary potential of *A. franciscana* (Kappas, 2001) (see also section 3.1 of this chapter). The inoculation project was initiated in 1982. Material of *A. franciscana* from San Francisco Bay (SFB), USA was used as a source population. Details on the culturing scheme can be found in Baert *et al.* (1997) and Quyhn and Nguyen (1987). Also, re-inoculation of the SFB strain have produced distinct year classes (strains Y1 to Y4) according to the number of culturing seasons (those strains were allowed to grow before harvesting). Allozyme screening (20 loci) has revealed that less than two decades after the introduction of *A. franciscana* in the area, at least two gene pools can be distinguished: the first represented by the initial inoculum and a second belonging to the originated Vietnamese strain (VC). A quite extraordinary fact though is that, even within four years of culturing, frequencies of genes at allozyme loci have changed to such a degree that the initial gene pool has been broken up to a series of genetic variability profiles. Two representative dendrograms (UPGMA method) of this condition are shown in Figure 1 (a, b). These data provide strong evidence for a minimal but detectable divergence in the form of changes in gene frequency and content. They also point to the fact that the high level of genetic variability in *A. franciscana* can rapidly respond to distinctly different environmental conditions. Finally, results like these justify the extensive use of the species for aquaculture purposes and cast further evidence on its remarkable phenotypic plasticity and extended gene pool.

On the other hand, a phenomenon of genome amplification seen in *A. franciscana* brings a new dimension to the understanding of *Artemia* speciation (Gajardo *et al.* 2001b). The increase in chromocentre numbers (see section 2.1 of this chapter) from *A. persimilis* to *A. franciscana* (and repetitive DNA content) could have occurred through amplification of heterochromatin or DNA content in *A. franciscana*, while the species kept a lower diploid number ($2n = 42$ instead of $2n = 44$ in *A. persimilis*). As mentioned in section 2.1, this should be regarded as an important evolutionary change since heterochromatin plays a dynamic role during speciation (Barigozzi *et al.* 1987), and chromosomal rearrangements like this (probably due to unequal crossing over according to Badaracco *et al.* 1987) can be directly adaptive to certain type of habitats and ecological niches. Interestingly, a north-south steady latitudinal decline of chromocentre number (which is positively correlated with repetitive DNA content of the Alu I type) is observed towards the equator in *A. franciscana* populations from the Northern Hemisphere. Chromocentre numbers increase from this point toward southern latitudes with a population in the Chilean Patagonia (around 52° latitude South) exhibiting the highest chromocentre frequency (Gajardo *et al.* 2001b).

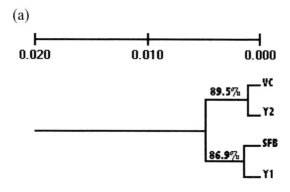

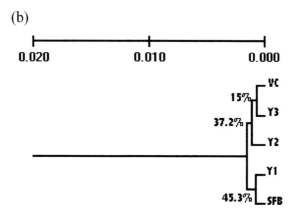

Figure 1. UPGMA dendrograms of Nei's (1972) genetic distance with bootstrap values (%) of topology support out of 1,000 permutations shown along branches. (a) Cyst samples raised at 80 ppt and 26 °C. (b) Live adults collected from field. SFB: San Francisco Bay (source population); Y1: Year one *i.e.* after one year in Vietnam; Y2: Year two (after two years in Vietnam); Y3: Year three (after three years in Vietnam); VC: Vihn Chau (*i.e.* the established Vietnamese strain of *A. franciscana* in the area, since 1986).

5. Reproductive Isolation

Although cross-fertility is said to be an inadequate trait for tracing the evolutionary story in a group, it is recognised as valuable for species recognition in *Artemia* (Browne and Bowen, 1991). Intrapopulation experimental crosses and cross-fertility tests have been performed in individual crosses of laboratory-reared populations of *A. franciscana* and *A. persimilis* (Gajardo *et al.* 2001a). The populations compared displayed significant variability in fecundity (total offspring, brood size) and in the ratio oviparity/ovoviviparity. Hybrid offspring were abundantly produced in cross-fertility tests and showed a pronounced switch to the encystment mode, particularly in crosses with *A. persimilis*. The production of laboratory hybrids between morphologically or genetically divergent allopatric populations appears to be a common phenomenon in some *Artemia* populations (Bowen et al. 1985; Pilla and Beardmore, 1994), and other members of the Order. Indeed, there are examples showing that morphologically distinct species that have been separated for long period of time are sexually compatible (Wiman, 1979; Maeda-Martinez et al. 1992).

This might be related with the fact that these populations have only completed the first stage of the allopatric speciation process (*i.e.* geographical separation) but have not completed the second stage required for the development of pre-mating isolating mechanisms (Mayr, 1969; Coyne, 1992). In addition, *A. franciscana* and *A. persimilis* are recently diverged according to Browne and Bowen (1991), and so are in the process of developing efficient barriers to gene exchange (Abreu-Grobois, 1987; Gajardo et al. 2001a).

These observations together with the fact that the BSC does not apply to parthenogenetic *Artemia*, highlight the need to rethink the application of this definition to the *Artemia* situation in general. A number of concepts maintain that reproductive isolation is not necessarily a key aspect of the Biological Species Concept, as many good species are able to maintain their genetic identity without substantial reproductive isolation (Cohesion Species Concept – Templeton, 1989). We have discussed in this chapter that *Artemia* species, including parthenogenetic forms, are indeed identified as good genetic entities by the genetic markers described in Table 2. In other words, the emphasis has been put in defining, though imperfectly (the example of *A. franciscana*), species in terms of genetic and phenotypic cohesion. In this way speciation should be regarded as the evolution of cohesion mechanisms, instead of isolation mechanisms. We leave this problem open for future experimental and intellectual work.

6. Final Remarks

The species concept and the mechanism by which species are formed (speciation) are still highly debated topics of evolutionary biology (Hey, 2001; Turelli et al. 2001). This is partially due to the need of integrating experimental and field studies from a vast array of species with varying life history strategies into a coherent theory that can explain the rules of gene pool transformation from one generation to another. Such rules require us, according to Lewontin (2000), to reconcile the genotypic and phenotypic spaces, two dimensions that are co-adapted (Gajardo and Beardmore, 2001), *i.e.* integrated via specific regulatory interactions at every level of the individual's organisational hierarchy. The external phenotype is the visible expression of such hierarchy, though the emphasis put lately on the molecular description of population variation has very often been left behind this level. And this is the primary object of the natural selection.

The *Artemia* work on the species and speciation reviewed in this chapter shows that an increasingly large data base with information produced at different levels of the biological hierarchy (morphological, karyological, allozyme, DNA-level) has been produced in the last decade. There is, however, the need to integrate all these results into a coherent body if progress on both topics is to be expected. The following aspects have been discussed:

1. *Artemia* has multiple advantages as an experimental organism for the study of speciation. From the list in Table 1, it is clear that the genus is suited with a range of evolutionary tools not very commonly seen at once in other organisms.
2. The use of different traits, either singly or combined, has helped to better identify *Artemia* species and has provided a more realistic view of the diversity of the genus. Seven bisexual species (see section 2 of this chapter) and several parthenogenetic types (with different ploidy levels) grouped unsuccessfully under the binomen *A. parthenogenetica*, are currently acknowledged. With regard to bisexual species, relative consensus exists to consider *A. persimilis* as a primitive species, *i.e.* more closely related to the original ancestor in the Mediterranean area from which originated, at different times, *A. salina*, *A. sinica* and *A. franciscana*.
3. *Artemia franciscana*, the best studied species so far, is an interesting case of an expanding species from an evolutionary viewpoint. It exhibits high genetic diversity at coding and non-coding genomic regions, part of which is correlated with key fitness traits as revealed by the work based on protein-coding loci (Gajardo and Beardmore, 1989; Zapata *et al.* 1990; Pilla and Beardmore, 1994; Kappas, 2001). At the DNA level, the species exhibits great interpopulation variation in the satellite I region (a repeat of 110 bp). This region is absent in Eurasian populations, and probably evolved through amplification of heterochromatin or DNA content by chromosomal rearrangements (unequal crossing over). Such rearrangements can be directly adaptive to environmental conditions, and so are highly relevant for *A. franciscana* speciation.
4. Two sets of gene pools are recognised in *Artemia*, the sexual (bisexual species) and asexual (parthenogenetic species) sets with qualitatively different phenotypic responses. Since the first affects the fertilisation system, it greatly influences the development of reproductive isolation, and hence speciation, whereas the asexual has a major role in survival. The phenotypic responses of the two genomes have been well documented in *Artemia* by both field and laboratory experiments.
5. The allopatric mode of speciation, or adaptive divergence, seems better to describe, what occurs in *Artemia*. However, the process turns out to be more complex, as many other elements participate in a non-exclusive manner. Noteworthy, differentiation can occur in spite of gene flow, or

by bottleneck effect due to permanent cycles of extinction and recolonisation. Similarly, chromosome rearrangement could play a relevant role as seen in *A. franciscana*. Despite the experimental advantages of *Artemia* (a sort of aquatic *Drosophila*), little laboratory work has been conducted in order to test the major predictions derived from the current speciation modes, and this is certainly an area that deserves special attention in the future.

6. There is a need to rethink the validity of the Biological Species Concept in *Artemia*, a genus where asexual reproduction is observed. Moreover, genetic differentiation is not always an indication of the development of reproductive isolation, while hybridisation seems to be more common in nature, contrary to what would be expected from Biological Species Concept (Arnold, 1997). The Cohesion Species Concept proposed by Templeton (1989) offers, we believe, better chances to integrate the work done so far in *Artemia*. The nature of the genetic differences (*i.e.* number and type of genes involved in reproductive isolation) (see Orr, 2001) should be approached from the perspective provided by the new scenario outlined in this chapter.

7. References

Abatzopoulos, T.J., Kastritsis, C.D. and Triantaphyllidis, C.D. (1986) A study of karyotypes and heterochromatic associations in *Artemia*, with special reference to two N. Greek populations. *Genetica* **71**, 3–10.

Abatzopoulos, T.J., Triantaphyllidis, C.D. and Kastritsis, C.D. (1987) Preliminary studies on some *Artemia* populations from northern Greece, in P. Sorgeloos, D.A. Bengtson, W. Decleir and E. Jaspers (eds.), *Artemia Research and its Applications*, Vol. 1, Universa Press, Wetteren, Belgium, pp. 107–113.

Abatzopoulos, T., Triantaphyllidis, C. and Kastritsis, C. (1993) Genetic polymorphism in two parthenogenetic *Artemia* populations from Northern Greece. *Hydrobiologia* **250**, 73–80.

Abatzopoulos, T.J., Zhang, B. and Sorgeloos, P. (1998) *Artemia tibetiana*: preliminary characterization of a new *Artemia* species found in Tibet (People's Republic of China). International Study on *Artemia*. LIX. *International Journal of Salt Lake Research* **7**, 41–44.

Abatzopoulos, T.J., Kappas, I., Bossier, P., Sorgeloos, P. and Beardmore, J.A. (2002) Genetic characterization of *Artemia tibetiana* (Crustacea: Anostraca). *Biological Journal of the Linnean Society* **75**, 333–344.

Abreu-Grobois F.A. (1983) Population genetics of *Artemia*, Ph.D. Thesis, University of Wales Swansea, Wales, UK.

Abreu-Grobois, F.A. (1987) A review of the genetics of *Artemia*. In P. Sorgeloos, D.A. Bengtson, W. Decleir and E. Jaspers (eds.), *Artemia Research and its Applications*, Vol. 1, Universa Press, Wetteren, Belgium, pp. 61–99.

Abreu-Grobois, F.A. and Beardmore, J.A. (1980) Genetic characterization of *Artemia* populations: an electrophoretic approach. In G. Persoone, P. Sorgeloos, O. Roels and E. Jaspers (eds.), *The Brine Shrimp Artemia*, Vol. 1, Universa Press, Wetteren, Belgium, pp. 133–146.

Abreu-Grobois, A. and Beardmore, J.A. (1982) Genetic differentiation and speciation in the brine shrimp *Artemia*. In C. Barigozzi (ed.), *Mechanisms of Speciation*, Alan R. Liss, New York, pp. 345–376.

Arnold, M.L. (1997) *Natural Hybridization and Evolution*, Oxford University Press, Oxford.

Artom, C. (1931) L'origine e l'evoluzione della partenogenesi attraverso I differenti biotopi di una specie collettiva (*Artemia salina* L.) con speciale riferimento al biotipo partenogenetico di Sete. *Memorie R. Accad. Ital. (Cl. Sci. fsi. Mat. Nat.)* **2**, 1–57.

Ayala, F. and Tracey, M. (1974) Genetic differentiation within and between species of the *Drosophila willistoni* group. *Proceedings of the National Academy of Sciences of the United States of America* **71**, 999–1002.

Avise, J.C. (1976) Genetic differentiation during speciation. In F.J. Ayala (ed.), *Molecular Evolution*, Sinauer Associates, Sunderland, MA, pp. 106–122.

Avise, J.C. (1994) *Molecular Markers, Natural History and Evolution*, Chapman and Hall, London.

Avivi, L. and Feldman, M. (1980) Arrangement of chromosomes in the interphase nucleus of plants. *Human Genetics* **55**, 281–295.

Badaracco, G., Baratelli, L., Ginelli, E., Meneveri, R., Plevani, P., Valsasnini, P. and Barigozzi, C. (1987) Variations in repetitive DNA and heterochromatin in the genus *Artemia*. *Chromosoma* **95**, 71–75.

Badaracco, G., Bellorini, M. and Landsberger, N. (1995) Phylogenetic study of bisexual *Artemia* using random amplified polymorphic DNA. *Journal of Molecular Evolution* **41**, 150–154.

Baert, P., Anh, N.T., Quynh, V.D. and Hoa, N.V. (1997) Increasing cyst yields in *Artemia* culture ponds in Vietnam: the multi-cycle system. *Aquaculture Research* **28**, 809–814.

Barigozzi, C. (1974) *Artemia*: A survey of its significance in genetic problems. *Evolutionary Biology* **7**, 221–251.

Barigozzi, C. (1989) Cytogenetics and speciation of the brine shrimp *Artemia*. *Atti della Accademia Nazionale dei Lincei Memorie* **19**, 57–96.

Barigozzi, C. and Tosi, M. (1959) New data on tetraploidy of amphigonic *A. salina* Leach and on triploids resulting from crosses between tetraploids and diploids. *Ricerca Sci.* **29**, 3–6.

Barigozzi, C. and Baratelli-Zambruni, L. (1983) New data on the chromosome number of the genus *Artemia*. *Atti della Accademia Nazionale dei Lincei, Rendiconti (Classe di Scienze fisiche, matematishe e naturali)* **73**, 139–143.

Barigozzi, C., Valsasnini, P., Ginelli, E., Badaracco, G., Levani, P. and Baratelli, L. (1987) Further data on repetitive DNA and speciation in *Artemia*. In P. Sorgeloos, D.A. Bengtson, W. Decleir and E. Jaspers (eds.), *Artemia Research and its Applications*, Vol. 1, Universa Press, Wetteren, Belgium, pp. 103–105.

Batuecas, B., Marco, R., Calleja, M. and Garesse, R. (1987) Molecular characterization of *Artemia* mitochondrial DNA: cloning, physical mapping, and preliminary gene organization. In W. Decleir, L. Moens, H. Slegers, E. Jaspers and P. Sorgeloos (eds.), *Artemia Research and its Applications*, Vol. 2, Universa Press, Wetteren, Belgium, pp. 355–367.

Beardmore, J.A. and Abreu-Grobois, F.A. (1983) Taxonomy and evolution in the brine shrimp *Artemia*. In G.S. Oxford and D. Rollinson (eds.), *Protein Polymorphism: Adaptive and Taxonomic Significance*, The Systematics Association Special Volume No. 24, Academic Press, London, pp. 153–164.

Belk, D. and Brtek, J. (1995) Checklist of the Anostraca. *Hydrobiologia* **298**, 315–353.

Booth, C.L., Woodruff, D.S. and Gould, S.J. (1990) Lack of significant associations between allozyme heterozygosity and phenotypic traits in the land snail *Cerion*. *Evolution* **44**, 210–213.

Bossart, J.L. and Prowell, D.P. (1998) Genetic estimates of population structure and gene flow: limitations, lessons and new directions. *Trends in Ecology & Evolution* **13**, 202–206.

Browne, R.A. and Bowen, S.T. (1991) Taxonomy and population genetics of *Artemia*. In R.A. Browne, P. Sorgeloos and C.N.A. Trotman (eds.), *Artemia Biology*, CRC Press, Boca Raton, Florida, pp. 221–235.

Bowen, S.T., Fogarino, E.A., Hitcher, K.N., Dana, G.L., Chow, H.S., Buoncristiani, M.R. and Carl, J.R. (1985) Ecological isolation in *Artemia*: population differences in tolerance of anion concentrations. *Journal of Crustacean Biology* **5**, 106–129.

Cai, Y. (1989) A redescription of the brine shrimp (*Artemia sinica*). *Wasmann Journal of Biology* **47**, 105–110.

Colihueque, N. and Gajardo, G. (1996) Chromosomal analysis in *Artemia* populations from South America. *Cytobios* **88**, 141–148.

Coyne, J.A. (1992) Genetics and speciation. *Nature* **355**, 511–515.

Coyne, J.A. and Orr, H.A. (2000) The evolutionary genetics of speciation. In R.M. Singh and C.B. Krimbas (eds.), *Evolutionary Genetics. From Molecules to Morphology*, Cambridge University Press, USA, pp. 532–569.

De los Rios, P. and Zuñiga, O. (2000) Biometric comparison of the frontal knob in American populations of *Artemia* (Anostraca, Artemiidae). *Revista Chilena de Historia Natural* **73**, 31–38.

De Meester, L. (1996) Local genetic differentiation and adaptation in freshwater zooplankton populations: patterns and processes. *Ecoscience* **3**, 385–399.

Dobzhansky, T. (1976) Organismic and molecular aspects of species formation. In F.J. Ayala (ed.), *Molecular Evolution*, Sinauer Associates, Sunderland, MA, pp. 95–105.

Freeman, J.A. and Chronister, R.B. (1988) Cell-specific endopolyploidy in developing *Artemia*. *Roux's Archives of Developmental Biology* **197**, 490–495.

Fryer, G.(1987) A new classification of the branchiopod Crustacea. *Zoological Journal of the Linnean Society* **91**, 357–383.

Gajardo, G.M. and Beardmore, J.A. (1989) Ability to switch reproductive mode in *Artemia* is related to maternal heterozygosity. *Marine Ecology Progress Series* **55**, 191–195.

Gajardo, G.M. and Beardmore, J.A. (1993) Electrophoretic evidence suggests that the *Artemia* found in the Salar de Atacama, Chile, is *A. franciscana* Kellogg. *Hydrobiologia* **257**, 65–71.

Gajardo, G. and Beardmore, J.A. (2001) Coadaptation: lessons from the brine shrimp *Artemia*, 'the aquatic *Drosophila*' (Crustacea, Anostraca). *Revista Chilena de Historia Natural* **74**, 65–72.

Gajardo, G.M., Wilson, R. and Zuñiga, O. (1992) Report on the occurrence of *Artemia* in a saline deposit of the Chilean Andes (Branchiopoda, Anostraca). *Crustaceana* **63**, 169–174.

Gajardo, G., da Conceicao, M., Weber, L. and Beardmore, J.A. (1995) Genetic variability and interpopulational differentiation of *Artemia* strains from South America. *Hydrobiologia* **302**, 21–29.

Gajardo, G., Colihueque, M., Parraguez, M. and Sorgeloos, P. (1998) International Study on *Artemia* LVIII. Morphologic differentiation and reproductive isolation of *Artemia* populations from South America. *International Journal of Salt Lake Research* **7**, 133–151.

Gajardo, G., Mercado, C., Beardmore, J.A. and Sorgeloos, P. (1999) International study on *Artemia*. LX. Allozyme data suggest that a new *Artemia* population in southern Chile (50° 29′ S; 73° 45′ W) is *A. persimilis*. *Hydrobiologia* **405**, 117–123.

Gajardo, G., Beardmore, J.A. and Sorgeloos, P. (2001a) Reproduction in the brine shrimp *Artemia*: evolutionary relevance of laboratory cross-fertility tests. *Journal of Zoology (London)* **253**, 25–32.

Gajardo, G., Beardmore, J.A. and Sorgeloos, P. (2001b) Genomic relationships between *Artemia franciscana* and *A. persimilis*, inferred from chromocentre numbers. *Heredity* **87**, 172–177.

Garesse, R., Carrodeguas, J.A., Santiago, J., Perez, M.L., Marco, R. and Vallejo, C.G. (1997) *Artemia* mitochondrial genome: molecular biology and evolutive considerations. *Comparative Biochemistry and Physiology B* **117**, 357–366.

Goldschmidt, E. (1952) Fluctuation in chromosome number in *Artemia salina*. *Journal of Morphology* **91**, 111–113.

Gosling, E.M. (1994) Speciation and wide-scale genetic differentiation. In A.R. Beaumont (ed.), *Genetics and Evolution of Aquatic Organisms*, Chapman and Hall, London, pp. 1–15.

Günther, R.T. (1890) Crustacea. In R.T. Günther (ed.), Contributions to the natural history of Lake Urmi, N.W. Persia and its neighbourhood. *Journal of the Linnean Society (Zoology)* **27**, 394–398.

Hanski, I. (1991) Single-species metapopulation dynamics: concepts, models and observations. *Biological Journal of the Linnean Society* **42**, 17–38.

Harris, H. (1966) Enzyme polymorphisms in man. *Proceedings of the Royal Society of London* B **164**, 298–310.
Hey, J. (2001) The mind of the species problem. *Trends in Ecology & Evolution* **16**, 326–329.
Hillis, D., Moritz, C. and Mable, B. (1996) *Molecular Systematics*, 2nd edition, Sinauer Associates, Sunderland, MA.
Hontoria, F. and Amat, F. (1992) Morphological characterization of adult *Artemia* (Crustacea, Branchiopoda) from different geographical origins. Mediterranean populations. *Journal of Plankton Research* **14**, 949–959.
Iwasaki, T. (1969) Chromosome number of *Artemia salina* obtained in the Great Salt Lake, Utah, USA. *Japanese Journal of Genetics* **44**, 105–106.
Kappas, I. (2001) Microevolution and genetic differentiation in brine shrimp *Artemia*, Ph.D. Thesis, University of Wales Swansea, Wales, UK.
Kellogg, V.A. (1906) A new *Artemia* and its life conditions. *Science* **24**, 594–596.
Kimura, M. (1983) *The Neutral Theory of Molecular Evolution*, Cambridge University Press, Cambridge, England.
Lenz, P.H. and Browne, R.A. (1991) Ecology of *Artemia*. In R.A. Browne, P. Sorgeloos and C.N.A. Trotman (eds.), *Artemia Biology*, CRC Press, Boca Raton, Florida, pp. 237–253.
Lewontin, R. (2000) The problems of population genetics. In R.M. Singh and C.B. Krimbas (eds.), *Evolutionary Genetics. From Molecules to Morphology*, Cambridge University Press, USA, pp. 5–23.
Lewontin, R.C. and Hubby, J.L. (1966) A molecular approach to the study of genic variation in natural populations. II. Amount of variation and degree of heterozygosity in natural populations of *Drosophila pseudoobscura*. *Genetics* **54**, 595–609.
Maeda-Martinez, A.M., Obregon-Barboza, H. and Dumont, H.J. (1992) *Branchinecta belki* n. Sp. (Branchiopoda: Anostraca), a new fairy shrimp from Mexico, hybridizing with *B. packardi* Pearse under laboratory conditions. *Hydrobiologia* **239**, 151–162.
Marco, R., Garesse, R., Cruces, J. and Renart, J. (1991) *Artemia* molecular genetics. In R.A. Browne, P. Sorgeloos and C.N.A. Trotman (eds.), *Artemia Biology*, CRC Press, Boca Raton, Florida, pp. 1–19.
Mayr, E. (1969) *Principles of Systematic Zoology*, McGraw-Hill, New York.
Mitrofanov, Y.A., Otradnova, V.V. and Val'vach, A.A. (1976) The karyotype of *Artemia salina*. *Tsitologiya* **18**, 233–237.
Mitrofanov, Y.A., Ivanofsky, Y.A. and Lesnikova, L.N. (1982) Chromosome numbers and karyotypes of some populations of *Artemia salina*. *Tsitologiya i Genetika* **4**, 11–14.
Mitton, J.B.(1997) *Selection in Natural Populations*, Oxford University Press, New York.
Moller, A.P. and Swaddler, J.P. (1997) *Asymmetry, Developmental Stability and Evolution*, Oxford University Press, Oxford.
Mura, G. (1990) *Artemia salina* (Linnaeus, 1758) from Lymington, England: frontal knob morphology by scanning electron microscopy. *Journal of Crustacean Biology* **10**, 364–368.
Naihong, X., Audenaert, E., Vanoverbeke, J., Brendonck, L., Sorgeloos, P. and De Meester, L. (2000) Low among-population genetic differentiation in Chinese bisexual *Artemia* populations. *Heredity* **84**, 238–243.
Nei, M. (1972) Genetic distance between populations. *American Naturalist* **106**, 283–292.
Orr, H.A. (2001) The genetics of species differences. *Trends in Ecology & Evolution* **16**, 343–350.
Orr, M.H. and Smith, T.B. (1998) Ecology and speciation. *Trends in Ecology & Evolution* **13**, 502–506.
Papeschi, A.G., Cohen, R.G., Pastorino, X.I. and Amat, F. (2000) Cytogenetic proof that the brine shrimp *Artemia franciscana* (Crustacea, Branchiopoda) is found in Argentina. *Hereditas* **133**, 159–166.
Parker Jr., E.D. and Niklasson, M. (2000) Genetic structure and evolution in parthenogenetic animals. In R.M. Singh and C.B. Krimbas (eds.), *Evolutionary Genetics. From Molecules to Morphology*, Cambridge University Press, USA, pp. 456–474.
Perez, M.L., Valverde, J.R., Batuecas, B., Amat, F., Marco, R. and Garesse, R. (1994) Speciation

in the *Artemia* genus: mitochondrial DNA analysis of bisexual and parthenogenetic brine shrimps. *Journal of Molecular Evolution* **38**, 156–168.

Piccinelli, M. and Prosdocimi, T. (1968) Descrizione tassonomica delle due species *Artemia salina* L. e *Artemia persimilis* n. sp. *Istituto Lombardo, Accademia di Scienze e Lettere, Rendiconti B* **102**, 170–179.

Pilla, E.J.S. (1992) Genetic differentiation and speciation in Old World *Artemia*, Ph.D. Thesis, University of Wales Swansea, Wales, UK.

Pilla, E.J.S. and Beardmore, J.A. (1994) Genetic and morphometric differentiation in Old World bisexual species of the brine shrimp *Artemia*. *Heredity* **73**, 47–56.

Queller, D.C., Strassmann, J.E. and Hughes, C.R. (1993) Microsatellites and Kinship. *Trends in Ecology & Evolution* **8**, 285–288.

Quynh, V.D. and Nguyen, N.L. (1987) Inoculation of *Artemia* in experimental ponds in central Vietnam: an ecological approach and a comparison of three geographical strains. In P. Sorgeloos, D.A. Bengtson, W. Decleir and E. Jaspers (eds.), *Artemia Research and its Applications*, Vol. 3, Universa Press, Wetteren, Belgium, pp. 253–269.

Rice, W.R. and Hostert, E.E (1993) Laboratory experiments on speciation: what have we learned in 40 years? *Evolution* **47**, 1637–1653.

Rieseberg, L.H. (2001) Chromosomal rearrangements and speciation. *Trends in Ecology & Evolution* **16**, 351–358.

Rodríguez Gil, S., Papeschi, A.G. and Cohen, R.G. (1998) Mitotic and meiotic chromosomes of *Artemia* (Branchiopoda) from populations of La Pampa Province, Argentina. *Journal of Crustacean Biology* **18**, 36–41.

Singh, R.M. (2000) Toward a unified theory of speciation. In R.M. Singh and C.B. Krimbas (eds.), *Evolutionary Genetics. From Molecules to Morphology*, Cambridge University Press, USA, pp. 570–604.

Sokal, R. and Rohlf, J. (1995) *Biometry*, 3rd edition, Freeman W.H. & Company, New York.

Sorgeloos, P. (1980) Life history of the brine shrimp *Artemia*. In G. Persoone, P. Sorgeloos, O. Roels and E. Jaspers (eds.), *The Brine Shrimp Artemia*, Vol. 1–3, Universa Press, Wetteren, Belgium, pp. XIX–XXIII.

Sorgeloos, P. and Beardmore, J.A. (1995) Editorial note: Correct taxonomic identification of *Artemia* species. *Aquaculture Research* **6**, 147.

Stefani, R. (1963) La digametia femminile in *A. salina* Leach e la constituzione del corredo cromosomico nei biotipi diploide amfigonico e diploide partenogenetico. *Caryologia* **16**, 625–636.

Sun, Y., Su, X. and Sun, G. (1995) Relationships between biological characteristics and chromosomal ploidy in *Artemia* from the coast of China. *Thai Journal of Aquacult. Science* **2**, 1–10.

Sun, Y., Zhong, Y., Song, W., Zhang, R. and Chen, R. (1999a) Detection of genetic relationships among four *Artemia* species using randomly amplified polymorphic DNA (RAPD). *International Journal of Salt Lake Research* **8**, 139–147.

Sun, Y., Song, W., Zhong, Y., Zhang, R., Abatzopoulos, T.J. and Chen, R. (1999b) Diversity of genetic differentiation in *Artemia* species and populations detected by AFLP markers. *International Journal of Salt Lake Research* **8**, 314–350.

Sweijd, N.A., Bowie, R.C.K., Evans B.S. and Lopata, A.L. (2000) Molecular genetics and the management and conservation of marine organisms. *Hydrobiologia* **420**, 153–164.

Tackaert, W. and Sorgeloos, P. (1991) Semi-intensive culturing in fertilized ponds. In R.A. Browne, P. Sorgeloos and C.N.A. Trotman (eds.), *Artemia Biology*, CRC Press, Boca Raton, Florida, pp. 287–315.

Templeton, A.R. (1989) The meaning of species and speciation: a genetic perspective. In D. Otte and J.A. Endler (eds.), *Speciation and its Consequences*, Sinauer Associates Inc., USA, pp. 3–27.

Thomas, K.M. (1995) Genetic variation and differentiation in Asian populations of *Artemia*, Ph.D. Thesis, University of Wales Swansea, Wales, UK.

Triantaphyllidis, G.V., Criel, G.R.J., Abatzopoulos, T.J. and Sorgeloos, P. (1997a) International Study on *Artemia*. LIV. Morphological study of *Artemia* with emphasis to Old World strains. II. Parthenogenetic populations. *Hydrobiologia* **357**, 155–163.

Triantaphyllidis, G.V., Criel, G.R.J., Abatzopoulos, T.J., Thomas, K.M., Peleman, J., Beardmore, J.A. and Sorgeloos, P. (1997b) International Study on *Artemia*. LVII. Morphological and molecular characters suggest conspecificity of all bisexual European and North African *Artemia* populations. *Marine Biology* **129**, 477–487.

Triantaphyllidis, G.V., Abatzopoulos, T.J. and Sorgeloos, P. (1998) Review of the biogeography of the genus *Artemia* (Crustacea, Anostraca). *Journal of Biogeography* **25**, 213–226.

Turelli, M., Barton, N.H. and Coyne, J.A. (2001) Theory and speciation. *Trends in Ecology & Evolution* **16**, 330–343.

Valverde, J.R., Batuecas, B., Moratilla, C., Marco, R. and Garesse, R. (1994) The complete mitochondrial DNA sequence of the crustacean *Artemia franciscana*. *Journal of Molecular Evolution* **39**, 400–408.

Ward, R.D. and Grewe, P.M. (1995) Appraisal of molecular genetic techniques in fisheries. In G.R. Carvalho and T.J. Pitcher (eds.), *Molecular Genetics in Fisheries*, Chapman & Hall, London, pp. 29–54.

White, M.J.D. (1978) *Modes of Speciation*, E. Freeman & Co., San Francisco.

Wiley, E.O. (1981) *Phylogenetics. The Theory and Practice of Phylogenetics Systematics*, John Wiley & Sons, New York.

Wiman, F.H. (1979) Mating patterns and speciation in the fairy shrimp genus *Streptocephalus*. *Evolution* **33**, 172–181.

Yang, Z. and Bielawski, J.P. (2000) Statistical methods for detecting molecular adaptation. *Trends in Ecology & Evolution* **15**, 496–503.

Zapata, C., Gajardo, G. and Beardmore, J.A. (1990) Multilocus heterozygosity and sexual selection in the brine shrimp *Artemia franciscana*. *Marine Ecology Progress Series* **62**, 211–217.

Zhang, L. and Lefcort, H. (1991) The effects of ploidy level on the thermal distributions of brine shrimp *Artemia parthenogenetica* and its ecological implications. *Heredity* **66**, 445–452.

Zuñiga, O., Wilson, R., Amat, F. and Hontoria, F. (1999) Distribution and characterization of Chilean populations of the brine shrimp *Artemia* (Crustacea, Branchiopoda, Anostraca). *International Journal Salt Lake Research* **8**, 23–40.

CHAPTER VI

APPLICATIONS OF *ARTEMIA*

JEAN DHONT and PATRICK SORGELOOS
Laboratory of Aquaculture & Artemia Reference Center
Faculty of Agricultural & Applied Biological Sciences
Ghent University, Rozier 44, B-9000 Gent
Belgium

1. Introduction

Even a cursory glance at the previous chapters on diverse aspects of *Artemia* will have convinced you of the astonishing convergence of unique features in this one organism, *Artemia*. It will come as no surprise then, that *Artemia* is cherished as a study object and a commodity by physiologists, biologists, biochemicists, ecotoxicologists, geneticists, aquaculturists and aquarists.

In this chapter we attempt to sketch the diverse applications of brine shrimp in all its aspects, starting with a short historic overview, up to the present time and projecting beyond this to offer some prospects on the supply of *Artemia* cysts.

2. *Artemia* Cyst Supply

2.1. A SHORT HISTORIC OVERVIEW

Artemia has been known to, and used locally by, humans probably for centuries. Hence its many popular names such as 'brine shrimp', 'Fezzan Wurm', 'Bahar-el-dud', and others. But its fame began to rise only in the 1930s when some investigators adopted it as a convenient replacement for the natural diet for fish larvae thus realizing the first break-through in the culture of commercially important fish species (Seale, 1933 and Rollefsen, 1939 in: Sorgeloos, 1980). In the 1950s *Artemia* cysts were still predominantly marketed for the aquarium and pet trade at costs as low as 10USD per kg. Commercial sources were few, basically there were two: the coastal salt works in the San Francisco Bay (SFB-California, USA) and the Great Salt Lake (GSL-Utah, USA). With fish and shrimp aquaculture developing from the early 1960s, new marketing opportunities were created for *Artemia* cysts. But by the mid seventies, increased demand, declining harvests from the Great Salt Lake, high import taxes in certain Third World countries and possibly artificial cyst shortage created by certain companies resulted in a severe price rise up to

50 or 100USD per kg *Artemia* cysts by the end of the eighties (Bengtson *et al.* 1991).

The dramatic impact of the cyst shortage on the expanding aquaculture industry invigorated research on the rationalization of the use of *Artemia* and exploration of new cyst resources. In that period the commercial exploitation of several other natural sources (Argentina, Australia, Canada, Colombia, France, PR China) and managed *Artemia* production sites (Brazil, Thailand) occurred. On the initiative of the *Artemia* Reference Center (Ghent University, Belgium) the International Study on *Artemia* (ISA) was established in order to co-ordinate a variety of different research initiatives (Sorgeloos, 1979). The cyst shortage also simultaneously invigorated the search for alternatives for *Artemia* such as micro-encapsulated diets (Jones *et al.* 1993; Samocha *et al.* 1999) in an attempt to abandon its use as live food in larval nutrition; a process that continues till today with slow but steady success.

With the development of improved techniques for cysts and nauplii applications (Léger *et al.* 1987a) and the exploitation of new natural resources, cyst prices returned to normal and market supply reached 50 metric tons and more by the 1980s. During the eighties improved techniques for harvesting from the open water and favorable hydrological and climatic conditions enabled a tenfold increase of the yields from GSL (> 200 t processed product) while the hatching quality was improved as well. This led to the precarious situation where, by the end of the eighties, cyst supply depended to more than 90% on one resource, namely GSL. But the Great Salt Lake proved – however big – to remain a natural ecosystem subject to climatic and other influences and has illustrated this by inflicting on us unpredictable and fluctuating cyst harvests (Figure 1). At this point it may be interesting to have a closer look at the characteristics of the Great Salt Lake.

2.2. GREAT SALT LAKE: ECOLOGICAL ASPECTS

Great Salt Lake is a desertic terminal lake located in the Great Basin, Utah (40°40′N, 112°20′W). The lake is 119 km long and 45 km wide, with a mean surface elevation of 1280 m above sea level. Like other terminal lakes depth, volume and salinity fluctuate widely in response to climatic cycles that determine rates of evaporation and precipitation (Wurtsbaugh and Smith Berry, 1990). Extremes in recent history were 1963, with the lowest lake surface and elevation and highest salinity, and 1986–1987 with the highest lake surface and elevation and lowest salinity (in the South arm) (Figure 2).

Besides climate, a major blow to the lake's stability was inflicted by the construction in 1959 of a solid-fill causeway replacing the existing wooden trestle for the Southern Pacific railway. This created two distinct bays that were named Gunnison Bay on the North, and Gilbert Bay on the South. Initially, the causeway was semi-permeable through two large culverts but over the years these became filled with debris thus allowing for little or no exchange between the northern and southern part. As virtually all surface inflow enters the

APPLICATIONS OF *ARTEMIA* 253

Figure 1. Harvest of a cyst 'streak' in the Great Salt Lake (Utah, USA) (courtesy Howard W. Newman, Desert Lake Technolgies, Klamath Falls, Oregon.

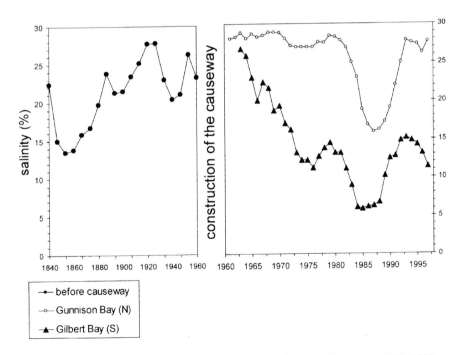

Figure 2. Salinity changes in the Great Salt Lake (Utah, USA) (data from USGS, 2001).

southern part of the lake the ecological and hydrological conditions between both parts have diverged rapidly. Basically, the salinity in Gilbert Bay decreased while it remained relatively constant in Gunnison Bay (Stephens, 1997).

Originally, *i.e.* before the divide, the lake biota primarily consisted of two species of benthic cyanobacteria, two species of brine fly, a green alga (*Dunaliella viridis*), some diatoms, protozoa and several species of halophilic bacteria (Stephens, 1974). After the divide, the salinity decline in the southern arm inevitably led to changes in species diversity and population dynamics. An invasion of several types of zooplankton other than *Artemia* brought about competition for food but also significant predation. *Artemia* populations that previously reached 5,000 to 18,000 individuals per m^3 dropped to less than 50 per m^3 in 1985–1986. Simultaneously, the commercial cyst harvest had to be stopped due to near depletion of the stocks (Figure 3). As a consequence of the reduced number of grazing *Artemia*, phytoplankton populations remained high, light penetration in the water column decreased, evaporation was slowed down and the effect of reduced salinity cascaded throughout the system (Stephens, 1997). In contrast, *Artemia* populations emerged in Gunnison Bay where they had always been very scarce due to traditionally high salinity. This allowed for modest cyst harvests as from 1983.

After 1987, salinities evolved back to normal and were followed by a considerable rise in cyst harvest. From 1995 till 1998 another salinity drop occurred (from 130 g/l to 85 g/l) this time causing a shift in phytoplankton composition from small chlorophytes and centric diatoms to a domination by pennate diatoms too large to be ingested by *Artemia* nauplii. In 1997, the commercial harvest was halted by the State at 'only' 2,700 tons of fresh product for fear of affecting the sustainability of the *Artemia* population (Stephens, 1999).

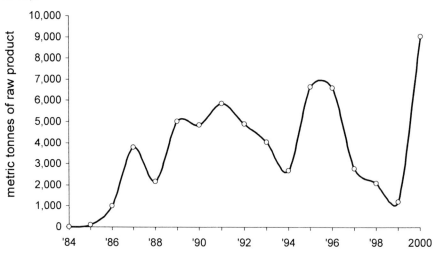

Figure 3. Cyst harvest (in metric tonnes raw product) from the Great Salt Lake (Utah, USA) (data from USGS, 2001).

2.3. TRENDS IN SUPPLY AND DEMAND

The unpredictable fluctuation of the cyst yields from Great Salt Lake has urged producers to explore new sites such as Lake Urmia in Iran, Aibi Lake in China, Bolshoye Yarovoye in Siberia, Kara Bogaz Gol in Turkmenistan, and several lakes in Kazakhstan (Lavens and Sorgeloos, 2000). In addition, numerous managed ponds and saltworks worldwide can provide small quantities (1–20 metric tons each). Although these sites are not believed to contribute substantially to the world supply of cysts, they provide interesting opportunities for local commercial development. Especially, in developing countries the local availability of cysts is an important asset for the development of a viable hatchery industry and a valued additional source of income.

New insights into hatching characteristics and nutritional essentials gave rise to the segregation of different cyst qualities. In the early 1990s cysts with good hatching quality could be purchased for as low as 20 USD per kg. However, prices of the small sized cysts with high eicosapentaenoic acid (EPA) levels, often referred to as AF-cysts, could reach over 100 USD per kg at times of short supply due to their critical role as starter food for marine fish larvae. With the severe cyst shortage in the mid 1990s and at the turn of the century (Figure 3) cyst prices inflated to levels around 100 USD per kg for GSL product and nearly 200 USD per kg for the so-called AF product. Following the super-harvest of over 9,000 metric tons of raw cysts from the Great Salt Lake in 2000–2001, prices started to come down again.

Over the last decade, cyst consumption increased exponentially as a consequence of the booming shrimp and marine fish industries. In 1997, some 6,000 hatcheries required over 1,500 metric tons of cysts annually. Some 80 to 85% of the total sales of *Artemia* went to shrimp hatcheries, mainly in China and South East Asia as well as Ecuador and a few other Latin-American countries, the remainder went to marine fish larviculture in Europe, China and Japan as well as to the pet fish producers. On the other hand, the rationalization of the use of *Artemia* in hatcheries (Sorgeloos *et al.* 1998; Sorgeloos *et al.* 2001) has enabled a dramatic reduction in the required amount of cysts per unit of produced fish or shrimp. For instance earlier a typical Mediterranean sea-bass and sea-bream hatchery would have been using some 150 kg cysts to produce 1 million fry while nowadays the required amount of cysts is brought down to 90 kg for bass and 70 kg for bream. Likewise, in shrimp hatcheries, the consumption of cysts dropped from 10 kg per million postlarvae to less than 5 kg.

Reliable estimates of future supplies remain difficult due to the lack of information on the ecology of new sites. Therefore, no stable cyst provision can be guaranteed and diversification of resources remains a most important issue along with the further rationalization of the use of *Artemia*.

3. *Artemia* and Solar Salt Production

3.1. IMPORTANCE OF SALT

Salt (NaCl or sodium chloride) has always been very precious to human societies, as an essential ingredient in our diet and as an indispensable tool for preserving food. Today, human consumption accounts for less than 10% of about 200 million tons produced annually, the bulk being consumed as crucial raw material for numerous chemical processes such as the chlorine-alkali industry (PVC, pulp and paper, organic chemicals, chloro-fluoro-carbons) soda ash, metallic sodium manufacture, salt for road de-icing, and many others (Salt Institute, 2000).

Equally important from another perspective is the role of salt as a source of income for the vast number of small farmers and their families making a living from artisanal salt production, mainly in developing countries.

3.2. SOLAR SALT PRODUCTION

Solar salt is typically produced by pumping seawater from one evaporation pond into another, allowing carbonates and gypsum to precipitate, and finally draining NaCl-saturated brine or 'pickle' into crystallizing ponds where sodium chloride precipitates (Figure 4). Before all NaCl has crystallized, the mother liquor, now called 'bittern', has to be drained off to reduce contamination of the sodium chloride with bromides and other salts that begin to precipitate

Figure 4. Typical lay-out of a integrated salt and *Artemia* farm (courtesy Patrick Sorgeloos).

at these elevated salinities. The technique of solar salt production thus involves fractional crystallization of salts in different ponds to obtain sodium chloride in the purest form possible, *i.e.* up to 99.7% on a dry weight basis.

The quantity and quality of salt produced is largely determined by the hydrobiological activity in a solar salt operation. Algal blooms are generally beneficial since they ensure increased solar heat absorption, resulting in faster evaporation and increased yields of salt. If they are not metabolized in time (*i.e.* consumed by brine shrimp), algal excretion and decomposition products, such as dissolved carbohydrates, act as chemical traps and consequently prevent early precipitation of gypsum, which will contaminate the sodium chloride and reduce salt quality. Furthermore, algal agglomerates and other organic impurities turn black upon oxidation and may contaminate the salt, reduce the size of the crystals and, hence, the salt quality. In the worst case, high water viscosity may completely inhibit salt formation and precipitation (Tackaert and Sorgeloos, 1993).

3.3. BENEFICIAL EFFECT OF *ARTEMIA* IN SALT PRODUCTION

Artemia enables the control of algal blooms and, through its metabolites and decaying animals, provides essential nutrients for the proliferation of *Halobacterium* in the crystallization ponds (Jones *et al.* 1981). High concentrations of these red halophilic bacteria reduce concentration of dissolved organics and promote heat absorption thereby accelerating evaporation.

Proper management of an *Artemia* population will not only lead to increased salt production and quality but will also provide opportunities for the harvest of a valuable by-product in the form of *Artemia* cysts and/or biomass. In small artisanal saltworks in (sub)tropical regions of South East Asia and Central and South America (mostly operated on a seasonal basis) yields of 20 kg DW cysts per ha and per month or 2 tons WW biomass per ha and per month can be achieved when they are managed for optimal *Artemia* production. This implies increasing the pond depth in order to prevent too high water temperatures and fertilizing regularly the lower salinity ponds with organic or inorganic products (Van Stappen and Sorgeloos, 1993). Recent field studies in Vietnam have proven that cyst yields in seasonal saltworks can be considerably improved when operating the system in a 'multi-cycle' way rather than the traditional continuous production (Baert *et al.* 1997).

3.4. AN EXAMPLE OF *ARTEMIA*, SALT AND SHRIMP INTEGRATION: THE BOHAI BAY, PR CHINA

Arguably the largest area for solar salt production in the world is located along the Bohai Bay in North East China where several hundred thousands of hectares of very labour intensive solar salt works are in operation. Yearly harvest of *Artemia* cysts amounts to several hundred tons fresh product, mainly collected in autumn.

Aside from cysts, huge quantities of *Artemia* biomass are collected by hundreds of saltworkers who lease the harvesting rights from the local salt companies. They sweep the pond with nets thus collecting over ten tons of live biomass per day (Figure 5). This is either used as live food in local shrimp and crab farms, or frozen and sold to aquaculture and aquarium pet farms or used as ingredient for local chicken feed (Tackaert and Sorgeloos, 1991a).

The saltworks provide unique conditions for the production of *Artemia*. As a result of the extreme eutrophication of the intake waters from the Bohai Bay, the evaporation ponds bloom with micro-algae in spring. Although this means sufficient food to sustain large populations of *Artemia*, the densities remain surprisingly low, especially at higher salinities. This is probably related to the particular climatic conditions of the Bohai Bay salterns as biotope: the main precipitation occurs in summer while winter and spring are mostly dry. The repopulation in spring by overwintering cysts is retarded because the hatching is constricted to the low salinity ponds. The higher salinity ponds become populated with *Artemia* only through the influx from low salinity ponds since the absence of rain prevents dilution of the upper brine layer down to salinity levels that permit hatching (Tackaert and Sorgeloos, 1991b).

Figure 5. Harvest of *Artemia* biomass in the Bohai bay (PR China) (courtesy Patrick Sorgeloos).

4. *Artemia* and Aquaculture

4.1. *ARTEMIA* AS INSTANT LIVE FOOD

The best feed from the farmer's viewpoint is seldom the food that the farmed animal prefers to eat. Likewise, a fish or shrimp farmer is looking for a cost-effective, easy-to-use, available-on-demand product while the fish or shrimp larvae would like to eat what its ancestors have been used to eating. So, when there appears to exist an organism that closely resembles the natural diet of most cultured fish and shrimp larvae and that – in one of its life forms – can be stored quasi indefinitely until one decides to activate it by a simple 24 h incubation: its success is guaranteed. Comprehensive literature reviews on the use of *Artemia* as live food in fish and shellfish larviculture have been published by Léger *et al.* (1986) and Sorgeloos *et al.* (1998, 2001). Today *Artemia* is used in the mass culture of different sea-bream species, sea-bass species, wolf-fish, cod, turbot, halibut, flounder species and other flatfish, milkfish, sturgeon, different carp and catfish species and whitefish species. The same is true for commercially important crustaceans such as several shrimp and prawn species, crawfish, several edible crab species and lobster.

However, one should not forget that any farmer will still switch to formulated feed as soon as it proves to be more cost-effective than *Artemia*. This switch will not only be triggered by the – constantly improving – quality of formulated feed but also by price and quality of *Artemia*. In general, most fish and shrimp larvae will be more likely to accept formulated feed as they grow bigger. This is not only a matter of size of mouth parts and particle size but also a matter of the developmental stage and efficiency of the digestive system. As a consequence, *Artemia* is essential only for those species that do need live food in their early life stages.

4.2. USE OF *ARTEMIA* NAUPLII

Nauplii in instar-I and instar-II stage are undoubtedly the most widely used forms of *Artemia* in aquaculture. They are also the easiest and earliest live food obtained from the cysts. Although the hatching process is relatively simple, several conditions need to be considered if one wishes to make optimal use of the cysts (Lavens and Sorgeloos, 1987). Primary critical factors are: light, temperature, salinity, oxygen level, pH and cyst density (Figure 6). Optimal values of these hatching conditions may vary among *Artemia* strains (Vanhaecke and Sorgeloos, 1983).

The quality of an *Artemia* strain is determined by both the hatching quality of the cysts and the nutritional value of the nauplii. There does not seem to be a correlation between these factors. Hatching quality of a batch of cysts (Figure 7) can be described by its hatching efficiency (number of nauplii per g cysts), hatching percentage (number of hatching cysts per 100 cysts), hatching synchrony (time between first and last hatching cysts) (Vanhaecke and

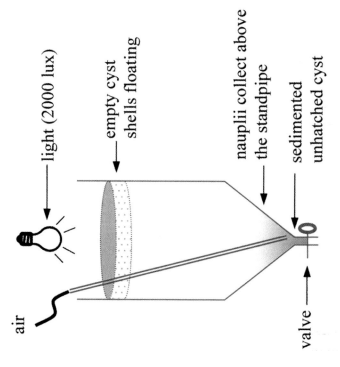

Figure 6. Hatching tank: small scale (left) and big scale (right).

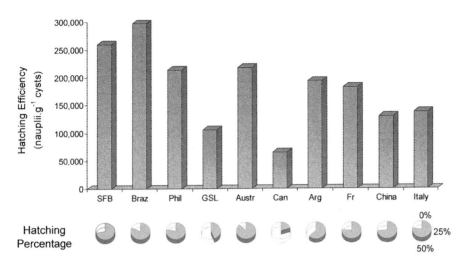

Figure 7. Hatching efficiency (in nauplii. g^{-1} cysts) and hatching percentage (in %) of *Artemia* cysts from various origin (modified from Vanhaecke and Sorgeloos, 1983).

Sorgeloos, 1982). Size and energy content are other important and strain-specific characteristics (Bengtson *et al.* 1978).

The nutritional value of nauplii is not at all a standard parameter that can be tagged to a specific batch of cysts. It can only be defined in relation to the species and stage of the larvae to which it is supposed to be fed, which also implies that the nutritional requirements of the latter is known. Even that is rarely the case. Needless to say, given a specific set of nutritional requirements, the suitability of nauplii differ among strains and even among seasonal harvests within the same strain (Léger *et al.* 1986). Indeed, cysts – thus nauplii – reflect the nutritional status of the parental material that in its turn is subject to ecological changes of its biotope (Table 1). The best documented nutritional component in this regard is undoubtedly the fatty acids.

4.3. THE MYSTERY OF THE 'UNIDENTIFIED FATTY ACIDS'

In the early 1970s several authors reported problems with the larviculture of shrimp, crab, prawn, lobster and marine fish larvae when using *Artemia* sources other than San Francisco Bay *Artemia*. Research teams far and wide conducted a frenetic quest for the hidden factor(s) responsible for those striking differences in culture results, supposedly related to the origin of the *Artemia* strain used (Bengtson *et al.* 1991). Extensive comparative studies with different strains of *Artemia* revealed differences in levels of specific polyunsaturated fatty acids (Kanazawa *et al.* 1979) (Table 2). Meanwhile it turned out that the fatty acid composition of a batch of cysts is largely environmentally determined. Millamena *et al.* (1988) and Lavens *et al.* (1989) have demon-

Table 1. Proximate composition of different developmental stages of Artemia from different origin (% on DW basis).

Artemia stage source (& reference)	Protein (%)	Lipid (%)	Carbohydrate (%)	Ash (%)	Fibre (%)	DW (µg)
Cysts						
GSL – USA [a]	55.8	11.2	6.9	5.9	–	4.83
SFB – USA [e]	45.2	3.9	36.3	5.2	–	–
Mexico [e]	41.4–50.2	0.3–1.0	36.4	5.8–12.6	–	–
Decapsulated cysts						
GSL – USA [a]	50.6	14.7	6.6	10.6	–	3.42
SFB – USA [d]	67.4	15.7	–	–	–	–
Nauplii						
GSL – USA [a]	56.2	17.0	3.6	7.6	–	2.31
GSL – USA [b]	41.6–47.2	20.9–23.1	10.5	9.5	–	1.65–2.70
GSL – USA [c]	61.9	14.4	10.6	7.1	5.9	–
P.R. China [b]	47.3	12.0	–	21.4	–	3.09
France [b]	55.7	12.4	–	15.4	–	2.7–3.1
SFB – USA [b, d]	41.9–59.2	15.9–27.2	11.2	8.17	–	1.45–2.87
Adults – wild population						
San Diego – USA [b]	64.0	12.0	–	20.6	–	
SFB – USA [b]	50.2–58.0	2.4–19.3	17.2	29.2	–	
Italy [g]	41.9	3.5	–	–	–	
Adults – cultured						
GSL – USA [b, c, f]	50.8–67.4	10.8–30.6	4.0–12.3	5.2–13.6	4.2	
France [b]	53.7	9.4	–	21.6	–	
SFB – USA [d]	39.4–64.0	4.5–12.1	–	–	–	
Italy [g]	55.4	4.0	20.0	20.6	–	

–: not mentioned
a: Garcia-Ortega et al. 1998
b: Léger et al. 1986
c: Lim et al. 2001
d: Dendrinos and Thorpe, 1987
e: Correa Sandoval et al. 1993
f: Correa Sandoval et al. 1994
g: Trotta et al. 1987

Table 2. Fatty acid composition of different developmental stages of *Artemia* from different origin (mg/g DW).

Artemia stage source (reference)	Palmitic acid (16:0)	Palmitoleic acid (16:1n-7)	Oleic acid (18:1n-9)	Linoleic acid (18:2n-6)	Linolenic acid (18:3n-3)	EPA (20:5n-3)	DHA (22:6n-3)
Cysts							
GSL – USA [a]	12.7	3.9	19.1	5.5	27.2	3.2	0.1
Decapsulated cysts							
GSL – USA [a, b]	16.1–25.7	5.0–8.1	24.2	6.9–11.1	34.2–49.4	3.9–4.7	0.1
France [b]	16.5	9.8	20.9	7.3	22.7	6.2	
Nauplii							
GSL – USA [a, c, d]	13.2–19.4	4.1–7.4	20.3–34.8	5.7–10.1	28.6–40.0	3.5–8.9	tr
SFB – USA [e]	3.6	0.1	6.1	1.8	8.2	0.6	
PR China [f, g, h]	12.6	6.7	17.8	11.0	3.6–39.3	1.4–7.5	0.0–0.4
Urmia, Iran [g]						2.7	0.4
A. parthenogenetica [g, h]	15.7	1.6	23.7	12.2	6.8	3.5–14.7	0.0–0.4
Madagascar [i]	18.6	20.2	21.2	7.4	6.25	24.5	0.0
A. persimilis – Argentina [h]	14.4	9.5	17.9	8.0	16.7	0.0	0.3
A. tibetiana – China [j]	23.6	2.4	40.4	6.2	tr	44.7	0.2
Adults							
GSL – USA [k]	3.9	1.7	7.4	15.4	1.7	1.7	0.8
GSL – USA [l]	9.1	4.3	18.3	15.9	0.0	2.8	0.0
SFB – USA [c]	4.5–14.5	0.2–1.0	9.0–17.1	1.7–30.5	3.5–13.5	2.5–3.9	0.1–0.4

a: Garcia-Ortega et al. 1998
b: Lavens et al. 1989a
c: Estevez et al. 1998
d: Han et al. 2000a
e: Dendrinos and Thorpe, 1987 (adults fed on yeast)
f: Dhert et al. 1993
g: Triantaphyllidis et al. 1995
h: Han et al. 2000b
i: Triantaphyllidis et al. 1996
j: Han et al. 1999
k: Lim et al. 2001 (adults fed on rice bran)
l: Lavens, 1989 (adults fed on soy pellets)

strated that the fatty acid profiles of adult *Artemia*, as well as the cysts they produce, strongly reflect the fatty acid profile of the diet they were fed.

From the work of Watanabe and his colleagues (1978) and later studies in the framework of the ISA (see also section 2 of this chapter) it became clear that the (n-3) series of fatty acids, especially eicosapentaenoic acid 20:5n-3 (EPA), determine the nutritional effectiveness more than any other single biochemical component. Later studies confirmed the correlation between EPA levels in the live food and culture success of several cultured fish and crustacean species (Watanabe *et al.* 1983; Léger *et al.* 1986) (Figure 8). More recently, the importance of another HUFA, docosahexaenoic acid 22:6n-3 (DHA), and of the ratio DHA/EPA have been demonstrated. Generally, aquatic animals seem to have a higher requirement for the n-3 series of fatty acids than terrestrial animals, for which the n-6 series is more important (Matty, 1989). These 'omega-3' fatty acids have a bewildering spectrum of biological activities amongst which is their role in the formation of biological membranes – having sweeping consequences for aquatic larvae – (Willis, 1987). Particularly cold water marine fish larvae require n-3 fatty acids in order to maintain the flexibility and permeability of their biomembranes at low temperatures. DHA seems to be specifically of paramount importance to the formation of neural, retina and brain tissue. A glance at any marine fish larvae with its relatively big eyes and head makes it easy to understand the importance of visual and neural functions for the survival of these sophisticated but fragile predators (Sargent, 1992).

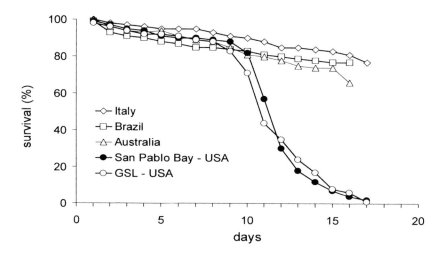

Figure 8. Survival of crab larvae (*Rhithropanopeus harrisii*) fed with *Artemia* nauplii from different origin (modified from Johns *et al.* 1980).

4.4. ENRICHMENT OR BIO-ENCAPSULATION TECHNIQUES

Commercially available cysts of various origin appeared to yield nauplii that lack essential nutrients such as essential fatty acids, vitamins, anti-oxidants etc. Fortunately, brine shrimp are non-selective filter feeders and this offers the possibility to manipulate their nutritional value in a fairly easy manner. As soon as brine shrimp moult to the second nauplius stage, that is 8 h or more after hatching, they will start filtering particles smaller than 25 µm irrespective of the particle's nature (Makridis and Vadstein, 1999; Gelabert Fernández, 2001). Taking advantage of this non-selective filter feeding, simple methods were developed to incorporate various kinds of products into nauplii prior to feeding them to predator larvae. This technique of 'bio-encapsulation' (Figure 9), commonly called *enrichment*, is widely applied in marine fish and crustacean hatcheries for enhancing the nutritional value of *Artemia* with essential fatty acids (Léger *et al.* 1987b), vitamins (Merchie *et al.* 1997), therapeutics (Chair *et al.* 1996), proteins (Tonheim *et al.* 2000), etc. A wide range of enrichment products has been developed such as unicellular algae, micro-encapsulated particles, coated yeast, self-emulsifying preparations (Léger *et al.* 1986). Basically this means that, if the nutritional requirements of the cultured species are sufficiently known, enrichment allows us to provide, rather easily, a live food that perfectly meets these requirements.

Currently, the highest enrichment levels (Table 3) are obtained using emulsified concentrates: freshly-hatched nauplii are transferred to an enrichment tank at a density of 100 to 300 nauplii/ml for a period of 24 h to 48 h. The enrichment is performed in seawater kept at 25 °C that has been disinfected with hypochlorite and neutralized before use. The enrichment emulsion is added in consecutive doses of 0.3 g/l every 12 h. Strong aeration using air-

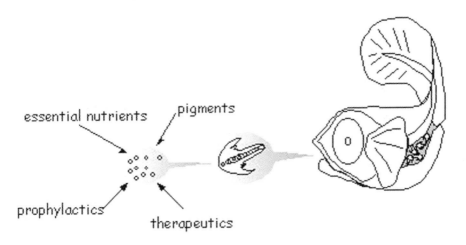

Figure 9. Schematic representation of bio-encapsulation.

Table 3. Overview of lipid levels obtained through enrichment of *Artemia franciscana* by various authors (in mg/g DW)

Stage Reference	Linolenic acid (LNA) 18:3n-3	Arachidonic acid (ARA) 20:4n-6	Eicosapentaenoic acid (EPA) 20:5n-3	Docosahexaenoic acid (DHA) 22:6n-3
Nauplii				
Devresse et al. 1992	0.5–0.8	0.0–1.3	9.5–18.3	22.7–35.7
Dhert et al. 1993	39.8	1.0	10.7–17.8	27.5–45.7
Triantaphyllidis et al. 1995	–	–	17.7	25.1
Evjemo et al. 1997	–	–	17.4	36.6
Estevez et al. 1998	20.6–21.7	2.1–14.3	10.0–25.1	5.2–16.6
Narciso et al. 1999	–	–	4.1–20.2	1.4–11.1
Han et al. 2000b	25.6–45.0	2.4–4.9	29.6–53.2	11.2–28.9
Adults and juveniles				
Han, 2001	14.9–26.0	13.4	23.1	4.6
Lim et al. 2001	1.8	–	5.6	1.9
Dhont et al. 1991	–	–	9.4–36.0	5.0–17.1

– : not mentioned

stones or pure oxygen is required to maintain dissolved oxygen levels above 4 ppm (Léger et al. 1987b).

Initially most attention was focused to the presence of ecosapentaenoic acid (EPA = 20:5n-3) in *Artemia* and its importance for the successful production of marine fish and crustacean larvae (Watanabe et al. 1983; Léger et al. 1985). In the late eighties and beginning nineties, more attention was paid to the level of docosahexaenoic acid (DHA = 22:6n-3); good survival appeared to be correlated with EPA, but DHA improved larval quality and growth (Lisac et al. 1986). Before long, several authors documented the importance of DHA, more particularly the requirement for high DHA/EPA ratios in promoting growth, stress resistance and pigmentation (Kraul et al. 1993; Mourente et al. 1993; Reitan et al. 1994; Lavens et al. 1995). Whereas previously satisfactory results were obtained with DHA/EPA ratios of less than 1, the emphasis evolved to consider levels of 2 and higher (Dhert et al. 1993). Since these values are not found naturally in *Artemia* (Dhert et al.1993; Triantaphyllidis et al. 1995), special enrichment formulations and *Artemia* with low DHA catabolizing activity had to be identified.

Contrary to rotifers, it is hard to maintain high DHA levels in enriched *Artemia franciscana* because of the inherent catabolism of DHA resulting in low DHA/EPA ratios as soon as enrichment is interrupted (Navarro et al. 1999). An interesting solution is expected from the capacity of some *Artemia* strains to reach high DHA levels during enrichment (Dhert et al. 1993; Velazquez, 1996) and to maintain them during subsequent starvation (Evjemo et al. 1997, Han et al. 2000a).

Recent studies by Koven et al. (2000) showed that arachidonic acid ARA

(20:4n-6) may improve larval growth and pigmentation in several marine fish species since it provides precursors for eicosanoid production (Castell et al. 1994; Estevez et al. 1997). However, the requirement of ARA in fish seems to depend on the fish species and larval development and needs to be dosed with extreme care since its effect seems to depend on the DHA concentration (Castell et al. 1994, Koven et al. 2000).

4.5. COLD STORAGE OF ARTEMIA NAUPLII

Even if hatching may be considered as a simple operation it still requires labor and infrastructure. In order to reduce hatching operations, methods were developed for storing hatched nauplii. Once again the robustness of brine shrimp would be exploited when it appeared that freshly-hatched nauplii could be stored at 4 °C in densities up to 8 million per litre for at least 24 h without marked mortality. Moreover, its reduced metabolic rate at such low temperatures prevented energy or weight losses and even losses of nutrients in the case of stored enriched nauplii (Léger et al. 1983) (Table 3). A recent study has shown that nauplii can be stored at 20 million per litre at 4 °C for at least 48 h with less than 20% mortality provided that disinfection of the hatching and nauplii suspension is applied (Anbaya Almalul, 2000).

Cold storage enables the farmer to reduce hatching efforts (less frequent hatching and harvest daily, fewer tanks, bigger volumes). Using cold storage also allows for more frequent and automatised distribution of nauplii to larvae. An additional advantage of using cold stored nauplii is their initial slower motility when transferred to the larval rearing tank. With poor hunters, such as the larvae of turbot, feeding cold-stored, less active Artemia resulted in a much more efficient food uptake (Léger et al. 1986).

4.6. USE OF DECAPASULATED CYSTS

Decapsulation is the process whereby the chorion that encysts the Artemia embryo is completely removed by a short exposure to a hypochlorite solution (Bruggeman et al. 1980). Some Artemia strains seem to be very sensitive to this harsh treatment, probably related to differences in chorion thickness or structure, and these require a modified decapsulation procedure (De Wolf et al. 1998). The use of decapsulated cysts is much more limited than the use of Artemia nauplii. Nevertheless, dried decapsulated Artemia cysts have proven to be an appropriate feed for larval rearing of various species like the freshwater catfish (Clarias gariepinus) and common carp (Cyprinus carpio), marine shrimp and milkfish larvae (Verreth et al. 1987; Stael et al. 1995).

The main advantages of using decapsulated cysts in larval rearing is that daily production of nauplii, a labour intensive job that requires additional facilities, is avoided; and decapsulation of non-hatching cysts means valorisation of an otherwise inferior product. Decapsulated cysts that have been dried before use have a high floating capacity and sink only slowly to the bottom of the culture tank. Finally, leaching of nutritional components (*e.g.* as in artificial diets) does not occur, since the outer cuticular membrane acts as a barrier for large molecules.

Decapsulated cysts on the other hand have the disadvantage that they are non-motile and thus less visually attractive to the predator. Moreover, decapsulated cysts dehydrated in brine sink rapidly to the bottom thus reducing their availability for fish larvae feeding in the water column. Older penaeid larvae on the contrary, are mainly bottom feeders and do not encounter this problem.

From the nutritional point of view, the gross chemical composition of decapsulated cysts is comparable to that of freshly-hatched nauplii (García-Ortega *et al.* 1998) (Table 3). In addition, their individual dry weight and energy content is on the average 30 to 40% higher than for instar-I nauplii (Figure 10). For example, for culture of carp larvae in the first 2 weeks, the use of decapsulated cysts constitutes a saving of over one third in amount of *Artemia* cysts, compared to the use of live nauplii (Vanhaecke *et al.* 1990).

4.7. USE OF JUVENILE AND ADULT *ARTEMIA*

Artemia juveniles and adults are believed to have a better nutritional value than cysts. They have a higher protein content (Table 3), are richer in essential amino acids (Lim *et al.* 2001) and can be enriched in similar way but faster than

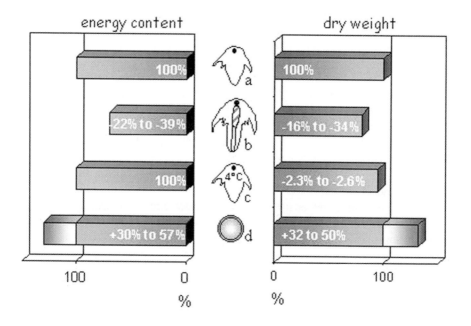

Figure 10. Energy content and dry weight of instar-I (a), instar-II (b), cold stored nauplii (c) and decapsulated cysts (d) (modified from Léger *et al.* 1987a).

nauplii (Dhont *et al.* 1991). They make a bigger prey thus resulting in a better energy balance per hunting effort for those predating fish or shrimp larvae that can ingest them. Besides nutritional and energetic advantages, the use of *Artemia* biomass for nursery feeding of fish and shrimp also results in improved economics as expenses for cysts and weaning diets can be reduced (Dhert *et al.* 1993). Live biomass has also proven to be an excellent maturation diet for penaeid shrimp (Naessens-Foucquart *et al.* 1996; Wouters *et al.* 1998), and is also very popular as live food for ornamental fish in the tropical aquarium industry (Lim *et al.* 2001). Finally the use of *Artemia* biomass has proven to improve stress resistance in *Penaeus monodon* (Dhert *et al.* 1993) and halibut larvae (Olsen *et al.* 1999).

Although the fresh-live form has the highest nutritive value, harvested *Artemia* can also be frozen, freeze-dried or acid-preserved (Abelin *et al.* 1991; Naessens *et al.* 1995) for later use, or made into flakes or other forms of formulated feed (Liying, 2000).

Juvenile and adult *Artemia*, often referred to as 'biomass', can be obtained through culturing (Dhert *et al.* 1992; Dhont and Lavens, 1996; Lim *et al.* 2001) or can be harvested from salt ponds or lakes (Baert *et al.* 1996). In China, thousands of tons are collected on an annual basis from the Bohai Bay salt ponds and used in the local culture of Chinese white shrimp, *Penaeus chinensis* (see also section 3.4 of this chapter). Although live biomass has a higher nutritive value, most of the 3,000 metric tons that harvested annually is marketed in frozen form. Part of it is also flaked, dried or incorporated in compound diets.

5. Applications of *Artemia* Pond Culture

5.1. VALORISATION OF UNICELLULAR ALGAE RESOURCES

The controlled production of micro-algae can be a rather complex and costly affair and will generally not pay off unless it yields valuable compounds or applications that guarantee a sufficient return. The cases where massive supplies of unicellular algae are available for brine shrimp culture are thus rare. Examples are the 'Artificial Upwelling Mariculture Project' on St. Croix, Virgin Islands (Tobias *et al.* 1979) where nutrient-rich deep ocean water was pumped into ponds at the surface where high density blooms of *Chaetoceros* were maintained. A simple flow-through culture system yielded as much as 20 kg WW per cubic meter in two weeks time.

5.2. INTEGRATED AQUACULTURE OPERATIONS RELYING ON *ARTEMIA* POND CULTURE

Various forms of integrated aquaculture operations are practiced of which some rely on the use of *Artemia* (Pudadera, 1988). Brine shrimp can act as converters of algal blooms into animal protein, blooms that may or may not have been induced through organic fertilizing, most often derived from chicken or pig farming. Other forms of integration rather rely on the sequential use of ponds for salt/*Artemia* production during dry season, followed by fish or shrimp during wet season (Jumalon and Ogburn, 1987).

Salt farmers who decide to incorporate *Artemia* in their operations find themselves rewarded with a *'brine shrimp biomass bonus'* (see section 6.3 of this chapter). An obvious further step is to expand their operation with a fish or a shrimp culture that can make use of that biomass rather than to market it. Equally important is the small but valuable harvest of cysts that can either yield an extra income or reduce the dependence on imported cysts.

5.3. VALORISATION OF ORGANIC WASTE AND USE OF *ARTEMIA* IN EFFLUENT TREATMENT

Growing environmental awareness has entailed the problem of waste treatment and disposal. Organic wastes can readily be tackled by bacterial oxidation or even by inducing algal growth but the resultant bacterial or algal sludge may require expensive chemical flocculation or mechanical separation. Techniques have been developed where brine shrimp are used as a convenient tool for harvesting the algae and/or bacteria at the same time yielding a high-profit product (McShan *et al.* 1974; Milligan *et al.* 1980). A most elegant application is the treatment of fish or shrimp pond effluents using *Artemia* that in its turn can be reused as valuable live food (Zmora and Shpigel, 2001).

6. Use of *Artemia* in Research and Education

6.1. THE CONSEQUENCES OF BEING 'EXTREMO-TOLERANT'

Because *Artemia* offers the rare opportunity of yielding live animals from a seemingly inert and storable product, it has been, and is being widely used by scientists in very diverse research fields. How to prove or quantify this popularity? Try a literature search on keywords [*Artemia* NOT aquaculture] and convince yourself. You will of course hit first and foremost on ecotoxicologists poisoning it with 1,001 different toxicants, but also on geneticists and molecular biologists in need of RNA, DNA, ribosomes and other cellular components; pharmacists mistreating it with excessive doses of their novel drug; biochemists and molecular biologists stripping it of its enzymes, haemoglobins or other required molecules; ecologists feeding it to their pet specimen while observing its behaviour; radiobiologists sending it into space to test various types of radiation on biological life forms, and many more uncanny treatments our unfortunate brine shrimp has to endure in the name of science. Finally, brine shrimp nauplii, being strong osmoregulators that can survive several hours in fresh water, have been used as live food for virtually any aquatic organism that has at one time or another been kept in a laboratory.

6.2. ARTEMIA IN ECOTOXICOLOGY

Acute toxicity tests with various aquatic species still form the cornerstone of methods to monitor and predict pollution effects of chemicals and effluents. Since the development and publication of the first pickle-jar bioassay in the 1940s, the number of available toxicity assays has grown dramatically (see review in: Janssen *et al.* 2000). While invertebrates are the most commonly used toxicity test organisms (Maltby and Calow, 1989 in: Janssen *et al.* 2000), increasing use is made of test-kits which contain a bio-indicator under the form of resting eggs from which the test organism can be hatched when needed (Persoone, 1998). Cyst-based toxicity tests have been developed for use in different environments but for the marine environment *Artemia* is undoubtedly the most used test-organism. Nevertheless the aptness of *Artemia* as a test organism is increasingly questioned due to its low ecological relevance for the marine ecosystem, but mostly for its lack of sensitivity for many of the common toxicants.

6.3. ARTEMIA AS A DIDACTIC TOOL

It is probably equally safe to claim that *Artemia* is frequently used in practical exercises by biology teachers but this is harder to prove. An example is found via the US National Association of Biology Teachers (NABT, 2000). It refers to an experiment where a brine shrimp bio-assay is used to identify potential medicinal plants and to determine anti-tumor activity.

Bearing in mind the adagio of experiential education: *'Tell me, and I will forget; Show me, and I may remember; Involve me and I will understand . . .'* a number of useful exercises can be devised for students to perform in order to contribute to a better understanding of biological and other principles. Think about population dynamics. A simple ecosystem fuelled by sunlight with unicellular algae and brine shrimp could beautifully illustrate the subtle balance between primary producers and consumers, the carrying capacity, overpopulation, and other factors. Koshida and Hiroki (1980) have developed and described a number of practical exercises on morphology, animal physiology (osmoregulation, enzyme activity), and animal behaviour. Other examples can easily be found in histology, environmental biology and many other fields.

6.4. ARTEMIA IN PERSONAL, SOCIAL AND SEX EDUCATION

Applications of *Artemia* are manifold indeed; we cite Tunnicliffe and Reiss (1999): 'Five hundred and seventy eight conversations were audio-taped while year 2 (6–7-year-old) or year 5 (9–10-year-old) pupils were observing meal worms (*Tenebrio molitar*) or brine shrimps (*Artemia salina*) during science activities in a UK primary school. Analysis revealed many comments on sex, reproduction, death, violence and the life history of the organisms. In addition, pupils made associations to humans, raised moral issues and expressed emotions and wonder. Significant differences were found between the comments of year 2 and year 5 pupils, between the comments of boys and girls, and between comments made when observing meal worms and when observing brine shrimps. These findings have implications both for the formal teaching of science and for the use of science as a vehicle for the teaching of sex education and other aspects of personal and social education (PSE).'

7. Acknowledgements

Over the past three decades, our studies with the brine shrimp, *Artemia*, have been supported by research contracts with the Belgian National Science Foundation (FWO), the Belgian Administration for Development Cooperation (ABOS), the Belgian Ministry of Science Policy, the European Commission, *Artemia* Systems NV and INVE Aquaculture NV, Belgium.

8. References

Abelin, P., Tackaert, W. and Sorgeloos, P. (1991) Ensiled *Artemia* biomass: a promising and practical feed for penaeid shrimp postlarvae. In P. Lavens, P. Sorgeloos, E. Jaspers and F. Ollevier (eds.), *Larvi'91 – Fish & Crustacean Larviculture Symposium, European Aquaculture Society, Special Publication Nr. 15*, Ostend, Belgium, pp. 125–127.

Anbaya Almalul, M.A. (2000) Effect of temperature and density on storage characteristics of *Artemia* nauplii, M.Sc. Thesis, Ghent University, Belgium.

Baert, P., Bosteels, T. and Sorgeloos, P. (1996) Pond production of *Artemia*. In P. Lavens and P. Sorgeloos (eds.), *Manual on the Production and Use of Live Food for Aquaculture*, FAO Fisheries Technical Paper 361, pp. 196–251

Baert, P., Anh, N.T.N., Vu Do Quynh and Hoa, N.V. (1997) Increasing cyst yields in *Artemia* culture ponds in Vietnam: the multi-cycle system. *Aquaculture Research* **28**, 809–814.

Bengtson, D.A., Beck, A.D. and Poxton, H.A. (1978) Comparative effects of live and artificial diets on growth and survival of juvenile Atlantic silverside, *Menidia menidia*. *Proceedings of the 9th Annual meeting WAS*, Atlanta, pp. 159–172.

Bengtson, D.A., Léger, P. and Sorgeloos, P. (1991) Use of *Artemia* as a food source for aquaculture. In R.A. Browne, P. Sorgeloos and C.N.A. Trotman (eds.), *Artemia Biology*, CRC Press, Boca Raton, Florida, pp. 255–285.

Bruggeman, E., Sorgeloos, P. and Vanhaecke, P. (1980) Improvements in the decapsulation technique of *Artemia* cysts. In G. Persoone, P. Sorgeloos, O. Roels and E. Jaspers (eds.), *The Brine Shrimp Artemia*, Vol. 3, Universa Press, Wetteren, Belgium, pp. 261–269.

Castell, J.D., Bell, J.G, Tocher, D.R., and Sargent, J.R. (1994) Effects of purified diets containing different combinations of arachidonic and docosahexaenoic acid on survival, growth and fatty acid composition of juvenile turbot (*Scophthalmus maximus*). *Aquaculture* **128**, 315–333.

Chair, M., Nelis, H.J., Léger, P., Sorgeloos, P. and De Leenheer, A. (1996) Accumulation of trimethoprim, sulfamethoxazole, and N-acetylsulfamethoxazole in fish and shrimp fed medicated *Artemia franciscana*. *Antimicrobial Agents and Chemotherapy* **40**, 1649–1652.

Correa Sandoval, F., Ramirez, L.F.B. and Lobina, D.V. (1993) The biochemical composition of the cysts of some Mexican populations of *Artemia franciscana* Kellogg, 1906. *Comparative Biochemistry and Physiology B* **104**, 163–167.

Correa Sandoval, F., Cordero Esquivel, B., Valenzuela-Espinoza, E. and Escobar Fernandez,

R. (1994) Biochemical composition of laboratory-cultured adults of *Artemia franciscana* Kellogg, 1906. *Rivista Italiana di Aquacoltura* **29**, 63–66.

Dendrinos, P. and Thorpe, J.P. (1987) Experiments on the artificial regulation of the amino acid and fatty acid contents of food organisms to meet the assessed nutritional requirements of larval, post-larval and juvenile Dover sole (*Solea solea* (L.)). *Aquaculture* **61**, 121–154.

Devresse, B., Léger, P., Sorgeloos, P., Murata, O., Nasu, T., Ikeda, S., Rainuzzo, J.R., Reitan, K. I., Kjorsvik, E. and Olsen, Y. (1992) Improvement of flat fish pigmentation through the use of DHA enriched rotifers and *Artemia*, *Book of Abstracts, 5th International Symposium on Fish Nutrition and Feeding*, Santiago, Chile, pp. 1–7.

De Wolf, T., Cirillo, A., Candreva, P., Deichmann, M. and Sorgeloos, P. (1998) Improvements in *Artemia* cyst decapsulation. In Aquaculture Europe '89: *Book of Abstracts, European Aquaculture Society, Special Publication n° 26*, Ostend, Belgium.

Dhert, P., Bombeo, R.B., Lavens, P. and Sorgeloos, P. (1992) A simple semi flow-through culture technique for the controlled super-intensive production of *Artemia* juveniles and adults. *Aquacultural Engineering* **11**, 107–119.

Dhert, P., Sorgeloos, P. and Devresse, B. (1993) Contributions towards a specific DHA enrichment in the live food *Brachionus plicatilis* and *Artemia* sp. In H. Reinertsen, L.A. Dahle, L. Jørgensen and K. Tvinnerheim (eds.), *Fish Farming Technology*, Balkema, Rotterdam, pp. 109–115.

Dhont, J. and Lavens, P. (1996) Tank production and use of ongrown *Artemia*. In P. Lavens and P. Sorgeloos (eds.), *Manual on the Production and Use of Live Food for Aquaculture*, FAO Fisheries Technical Paper 361, pp. 164–195.

Dhont J., Lavens, P. and Sorgeloos, P. (1991) Development of a lipid enrichment technique for *Artemia* juveniles produced in an intensive system for use in marine larviculture. In P. Lavens, P. Sorgeloos, E. Jaspers and F. Ollevier (eds.), *Larvi'91 – Fish & Crustacean Larviculture Symposium, European Aquaculture Society, Special Publication Nr. 15*, Ostend, Belgium, pp. 51–55.

Estévez, A., Ishikawaq, M. and Kanazawa, A. (1997) Effects of arachidonic acid on pigmentation and fatty acid composition of Japanese flounder, *Paralichthys olivaceus* (Temminck and Schlegel). *Aquaculture Research* **28**, 279–289.

Estévez, A., McEvoy, L.A., Bell, J.G. and Sargent, J.R. (1998) Effects of temperature and starvation time on the pattern and rate of loss of essential fatty acids in *Artemia* nauplii previously enriched using arachidonic acid and eicosapentaenoic acid-rich emulsions. *Aquaculture* **165**, 295–311.

Evjemo, J.O., Coutteau, P., Olsen, Y. and Sorgeloos, P. (1997) The stability of docosahexaenoic acid in two *Artemia* species following enrichment and subsequent starvation. *Aquaculture* **155**, 135–148.

García-Ortega, A., Verreth, J.A.J., Coutteau, P., Segner, H., Huisman, E.A. and Sorgeloos, P. (1998) Biochemical and enzymatic characterization of decapsulated cysts and nauplii of the brine shrimp *Artemia* at different developmental stages. *Aquaculture* **161**, 501–514.

Gelabert Fernández, R. (2001) *Artemia* bioencapsulation I. Effect of particle sizes on the filtering behavior of *Artemia franciscana*. *Journal of Crustacean Biology* **21**, 435–442.

Han, K. (2001) Use of lipid emulsions for the bio-encapsulation of highly unsaturated fatty acids in the brine shrimp *Artemia*, Ph.D. Thesis, Ghent University, Belgium.

Han, K., Geurden, I. and Sorgeloos, P. (1999) Enrichment of the nauplii of two *Artemia* species with docosahexaenoic acid. *Crustacean Issues* **12**, 599–604.

Han, K., Geurden, I. and Sorgeloos, P. (2000a) Comparison of docosahexaenoic acid (22:6n-3) levels in various *Artemia* strains during enrichment and subsequent starvation. *Journal of the World Aquaculture Society* **31**, 469–475.

Han, K., Geurden, I. and Sorgeloos, P. (2000b) Enrichment strategies for *Artemia* using emulsions providing different levels of n-3 highly unsaturated fatty acids. *Aquaculture* **183**, 335–347.

Janssen, C.R., Vangheluwe, M. and Van Sprang, P. (2000) A brief review and critical evaluation of the status of microbiotests. In G. Persoone, C. Janssen and W. De Coen (eds.), *New*

Microbiotests for Routine Screening and Biomonitoring, Kluwer Academic Press, New York, pp. 27-37.

Johns, D.M., Peters, M.E. and Beck, A.D. (1980) International Study on *Artemia* VI. Nutritional value of geographical and temporal strains of *Artemia*: effects on survival and growth of two species of Brachyuran larvae. In G. Persoone, P. Sorgeloos, O. Roels and E. Jaspers (eds.), *The Brine Shrimp Artemia*, Vol. 3, Universa Press, Wetteren, Belgium, pp. 291-304.

Jones, A.G., Ewing, C.M. and Melvin, M.V. (1981) Biotechnology of solar saltfields. *Hydrobiologia* **82**, 391-406.

Jones, D.A., Kamarudin, M.S. and Le Vay, L. (1993) The potential for replacement of live feeds in larval culture. *Journal of the World Aquaculture Society* **24**, 199-210.

Jumalon, N.A. and Ogburn, D.M. (1987) Nutrient flow and physicochemical profile studies of an integrated poultry-salt-*Artemia*-milkfish-sea bass-shrimp pond production system. In P. Sorgeloos, D.A. Bengtson, W. Decleir and E. Jaspers (eds.), *Artemia Research and its Applications*, Vol. 3, Universa Press, Wetteren, Belgium, pp. 239-251.

Kanazawa, A., Teshima, S.I. and Ono, K. (1979) Relationship between essential fatty acid requirements of aquatic animals and the capacity for bioconversion of linolenic acid to highly unsaturated fatty acids. *Comparative Biochemistry and Physiology B* **63**, 295-298.

Koshida, Y. and Hiroki, M. (1980) *Artemia* as a multipurpose biomaterial for biology education. In G. Persoone, P. Sorgeloos, O. Roels and E. Jaspers (eds.), *The Brine Shrimp Artemia*, Vol. 1, Universa Press, Wetteren, Belgium, pp. 289-298.

Koven, W., Barr, Y., Lutzky, S., Ben-Atia, I., Harel, M., Behrens, P., Weiss, R. and Tandler, A. (2000) The effect of dietary arachidonic acid (20:4n-6) on growth and survival prior to and following handling stress in the larvae of gilthead seabream (*Sparus aurata*). *Aqua 2000: Book of Abstracts, European Aquaculture Society, Special Publication Nr. 28*, pp. 248.

Kraul, S. (1993) Larviculture of the mahimahi (*Coryphaena hippurus*) in Hawaii, USA. *Journal of the World Aquaculture Society* **24**, 410-421.

Lavens, P. (1989) Intensive production and quality evaluation of *Artemia* adults and their offspring (in Dutch), Ph.D. Thesis, Ghent University, Belgium.

Lavens, P. and Sorgeloos, P. (1987) The cryptobiotic state of *Artemia* cysts, its diapause deactivation and hatching: a review. In P. Sorgeloos, D.A. Bengtson, W. Decleir and E. Jaspers (eds.), *Artemia Research and its Applications*, Vol. 3, Universa Press, Wetteren, Belgium, pp. 27-63.

Lavens, P. and Sorgeloos, P. (2000) The history, present status and prospects of the availability of *Artemia* cysts for aquaculture. *Aquaculture* **181**, 397-403.

Lavens, P., Léger, P. and Sorgeloos, P. (1989) Manipulation of the fatty acid profile in *Artemia* offspring produced in intensive culture systems. In N. De Pauw, E. Jaspers, H. Ackefors and N. Wilkins (eds.), *Aquaculture – a Biotechnology in Progress*, European Aquaculture Society, Bredene, Belgium, pp. 731-739.

Lavens, P., Coutteau, P. and Sorgeloos, P. (1995) Laboratory and field variation in HUFA enrichment of *Artemia* nauplii. In P. Lavens, E. Jaspers and I. Roelants (eds.), *Larvi'95 – Book of Abstracts, European Aquaculture Society, Special Publication n° 24*, Gent, pp. 137-140.

Léger, P., Vanhaecke, P. and Sorgeloos, P. (1983) International Study on *Artemia*. XXIV. Cold storage of live *Artemia* nauplii from various geographical sources: Potentials and limits in aquaculture. *Aquacultural Engineering* **2**, 69-78.

Léger, P., Sorgeloos, P., Millamena, O.M. and Simpson, K.L. (1985) International Study on *Artemia*. XXV. Factors determining the nutritional effectiveness of *Artemia*: the relative impact of chlorinated hydrocarbons and essential fatty acids in San Francisco Bay and San Pablo Bay *Artemia*. *Journal of Experimental Marine Biology and Ecology* **93**, 71-82.

Léger, P., Bengtson, D.A., Simpson, K.L. and Sorgeloos, P. (1986) The use and nutritional value of *Artemia* as a food source. *Oceanography and Marine Biology: an Annual Review* **24**, 521-623.

Léger, P., Bengtson, D.A., Sorgeloos, P., Simpson, K.L. and Beck, A.D. (1987a) The nutritional value of *Artemia*: a review. In P. Sorgeloos, D.A. Bengtson, W. Decleir and E. Jaspers (eds.), *Artemia Research and its Applications*, Vol. 3, Universa Press, Wetteren, Belgium, pp. 357–372.

Léger, P., Naessens-Foucquaert, E. and Sorgeloos P. (1987b) International Study on *Artemia* XXXV. Techniques to manipulate the fatty acid profile in *Artemia* nauplii, and the effect on its nutritional effectiveness for the marine crustacean *Mysidopsis bahia* (M.). In P. Sorgeloos, D.A. Bengtson, W. Decleir and E. Jaspers (eds.), *Artemia Research and its Applications*, Vol. 3, Universa Press, Wetteren, Belgium, pp. 411–424.

Lim, L.C., Soh, A., Dhert, P. and Sorgeloos, P. (2001) Production and application of ongrown *Artemia* in freshwater ornamental fish farm. *Aquaculture Economics and Management* **5**, 211–228.

Lisac, D., Franicevic, V., Vejmelka, Z., Buble, J., Léger, P. and Sorgeloos, P. (1986) International Study on *Artemia*. XLIII. The effect of live food fatty acid content on growth and survival of sea bream (*Sparus aurata*) larvae. *Conference of Ichthyopathology in Aquaculture*, Dubrovnik, pp. 1–10.

Liying, S. (2000) Use of *Artemia* biomass in practical diets and decapsulated cysts as food source for common carp (*Cyprinus carpio* L.), M.Sc. Thesis, Ghent University, Belgium.

Makridis, P. and Vadstein, O. (1999) Food size selectivity of *Artemia franciscana* at three developmental stages. *Journal of Plankton Research* **21**, 2191–2201.

Matty, A.J. (1989) Fatty acid confusion in fish diets. *Fish Farming International* **16**, 13–14.

McShan, M., Trieff, N.M. and Grajcer, D. (1974) Biological treatment of wastewater using algae and *Artemia*. *Journal of the Water Pollution Control Federation* **46**, 1742–1750.

Merchie, G., Lavens, P. and Sorgeloos, P. (1997) Optimization of dietary vitamin C in fish and crustacean larvae: a review. *Aquaculture* **155**, 165–181.

Millamena, O.M., Bombeo, R., Jumalon, N.A. and Simpson, K.L. (1988) Effects of various diets on the nutritional value of *Artemia* sp. as food for the prawn *Penaeus monodon*. *Marine Biology* **98**, 217–222.

Milligan, D.J., Quick, J.A., Hill, S.E., Morris, J.A. and Hover, R.J. (1980) Sequential use of bacteria, algae and brine shrimp to treat industrial wastewater at pilot plant scale. In G. Persoone, P. Sorgeloos, O. Roels and E. Jaspers (eds.), *The Brine Shrimp Artemia*, Vol. 3, Universa Press, Wetteren, Belgium, pp. 193–206.

Mourente, G., Rodriguez, A. and Sargent, J.R. (1993) Effects of dietary docosahexaenoic acid (DHA; 22:6n-3) on lipid and fatty acid composition and growth in gilthead sea bream (*Sparus aurata* L.) larvae during first feeding. *Aquaculture* **112**, 79–98.

NABT (2000) National Association of Biology Teachers, http://www.nabt.org

Naessens, E., Pedrazzoli, A., Vargas, V., Townsend, S., Cobo, M.L. and Dhont, J. (1995) Evaluation of preservation methods for *Artemia* biomass and application in postlarval rearing of *Penaeus vannamei*. In P. Lavens, E. Jaspers and I. Roelants (eds.), *Larvi'95 – Book of Abstracts*. European Aquaculture Society, Special Publication n° 24, Ostend, Belgium, pp. 338–341.

Naessens, E., Lavens, P., Gomez, L., Browdy, C.L., McGovern-Hopkins, K., Spencer, A.W., Kawahigashi, D. and Sorgeloos, P. (1997) Maturation performance of *Penaeus vannamei* co-fed *Artemia* biomass preparations. *Aquaculture* **155**, 87–101.

Narciso, L., Poussão-Ferreira, P., Passos, A. and Luís, O. (1999) HUFA content and DHA/EPA improvements of *Artemia* sp. with commercial oils during different enrichment periods. *Aquaculture Research* **30**, 21–24.

Navarro, J.C., Henderson, R.J., McEvoy, L.A., Bell, M.V. and Amat, F. (1999) Lipid conversions during enrichment of *Artemia*. *Aquaculture* **174**, 155–166.

Olsen, A.I., Attramadal, Y., Jensen, A. and Olsen, Y. (1999) Influence of size and nutritional value of *Artemia franciscana* on growth and quality of halibut larvae (*Hippoglossus hippoglossus*) during the live feed period. *Aquaculture* **179**, 475–487.

Persoone, G. (1998) Development and validation of Toxkit microbiotests with invertebrates, in particular crustaceans. In P.G. Wells, K. Lee and C. Blaise (eds.), *Microscale Testing in Aquatic Toxicology*, CRC Press, pp. 437–449.

Pudadera, B.J. (1988) Integrated brackish water aquaculture systems in the Asia-pacific region, Manuscript for FAO.
Reitan, K.I., Rainuzzo, J.R. and Olsen, Y. (1994) Influence of lipid composition of live feed on growth, survival and pigmentation of turbot larvae. *Aquaculture International* **2**, 33–48.
Salt Institute (2000): http://www.saltinstitute.org
Samocha, T.M., Matsumoto, T., Jones, E.R. and Torano, M. (1999) Use of artificial diets to reduce *Artemia* nauplii requirements for production of *Litopenaeus vannamei* postlarvae. *The Israeli Journal of Aquaculture-Bamidgeh* **51**, 157–168.
Sargent, J. (1992) New developments in the omega-3 story from man to fish and heart to brain, *Aquaculture News*, July, 4–5.
Sorgeloos, P. (1979) The brine shrimp, *Artemia salina*: A bottleneck in mariculture? In T.V.R. Pillay and W.A. Dill (eds.), *FAO Technical Conference on Aquaculture*, Fishing News Books Ltd, Farnham, pp. 321–324.
Sorgeloos, P. (1980) Life history of the brine shrimp *Artemia*. In G. Persoone, P. Sorgeloos, O. Roels and E. Jaspers (eds.), *The Brine Shrimp Artemia*, Vol. 1–3, Universa Press, Wetteren, Belgium, pp. XIX–XXIII.
Sorgeloos, P., Coutteau, P., Dhert, P., Merchie, G. and Lavens, P. (1998) Use of brine shrimp, *Artemia* spp., in larval crustacean nutrition: a review. *Reviews in Fisheries Science* **6**, 55–68.
Sorgeloos, P., Dhert, P. and Candreva, P. (2001) Use of brine shrimp, *Artemia* spp., in marine fish larviculture. *Aquaculture* **200**, 147–159.
Stael, M., Sanggontanagit, T., Van Ballaer, E., Puwapanich, N., Tunsutapanich, A. and Lavens, P. (1995) Decapsulated cysts and *Artemia* flakes as alternative food sources for the culture of *Penaeus monodon* postlarvae. In P. Lavens, E. Jaspers and I. Roelants (eds.), *Larvi'95 – Book of Abstracts, European Aquaculture Society*, Special Publication n° 24, Gent, Belgium, pp. 342–345.
Stephens, D.W. (1974) A summary of biological investigations concerning the Great Salt Lake, Utah (1861–1973). *Great Basin Naturalist* **34**, 221–229.
Stephens, D.W. (1997) Salinity-induced changes in the aquatic ecosystem of Great Salt Lake, Utah. *Utah Geological Association Guidebook* **26**, 1–7.
Stephens, D.W. (1999) Dynamics of the *Artemia franciscana* population in Great Salt Lake during the boom and bust period of 1995–1998, *Book of Abstracts of the 7th International Conference on Salt Lakes*, Death Valley, California, p. 14.
Tackaert, W. and Sorgeloos, P. (1991a) Salt, *Artemia* and shrimp: Integrated production of salt, *Artemia* and shrimp in the Peoples' Republic of China: The Tang Gu saltworks. *World Aquaculture (Magazine)* **22**(3), 11–17.
Tackaert, W. and Sorgeloos, P. (1991b) Biological management to improve *Artemia* and salt production at the Tang Gu saltworks in the Peoples' Republic of China. *Proc. Int. Symp. Biotechn. Saltponds, Sept 1990*, Tanggu, PR China, pp. 78–83.
Tackaert, W. and Sorgeloos, P. (1993) The use of brine shrimp *Artemia* in biological management of solar salt works, *Proc. 7th International Symposium on Salt*, Kakihana, Japan, pp. 617–622.
Tobias, W.J., Sorgeloos, P., Bossuyt, E. and Roels, O.A. (1979) The technical feasibility of mass-culturing *Artemia salina* in the St. Croix 'artificial upwelling' mariculture system. In J.W. Avault (ed.), *Proceedings 10th Annual Meeting of the World Mariculture Society*, Baton Rouge, USA, pp. 203–214.
Tonheim, S.K., Koven, W. and Rønnestad, I. (2000) Enrichment of *Artemia* with free methionine. *Aquaculture* **190**, 223–235.
Triantaphyllidis, G., Coutteau, P. and Sorgeloos, P. (1995) The stability of (n-3) highly unsaturated fatty acids in various *Artemia* populations following enrichment and subsequent starvation. In P. Lavens, E. Jaspers and I. Roelants (eds.), *Larvi'95 – Book of Abstracts, European Aquaculture Society*, Special Publication n° 24, Gent, Belgium, pp. 149–153.
Triantaphyllidis, G.V., Abatzopoulos, T.J., Miasa, E. and Sorgeloos, P. (1996) International Study on *Artemia*. LVI. Characterization of two *Artemia* populations from Namibia and Madagascar:

cytogenetics, biometry, hatching characteristics and fatty acid profiles. *Hydrobiologia* **335**, 97–106.

Trotta, P., Villani, P., Palmegiano, G.B., Forneris, G. and Sarra, C. (1987) Laboratory-grown *Artemia* as reference food for weaning fish fry and shrimp postlarvae. In P. Sorgeloos, D.A. Bengtson, W. Decleir and E. Jaspers (eds.), *Artemia Research and its Applications*, Vol. 3, Universa Press, Wetteren, Belgium, pp. 459–463.

Tunnicliffe, S.D. and Reiss, M.J. (1999) Opportunities for sex education and personal and social education (PSE) through science lessons: the comments of primary pupils when observing meal worms and brine shrimps. *International Journal of Science Education* **21**, 1007–1020.

USGS (2001) United States Geological Survey – Great Salt Lake: http://ut.water.usgs.gov/greatsaltlake/

Van Stappen, G. and Sorgeloos, P. (1993) The cosmopolitan brine shrimp. *Infofish International* **4**, 45–50.

Vanhaecke, P. and Sorgeloos, P. (1982) International Study on *Artemia*. XVIII. The hatching rate of *Artemia* cysts – A comparative study. *Aquacultural Engineering* **1**, 263–273.

Vanhaecke, P. and Sorgeloos, P. (1983) International Study on *Artemia*. XIX. Hatching data for ten commercial sources of brine shrimp cysts and re-evaluation of the 'hatching efficiency' concept. *Aquaculture* **30**, 43–52.

Vanhaecke, P., De Vrieze, L., Tackaert, W. and Sorgeloos, P. (1990) The use of decapsulated cysts of the brine shrimp *Artemia* as direct food for carp *Cyprinus carpio* L. larvae. *Journal of the World Aquaculture Society* **21**, 257–262.

Velazquez, M.P. (1996) Characterization of *Artemia urmiana* Günther (1900) with emphasis on the lipid and fatty acid composition during and following enrichment with highly unsaturated fatty acids, M.Sc. Thesis, Ghent University, Belgium.

Verreth, J., Segner, H. and Storch, V. (1987) A comparative study on the nutritional quality of decapsulated *Artemia* cysts, micro-encapsulated egg diets and enriched dry feeds for *Clarias gariepinus* (Burchell) larvae. *Aquaculture* **63**, 269–282.

Watanabe, T., Oowa, F., Kitajima, C. and Fujita, S. (1978) Nutritional quality of brine shrimp, *Artemia salina*, as a living feed from the viewpoint of essential fatty acids for fish. *Bulletin of the Japanese Society of Scientific Fisheries* **44**, 1115–1121.

Watanabe, T., Tamiya, T., Oka, A., Hirata, M., Kitajima, C. and Fujita, S. (1983) Improvements of dietary value of live foods for fish larvae by feeding them on (n-3) highly unsaturated fatty acids and fat-soluble vitamins. *Bulletin of the Japanese Society of Scientific Fisheries* **49**, 471.

Willis, A.L. (1987) *Handbook of Eicosanoids: Prostaglandins and Related Lipids, Volume 1 – Part B. Chemical and Biochemical Aspects*, CRC Press, Boca Raton, Florida.

Wouters, R., Gomez, L., Lavens, P. and Calderon, J. (1999) Feeding enriched *Artemia* biomass to *Penaeus vannamei* broodstock: its effect on reproductive performance and larval quality. *Journal of Shellfish Research* **18**, 651–656.

Wurtsbaugh, W.A. and Smith Berry, T. (1990) Cascading effects of decreased salinity on the plankton, chemistry, and physics of the Great Salt Lake (Utah). *Canadian Journal of Fisheries and Aquatic Sciences* **47**, 100–109.

Zmora, O. and Shpigel, M. (2001) Filter feeders as biofilter in marine land-based systems. In S. Deroe (ed.), *Aquaculture 2001 – Book of Abstracts*, Orlando, Florida, USA, p. 716.

INDEX

A. *franciscana*, 40, 141, 147, 176–182, 184, 191, 197–199, 226, 228–232, 234–236, 238–245
A. *monica*, 177, 181, 199, 226, 228
A. *parthenogenetica*, 184, 227, 244, 263
A. *persimilis*, 182, 184, 197, 199, 228, 229, 231, 232, 239–244, 263
A. *salina*, 182, 184, 195, 199, 228, 229, 232, 236, 244
A. *sinica*, 182, 184, 185, 196, 199, 228, 232, 234, 238, 244
A. *tibetiana*, 178, 182, 196, 199, 228, 263
A. *tunisiana*, 227
A. *urmiana*, 182, 184–186, 199, 228, 229, 233
abdominal segments, 1, 6, 19, 114
accessory glands, 24, 29, 116
accessory nuclei, 56–58, 60, 63, 66
acetylated tubulin, 11, 32
acid phosphatase, 11, 57
acrosome, 27, 78, 81, 84
α-crystallin, 39, 42, 138
actin, 23, 27, 96
adaptation, 129, 141, 142, 146, 149, 150, 158, 176–178, 180, 189, 231, 233, 238
AFLPs, 231, 232
algal blooms, 191, 257, 270
alimentary tract, 6, 9
allopatric, 182, 227, 240, 243, 244
allopatry, 227
allozyme, 196, 225, 227, 230–233, 237, 239, 241, 244
allozyme loci, 233, 241
Alu I, 229, 232, 241
anaphase, 24, 30, 68, 74, 92, 93, 95, 97, 101, 102, 130
Anas sp., 187, 188
aneuploidy, 228
annulate *lamellae*, 54–56
annulate membranes, 69
anostracan, 56, 91, 115, 176, 179, 180, 188, 190, 193
anoxia, 40, 42, 107, 131, 135–140, 145, 146, 148
antennae, 1, 3–5, 9, 26, 29, 32, 115, 189

antennal, 10, 11, 13, 113, 114, 116, 118
antennal glands, 11, 113
Antennulae, 3
antibody, 23, 29, 30, 41–47, 115
antioxidants, 134
Aphanius, 176
apocrine, 7, 117
apolysis, 111, 117
apomixis, 128
apoptosis, 42, 56, 62
apoptotic stress, 42
aquaculture, 171, 173, 189, 193, 197, 215, 225, 231, 234–236, 241, 251, 252, 258, 259, 269, 270, 272
Artemia biomass, 187, 258, 269
Artemia juveniles, 268
Artemia Reference Center, 171, 251, 252
Artemia sp., 174, 195, 228
artemin, 135, 138, 140
arthrodial membranes, 116, 117
ascorbic acid-2-sulphate, 135
asexual, 182, 185, 226, 228, 232, 236, 237, 244, 245.
asexual gene pool, 236, 237
ATP, 52, 138, 145, 146, 157
α-tubulin, 23, 30, 32, 46, 47, 50, 51
australia, 154, 156, 171, 173, 175, 179, 182, 189–194, 197, 200, 228, 252
automixis, 229

β-tubulin, 46, 47, 50, 51
benthic cyanobacteria, 254
bicarbonate, 47, 51, 157
bioconservation, 193
biodiversity, 193
bio-encapsulation, 265
bio-indicator, 271
biological diversity, 239
Biological Species Concept, 226, 243, 245
biotic elements, 179
biotope, 174, 175, 177–180, 182, 186, 187, 193, 196, 197, 258, 261
birds, 175, 176, 181, 186–188, 196–198, 234, 240
Birgus latro, 76

bisexual species, 39, 182, 184–186, 196, 227, 228, 231–233, 236, 244
bittern, 256
Black Sea basin, 171, 195
blood cell forming organs, 9
blood cells, 9–11, 15, 112, 114, 118
blood platelets, 133
Boeckella, 179, 191
Bombyx, 93
Brachionus, 180.
brain, 12, 13, 16–18, 113, 114, 118, 264
Branchinecta, 179
Branchinectella, 179
Branchinella spinosa, 176
branchiopods, 80, 81, 100, 101, 226
brine, 41, 45, 129, 131, 147, 151, 152, 156–158, 171, 174–176, 178–180, 182–189, 195–197, 215, 225–227, 230, 234, 239, 251, 254, 256–258, 265, 267–272
brine fly, 158, 180, 254
bromides, 256

Caeca, 6
Calidris sp., 187
Cancer salinus, 227
carbonate habitat, 177
carbonate lake, 178
carotenoids, 134
casein kinase II, 146
caudal *ostium*, 9
cavity receptor organ, 16–18
centriole, 67, 69, 70, 72–76, 80, 81, 84, 87, 89, 93
centromere, 68, 229
centromeric constriction, 229
Chaetoceros, 269
Chanos sp., 176
chaperone, 138–141
chitin, 1
Chlidonias sp., 187
chloride cells, 150, 152
chlorophytes, 254
chromatin, 12, 27, 54, 61–63, 67, 69, 71, 72, 74, 79, 80, 87–89, 92, 95, 97
chromocentre, 228, 229, 232, 240, 241
chromosome, 66–68, 70, 73, 74, 79, 91, 93, 97, 99, 106, 228–230, 245
Cladocera, 108, 179, 188
cladoceran, 188
Clarias gariepinus, 267
clasper, 2–4
clonal diversity, 127, 137
coexistence, 179–183, 185, 186
Cohesion Species Concept, 243, 245

cold storage, 267
Coleoptera, 176
colonisation, 178, 180, 184, 193, 235
commercial exploitation, 173, 193, 235, 252
competition, 173, 175, 183, 184, 192, 193, 198, 238, 254
compound eyes, 1, 3, 14
conchostracan, 188
conspecific population, 227, 231
conspecific/s, 227, 231, 238
contamination, 147, 178, 188, 256
copepods, 76, 176, 179, 180, 191
cortical granules, 58, 100
Cretaceous, 226
cross-breeding, 180, 195, 197, 227, 239
cross-fertility, 181, 242, 243
cryptobiotic cyst, 106
crystalline cones, 14, 15
crystallizing ponds, 151, 152, 256
cuticle, 1–3, 6, 9, 12, 14, 16, 17, 27, 48, 91, 103, 105, 106, 110, 111, 115, 116, 130, 131, 158
Cyclops spp., 176
Cyprinus carpio, 267
Cyst, 10, 21, 39, 40, 42–44, 47, 48, 51, 52, 103, 105–107, 109, 112, 129–132, 134–146, 148, 149, 156–158, 173, 177, 178, 181, 184–191, 193–199, 226, 230, 234–236, 239, 242, 251–255, 257–259, 261–265, 267–269, 270, 271
cyst bank, 214
cyst harvests, 193, 252, 254
cyst shortage, 199, 251, 252, 255
cyst supply, 251, 252
cytochrome oxidase, 97, 143, 151
cytogenetic, 197, 227, 228, 230, 239

Daphnia, 101
Daphniopsis sp., 179
decapsulated cysts, 196, 262, 263, 267, 268
decapsulation, 267
dehydration (see also desiccation), 42, 103, 106, 107, 132, 145
dendrites, 17
desiccation, 42, 52, 131–133, 135, 138, 145, 184, 198
detyrosinated tubulin, 11, 46, 48
Deuterocerebrum, 12, 112
devonian, 226
DHA, 70, 263, 264, 266, 267
DHA/EPA, 264, 266
diakinesis, 71, 72
diapause, 40, 42, 44, 46, 47, 51, 103, 112,

INDEX

130–133, 135, 137–140, 142, 144, 145, 148, 151, 157, 226
Diaptomus spp., 176
diatoms, 190, 254
differentiation, 19, 26, 42, 52, 55, 62, 67, 113, 115–119, 143, 181, 182, 190, 196, 197, 227, 229, 231, 233, 237, 238, 240, 241, 244, 245
diploid, 93, 182, 183, 185, 199, 228, 237, 240, 241
diploidy, 225, 228
diplotene, 53, 71
dispersal, 181, 186–188, 191–193, 197, 226, 231, 238, 239
dispersal mechanism, 181
distribution, 11, 32, 71, 74, 101, 103, 107, 171–177, 179, 181–183, 186, 188, 189, 191–193, 196, 228, 229, 238, 267
distribution vectors, 175
divergence, 154, 189, 230, 231, 233–241, 244
diversity, 18, 47, 141, 174, 180, 196, 225, 227, 233, 237, 244, 254
DNA, 40, 42, 45, 53, 55, 143, 196, 225–227, 229–232, 234, 241, 244, 270
DNA markers, 226, 230, 232
docosahexaenoic acid (see also DHA), 264, 266
Dolichopodidae, 179
dormancy, 44, 106, 130, 137
dorsal frontal organs, 17
Drosophila, 119, 226, 230, 238, 240, 245
Dunaliella salina, 181
Dunaliella sp., 174
Dunaliella viridis, 181, 254

ecdysteroid, 19, 64
ecological barrier, 181
ecological isolation, 177, 180, 226
ecology, 129, 133, 149, 183, 192, 233, 255
ectoderm, 110–114, 116–119
ectodermal, 6, 110–114, 116, 118
effective stroke, 6
eicosapentaenoic acid (see also EPA), 255, 264, 266
electrophoresis, 41, 47, 50, 51, 153, 154, 186, 196, 230, 232, 233, 239
elevated ploidy hypothesis, 237
embryonic cuticle, 40, 45, 48, 51, 112, 116, 144, 148, 149, 151, 157
emergence, 10, 40, 45, 51, 112, 116, 144, 148, 149, 151, 157
endites, 5, 117
endocuticle, 2
endocytosis, 60

endoderm, 110, 112, 113, 117, 119
endodermal, 7, 25, 32, 110, 112, 113
endogenous yolk, 55–57, 59, 99
endogenous yolk formation, 56, 57
endoplasmic reticulum, 54, 56, 58,–60, 62, 67, 69, 97, 101, 107
engrailed gene, 115
enrichment, 23, 265, 266
environmental variation, 226
enzyme loci, 233
EPA, 255, 263, 264, 266
Ephydra, 158, 174, 180
Ephydra sp., 174
Epicuticle, 2, 116
Ereunetes sp., 187
Ergasilus, 76
Erolia sp., 187
Euploid, 228
Eurycestus, 188
evaporation, 175, 178, 252, 254, 256–258
excretion, 11, 257
excretory glands, 11
excretory organs, 3
excystment, 42
exites, 6, 117
exogenous yolk, 60, 61
exogenous yolk formation, 60
exopodites, 2
exoskeleton, 1, 4, 116, 117, 158
exploration, 171, 173, 191, 193, 194, 196, 252

faecal pellets, 8
fatty acid, 261, 263–265
feeding behaviour, 187, 190
feral strains, 174
fertilisation, 39–42, 63, 75, 77, 78, 82, 84, 85, 87, 91, 93, 97, 100–103, 105, 106, 109, 110, 237, 244
fertilisation membrane, 84, 97, 100–103, 105, 106, 109, 110
filter feeder, 190, 265
filter system, 6
filter-feeding, 5
filtering apparatus, 5
filtering organs, 5
first antennae, 1, 3
fish and shrimp larvae, 259
fitness-related traits, 234, 236
Flagellum, 78
Flamingolepis, 188
food competitors, 180, 192
food groove, 5, 6
formulated feed, 259, 269
frontal knobs, 3, 4, 240

frozen niche hypothesis, 237
furca, 1
furcal *rami*, 6
fusome, 52, 53, 68, 72, 107

γ-tubulin, 30, 32, 93
G+C content, 230
gametogenesis, 93, 237
gastric caeca, 6
gastrula, 40, 44, 116, 129–131, 133, 142–144
gastrulation, 110, 112, 117
Gecarcinus, 52
gene bank, 198
gene flow, 185, 197, 231, 238, 240, 244
gene pool, 174, 197–199, 235–237, 239, 241, 243, 244
general purpose genotype hypothesis, 237
genetic distance, 186, 228, 231, 232, 238, 240, 242
genetic diversity, 196, 244
genetic variation, 183, 196, 233, 234, 236
genetics population, 226, 239, 240
genome amplification, 231, 241
genotype, 198, 234, 235, 237
geographical distribution, 174, 175, 182, 183
germinal vesicle breakdown, 63, 64
Gilbert Bay, 252, 254
glands, 3, 5, 6, 11, 13, 19, 21, 23, 24, 29, 113, 116
glycerol, 51, 52, 133, 141, 149, 151
glycocalyx, 8
glycogen, 8, 9, 15, 51, 131, 133, 151
gnathobase, 5
Golgi, 8, 55, 59, 60, 64, 69, 80, 97, 101, 102, 106, 107
gonopods, 1, 6
Gp4G, 146–148, 151
gravity sensor, 140
Great Salt Lake (see also GSL), 129, 140, 156, 173, 191, 193, 194, 198, 202, 234, 236, 251–255
green algae, 174, 254
GSL, 194, 251, 252, 255, 262, 263
GTP, 138, 146
Gunnison Bay, 252, 254

habitat, 141, 155, 173, 174, 177–181, 183–186, 188, 191–197, 199, 226, 228, 231, 234, 235, 238, 239, 241
haemocytes (see also blood cells), 9–11
haemoglobin, 152, 153, 157, 182, 183, 189, 190, 193, 270
haemoglobin antiquity, 158
haemoglobin characterisation, 153

haemoglobin gene expression, 155
haemoglobin role, 157, 158
haemoglobin structure, 154
Haliphthoros milfordensis, 188
Halobacterium, 257
halobiont, 179, 193
halophilic bacteria, 254, 257
haploid genome, 226, 229, 230
hatcheries, 255, 265
hatching, 11, 106, 111, 114, 116, 118, 136, 140, 142, 144, 146, 148, 149, 151, 155–157, 177, 178, 184, 188, 190, 194, 230, 252, 255, 258–261, 265, 267
hatching characteristics, 255
hatching conditions, 259
hatching efficiency, 259, 261
hatching percentage, 259, 261
hatching quality, 194, 252, 255, 259
hatching synchrony, 259
heart, 9, 14, 119
heat shock, 39, 40, 42, 119, 138–141, 145
heat shock proteins (see also hsp), 39, 42, 138, 139, 140, 141
heat shock response, 140
Hemiptera, 176
heterochromatic, 229
heterochromatin, 67, 72, 226, 230, 241, 244
heteroploidy, 228
Hindgut, 6, 7, 9, 14, 25, 113, 116, 117
holocene, 180
holocrine, 7
holokinetic chromosomes, 68
hsp70, 139
hsp90, 139
HUFA (see also fatty acids), 264
hybridisation, 46, 182, 245
Hymenolepis, 188
hypochlorite, 105, 265, 267
hypo-osmotic regulation, 190

immunofluorescent staining, 11, 23–25, 30, 41
incipient species, 180, 226, 240
infestation, 188, 189
inland lakes, 12, 173, 186, 195
inoculation, 178, 187, 197, 198, 234–236, 241
instar-I, 26, 40, 44–48, 138, 140, 150, 196, 230, 259, 268
instar-II, 30, 40, 44–46, 48, 259, 268
intercellular bridges, 56, 62, 67, 68, 72
internal osmotic pressure, 51
International Study on Artemia, 252
interphase, 32, 72, 230
interpopulation, 197, 226, 231, 238, 244
intracellular pH (see also pHi), 40, 42, 136

intrapopulation, 242
intrapopulation diversity, 196
ionic composition, 86, 177, 180, 181, 190, 192, 226, 234

karyogram, 229
karyotype, 229
kinetochore, 68, 93, 97

Labidocera, 76
Labrum, 1, 3, 6, 13, 26, 113, 114, 116, 118, 189
lagoons, 155, 171, 174, 179, 196, 197, 209
Larus, 187, 188
larval epidermis, 3
larval salt gland, 23, 100, 151
larval stages, 1, 114, 115, 134, 135, 150
larviculture, 255, 259, 261
latitude, 182, 183, 185, 228, 229, 237, 241
L-DOPA, 11
life history traits, 174, 181, 195, 233, 235, 238
light intensity, 175
live biomass, 258, 259, 264, 265, 269, 270
live food, 252, 258, 259, 264, 265, 269, 270
locomotion, 5

macro-evolution, 230
male clasper, 2, 4
mandible, 1, 5, 6, 10, 11, 111–113, 115
maxillae, 1, 5, 118
maxillary gland, 5, 11–13, 152
maxillulae, 1
median eye, 3, 16, 17
Mediterranean, 171, 173, 176, 182, 188, 189, 195, 225, 227, 228, 244, 255
medulla terminalis, 17
meiosis, 63, 67, 70–72, 93, 107, 108, 228
meiotic spindle, 66, 92, 93, 97
membranous system, 72, 73, 75–77, 80–82, 87, 91
mesoderm, 10, 110–114, 117–119
mesodermal cells, 25, 110, 112, 113, 117, 118
Mesozoic, 189
metabolic rate depression, 136, 146
metabolic regulation, 145, 148
metanaupliar mesoderm, 110–114, 118
metanauplii, 8, 176
metaphase, 24, 30, 63, 66–68, 72–74, 91–94, 97, 230
metaphase I, 63, 72, 91, 94, 97
metaphase II, 73
metaphase tetrads, 93
meta-population, 238
metepipodites, 2, 152

microclimates, 175
micro-encapsulated diets, 252
micro-evolution, 231, 241
microfilaments, 27, 30, 32, 86–88, 95, 96, 111, 116, 117
microtubule organising centre, 93
microtubules, 11, 27, 29, 32, 67, 68, 72, 73, 75–79, 82, 93, 97, 100, 105, 111, 117
microvilli, 7, 8, 55, 60, 62, 96
midgut, 6, 7, 9, 14, 25, 112, 113, 116, 117, 208
midgut epithelium, 7
mitochondria, 7–10, 12, 55, 56, 61–63, 67, 69, 72, 73, 75, 77, 78, 80, 87, 89, 91, 93, 96, 97, 101–103, 143
mitochondrial DNA, 225–227, 230
Mitosis, 40, 46, 57, 68, 72, 87, 110–113, 117, 228
mitotic apparatus, 32
mitotic spindle, 30, 32, 52, 67–69, 72
mixed population, 185, 186
molecular markers, 231
Mono Lake, 177, 180, 181, 187, 201, 228
morphological traits, 228, 232
moult, 1, 3, 19, 21, 27, 32, 63, 64, 114, 115, 265
moulting, 1, 19, 63, 115, 148
MTOC, 93
mucopolysaccharides, 7, 29
Mugil sp., 176
multivesicular bodies, 56, 80, 84, 87, 102
Myticola, 76

Na,K-ATPase, 150–152
Nannochloris, 181
Naobranchia cygniformis, 76
natural distribution, 175
natural diversity, 227
natural resources, 252
naupliar mesoderm, 110, 113
nauplii, 3, 9, 18, 23, 40, 44, 45, 48, 51, 52, 103, 112–114, 116, 140, 145, 147, 148, 150, 151, 153, 176, 180, 196, 197, 226, 228, 230, 234, 235, 239, 252, 254, 259, 261–270
nauplius eye (see also median eye), 12, 16, 17, 112–114
nerves, 3, 9, 12–14, 18
neurohaemal organ, 18
neurosecretory cells, 18
neurosecretory system, 18, 113
niche overlap, 184
niche partitioning, 183, 184, 237
notostracan, 188
nuage, 54, 55, 57, 58, 61, 63, 64, 77
nuclear body, 74, 80, 81, 87, 89

nuclear membrane, 11, 55, 71–73, 76, 80, 82, 85, 86, 88, 89, 92, 99
nucleolus, 54, 67, 70, 71
nucleotides, 146, 147
nurse cells, 20, 21, 52, 55, 56, 61–66
nutrition, 5, 252
nutritional composition, 196
nutritional requirements, 196
nutritional value, 259, 261, 265, 268

ocellus, 16
Odonata, 176
oesophagus, 6, 7, 9, 13, 113, 116, 118
ommatidia, 3, 14, 18
ommatidium, 15
oocyte, 19–21, 39, 40, 52–66, 74, 82, 87, 91, 100, 102, 107, 108, 134
oogenesis, 19, 21, 52
oogonia, 19, 52
Orchestia, 58
ornamental fish, 269
ornithine decarboxylase, 151
osmoregulation, 5, 150, 151, 152, 158, 177, 181, 271
ostracod, 188
outer cuticular membrane, 48, 105, 106, 268
ovaries, 19, 22, 52, 63, 67, 108
ovary, 19, 20, 52, 56, 60, 62, 63, 65–67, 86, 91
oviducts, 9, 22, 23, 64, 85, 92, 93
oviparity, 23, 226, 242
oviparous, 19, 39–41, 47, 103, 130
ovisac, 1, 3, 6, 19, 22, 23, 64, 82, 84, 86, 91, 93, 101, 106, 110, 116
ovoviviparity, 23, 130, 226, 235, 242
ovoviviparous, 19, 39–41, 103, 106, 112, 130, 192, 199
ovulation, 1, 19, 57, 62, 64, 65, 101
oxygen consumption, 153
oxygen transport, 153, 156, 190

P. contracta, 192
p26, 39–46, 135, 138, 140, 141, 145, 148, 151
p26 mRNA, 39, 40, 44
pachytene, 53, 54, 68, 71, 72, 74, 75
paired penes, 1, 6
Palaemonetes, 176
pancreatic islets, 132
parallel evolution, 190
Parartemia, 154, 155–158, 173, 175, 179, 189, 190–193
Parartemia minuta, 191.
Parartemia zietziana, 154, 156, 192
parasites, 188

parasitism, 188, 189
parthenogenesis, 91, 183, 184, 186, 225, 232, 237
parthenogenetic population, 12, 39, 181, 183, 184, 186, 191, 196, 199, 229
parthenogenetic strain, 39, 93, 134, 183–186, 191, 195, 196, 199
Penaeus chinensis, 269
Penaeus monodon, 269
penis, 22, 27, 29
pentaploid, 183, 237
pentaploidy, 228
pericentriolar material, 71, 73, 74, 79, 80, 93
peritrophic membrane, 8, 117
peroxidase, 58, 60, 102, 134
phagocytosis, 11
Phalaropus sp., 187
phenotypic variation, 240
Phoenicopterus, 187
phyllopods, 5
phylogenetic, 190, 227, 231
phytoplankton, 180, 181, 187, 190, 254
pigment cells, 16, 17, 112
pinocytosis, 60, 63
ploidy levels, 199, 228, 237, 239, 244
Podiceps, 187
polar body, 87, 91, 93, 95–97, 99, 103, 109
polymorphic, 196, 232, 233
polymorphic loci, 233
polymorphism, 196, 230, 232
polyploid, 61, 173, 181, 183, 185, 228, 229, 237
polyploidy, 183, 225
polyribosomes, 55, 69
population size, 178, 226, 234
post-diapause development, 42, 44, 46, 142, 144, 145
post-mating isolation, 233
post-oesophageal ganglia, 13
post-zygotic isolation, 238
precipitation, 175, 252, 257, 258
predation, 176, 177, 186, 187, 193, 254
pre-mating isolation, 227
pre-nauplius, 47, 51, 149
previtellogenesis, 19, 54–56
primary spermatocytes, 26, 67, 68, 70, 77
proctodeum, 6, 112–114
procuticle, 2, 116
pronucleus, 84, 87–89, 91, 92, 97, 100, 101
prophase, 63, 73, 79, 97, 230
protease inhibitors, 148
proteases, 148
protein genes, 230
protein-coding loci, 233, 244

protocerebrum, 12, 16, 112, 114
protopodite, 3–5, 11
protozoa, 187, 254
proximal thoracopodal glands, 3
purines, 147

RAPDs, 231, 232
receptaculum seminis, 64, 82, 86
recombination nodules, 71, 73
Recurvirostra, 187, 188
repetitive DNA, 226, 229, 230, 241
reproduction mode, 171
reproductive isolation, 180, 182, 185, 226–228, 233, 236–238, 240, 242–245
RER, 54, 55, 102
respiration, 5, 142, 151–153, 193
retinula cells, 14–16, 112
RFLPs, 231, 232
RNA, 41, 44, 50, 56, 137, 142, 148, 270
RRNA, 61, 230

S6 kinase, 142.
salinity, 130, 141, 142, 150, 152, 153, 155, 174–181, 184–186, 190, 192, 193, 195, 197, 198, 225, 235, 237, 252–254, 257–259
salt, 23, 26, 32, 41, 45, 46, 113, 116, 129, 134, 140, 150–152, 156, 157, 171, 173–183, 185–188, 191–198, 200, 202, 210, 234, 236, 251–258, 269, 270.
salt gland, 23, 26, 32, 41, 45, 46, 113, 116, 150, 151
salt lakes, 152, 156, 171, 175, 176, 179, 180, 182, 185–187, 192, 194–196, 198, 200
salt operations, 174, 182
salt quality, 257
saltworks, 171, 174, 179, 180, 187, 191, 194–199, 210, 227, 255, 257, 258
San Francisco Bay (see also SFB), 129, 177, 178, 180, 184, 187, 191, 194, 202, 230, 235, 241, 242, 251, 261
satellite DNA, 232
sea-bass, 255, 259
sea-bream, 255, 259
second antennae, 1, 3, 4
segmentation, 110, 115–118
semispecies, 180
sensillae, 3
sensory function, 3, 16
SER, 54
seta, 29, 32, 116
setae, 4–6, 26, 29, 32, 113, 116
setal cell, 27, 29, 30, 32, 116
sexual, 52, 141, 182, 183, 185, 226, 236, 237, 244.

sexual gene pool, 236, 237
SFB, 141, 178, 184, 197, 235, 241, 242, 251, 262, 263
shell, 19–21, 23, 24, 47, 48, 51, 52, 59, 86, 103, 105, 109, 116, 134, 135, 138, 142, 149, 150, 157, 158
shell glands, 19–21, 23, 48, 86, 105, 116
sibling species, 182, 240
sodium chloride, 256, 257
solar salt production, 256, 257
sperm aster, 89, 97, 99
spermatids, 27, 67, 74, 76, 77, 80, 81
spermatocytes, 26, 27, 67, 68, 70, 72, 73, 77, 78
spermatogonia, 26, 64, 67, 68
spermatozoa, 27, 64, 67, 74, 76–78, 80–82, 84–86
stomodeum, 6, 112–114
strains, 12, 39, 93, 101, 153, 173, 174, 177, 178, 180–186, 191, 193–198, 235, 241, 259, 261, 266, 267
stress proteins, 137, 139, 140
superspecies, 181, 226, 240
sympatric, 183
sympatry, 127, 237, 240
synaptonemal complex, 53, 70–72

Tadorna tadorna, 187
Tanymastix, 101, 105
tegumental glands, 3
telophase, 24, 68, 73, 92, 96, 97, 99
telson, 1, 6, 114
tendon cell, 9, 23, 29, 46, 110–112, 114, 116–118
Tenebrio molitar, 271
tertiary envelope, 48
testis, 22, 25–27, 64, 67, 73, 77, 79, 81, 82
tetraploid, 116, 181, 185, 237
tetraploidy, 228
thermal resistance, 139
thoracal blood organs, 10
thoracic segments, 3, 5, 13, 19, 114–116
thoracopod, 2, 111, 116, 117, 119
thoracopod morphogenesis, 117
Tilapia, 176
Tisbe, 76
tolerance levels, 176
toxicity tests, 271
trehalase, 144, 145
trehalose, 51, 52, 132, 133, 137, 138, 140, 141, 144, 145, 151
trehalose phosphorylase, 145
Trichocorixa spp., 176
triploid, 199, 237

triploidy, 228
tritocerebrum, 13
tubulin, 11, 23, 25, 27, 29, 30, 32, 46–51, 72, 93
tubulin mRNA, 47, 50
tyrosinated tubulin, 32

urine, 11, 12, 152

vas deferens, 22, 27, 29, 64, 77, 78, 82, 83, 85
ventral frontal organs, 16
Vietnam, 141, 174, 194, 198, 235, 241, 242, 257
vitelline envelope, 100

vitellogenesis, 11, 19–21, 55–58, 60–62, 65, 101
vitellogenin, 60
vitrification, 132, 133, 135

Water Replacement Hypothesis, 132

yeast, 141, 263, 265
yolk droplets, 10, 57–59, 61–64, 107
yolk nucleus, 57, 58, 60, 99, 104
yolk platelets, 57–62, 97, 102, 107, 113, 114, 131, 143, 144, 148

zooplankton, 175, 176, 179, 195, 254